THE HISTORY OF ALCHEMY

Influences on Culture, Science and Society

THE HISTORY OF ALCHEMY

Influences on Culture, Science and Society

Kurt Almqvist Tobias Churton Per Faxneld Hjalmar Fors Peter J. Forshaw Mattias Fyhr
Georgiana D Hedesan John M. MacMurphy Christopher McIntosh Mark S. Morrisson Fabrizio Pregadio
Salam Rassi Cristina Viano Andreas Winkler Dagmar Wujastyk

Edited by Carl Philip Passmark

BOKFÖRLAGET STOLPE

CONTENTS

ALCHEMY IN CONTEMPORARY SPIRITUAL, INTELLECTUAL AND ARTISTIC MOVEMENTS

FOREWORD

In modern times, alchemy has been dismissed as a quasi-scientific tradition tainted by mystical and occult tendencies. This view has caricatured alchemy as a singular pursuit, the transmutation of base metals into gold – a misconception that has long obscured its true complexity and diversity across cultures and epochs.[1] For much of modern history, this perspective reduced alchemy to an archaic curiosity, hindering a more nuanced understanding of its significance.[2] However, recent scholarship has illuminated the profound depth of alchemical thought and its intricate connections to religion, philosophy, science and the human quest for transcendence.

A key turning point was Antoine Faivre's conceptualisation of Western esotericism in the 1990s, which provided a pivotal framework for re-evaluating alchemy. Faivre proposed that alchemy should be understood as a constellation of traditions unified by an esoteric worldview, rather than a uniform practice.[3]

Central to this worldview is the principle of correspondence, wherein the macrocosm (the cosmos) and the microcosm (the individual) reflect and influence one another. From this perspective, alchemy functions as a symbolic system, bridging the material and the spiritual, the earthly and the divine, by revealing hidden wisdom embedded in the natural world.[4] This understanding challenges the reductionism of the 19th century and instead positions alchemy as a complex system of thought that engages both metaphysical and practical questions.

This anthology seeks to contribute to an in-depth exploration of its multifaceted nature. Through contributions from leading scholars, it traces the development of alchemy across different cultures and historical contexts, revealing a rich tradition that defies simplistic categorisation.

The essays in this volume encompass a broad range of alchemical traditions. With Chinese alchemy as a starting point, Fabrizio Pregadio traces its development as a dual practice, encompassing external (*wai-dan*) and internal (*nei-dan*) methodologies. Andreas Winkler, meanwhile, investigates the interwoven nature of alchemy, astrology and magic in late Roman Egypt. Tobias Churton continues the exploration of the Hellenistic and Egyptian traditions, highlighting the contributions of Maria the Jewess and the foundational concept of *prima materia*. Collectively, they illustrate how alchemy functioned as both a practical and symbolic endeavour, shaping later philosophical traditions.

The volume further expands its scope with works such as Dagmar Wujastyk's essay on *rasavāda*, the Indian alchemical tradition, which pursued both material and spiritual perfection through its integration of Ayurvedic and yogic practices. Wukastyk thus situates Indian alchemy within a broader cross-cultural context, linking it to Chinese and Islamic traditions and drawing parallels with European alchemy.

In the first of his three essays, Peter J. Forshaw explores the *Emerald Tablet* and its central axiom, 'As above, so below', emphasising its enduring role in alchemical thought across civilisations. He also examines medieval Latin alchemists such as Albertus Magnus and Roger Bacon, who merged alchemy with Christian theology, portraying the philosophers' stone as both a material and spiritual symbol of transformation.

The anthology also turns to early modern European alchemy, where its impact on science becomes particularly evident. Georgiana D Hedesan reveals alchemy's contributions to the rise of pharmaceutical medicine and its role in laying the groundwork for modern scientific enquiry. Hjalmar Fors challenges the notion of a strict divide between alchemy and chemistry, demonstrating how alchemical techniques and methodologies directly influenced the evolution of modern science. Both essays underscore alchemy's lasting legacy.

The volume concludes with an examination of alchemical imagery in literature and surrealist art. From Dante's *Divine Comedy* to the works of modern surrealists like Max Ernst, alchemy has functioned as a powerful metaphor for creativity, transformation and the exploration of the human condition. Contributions by Mattias Fyhr and Per Faxneld underline the enduring resonance of alchemical themes in cultural and artistic expressions, highlighting their capacity to inspire new ways of thinking and imagining the world.

Far from being a singular pursuit, alchemy emerges as a global phenomenon that intersects with religion, philosophy and science in diverse and significant ways. This anthology invites the reader to engage with the history of alchemy as a sophisticated intellectual and spiritual endeavour – one that has sought to unlock the mysteries of existence and the interconnectedness of all things.

Whether practised in ancient laboratories, reinterpreted by Renaissance philosophers or reimagined in modern art and psychology, alchemy remains a testament to humanity's enduring quest for transcendence.

Carl Philip Passmark
Stockholm, March 2025

坤鉛乾汞金丹
訣曰
二氣壺中結。三華頂上浮。
要知鉛汞理。紅黑大丹頭。

EARLY ALCHEMICAL THOUGHT IN THE ANCIENT WORLD

ELIXIRS AND IMMORTALITY

A Short History of Chinese Alchemy

FABRIZIO PREGADIO

The history of Chinese alchemy is documented for more than two millennia, from the second century BCE to the present day. The tradition is divided into two main branches, known as Waidan, or External Alchemy, and Neidan, or Internal Alchemy. External Alchemy, which developed earlier, is based on the manipulation of different substances – primarily minerals and metals – which, when submitted to the action of fire, release their essences. In Internal Alchemy, which, as far as we know, developed from around 700 CE, the elixir is compounded within the alchemist's own person, according to two main models. In the first model, the ingredients are the primary components of the cosmos and the human being; in the second, the elixir is achieved through the purification of the mind from attachments, desires, emotions and similar phenomena. Beyond a large variety of doctrinal formulations and models of practice, both Waidan and Neidan intend to provide knowledge of the Dao ('Way'), the principle that, according to the Taoist view, constantly gives birth to the cosmos and to the 'ten thousand things' (*wanwu*), i.e. all entities and phenomena that live and occur in the cosmic domain.

At the centre of both Waidan and Neidan is the *dan*, translated as 'elixir'. In the context of alchemy, this word refers to a material or immaterial substance – in Waidan and Neidan respectively – that grants access to the Dao. The term itself, however, is quite complex. Chinese and Japanese dictionaries give four main meanings: (1) the colour cinnabar red; (2) the mineral cinnabar; (3) sincerity; and (4) the 'medicine of immortality'. As a colour, *dan* is part of a range of words that denote different shades of red, including vermilion (*zhu*), scarlet (*chi*) and crimson (*jiang*). In the second meaning, *dan* appears for instance in the compound words *dansha* and *zhusha*, literally meaning 'cinnabar powder' and 'vermilion powder'. The third meaning is present in the term 'cinnabar heart' (*danxin*) and other analogous expressions, such as 'sincere like cinnabar' (*dancheng*) and 'genuine like cinnabar' (*dankuan*). The meaning of 'elixir' appears to have originated later, based on the other meanings of the word *dan*. According to a major Chinese philologist (Duan Yucai, 1735–1815), '*dan* is the essence of the stones, therefore the essence of all medicinal substances is called *dan*'. These definitions and related terms show that the semantic field of the word evolves from a root meaning of 'essence'; its connotations

Silk scroll painting of the Eight Immortals on their mythical island. Unknown artist, undated.

Previous vignette:
Detail of a page from an early Qing dynasty album depicting Taoist alchemy. 17th century.

Ink drawing on paper depicting Laozi, delivering the *Daode jing*. Ming dynasty, 16th century.

include the real nature of an entity or its essential part, and by extension the cognate concepts of reality, authenticity and truth. These concepts are associated with the colour red; as the late French sinologist Jean-Pierre Diény remarked, red in China is traditionally '*la couleur du sacré*' ('the color of the sacred').[1]

External Alchemy (Waidan)

Virtually nothing is known about the historical origins of Waidan. Early sources consist of tales on the search of a 'medicine of deathlessness' (*busi zhi yao*), often said to be found on the mythical islands where the Taoist immortals dwell, or attribute doctrines and methods to deities who transmitted them to one another in the heavens and later revealed them to humanity.[2] In other cases, Laozi – supposed to be the author of the *Daode jing* or *Book of the Way and Its Virtue*, the work that Taoists place at the origins of their entire tradition – is deemed to be the first master to transmit alchemical teachings, which he in turn had received from his mother.[3]

The first historical document on Waidan associates its origins with the *fangshi* (literally 'masters of methods'), the general designation of practitioners of different techniques who were often admitted to court by emperors and local rulers. Their main areas of expertise included astrology, divination, exorcism and medicine.[4] Around 133 BCE, a *fangshi* named Li Shaojun suggested to Emperor Wu that he should follow the example of the mythical Yellow Emperor (Huangdi), who had performed an alchemical method at the beginning of human history. Emperor Wu, said Li Shaojun, should make offerings to an alchemical stove in order to summon supernatural beings, in whose presence cinnabar would transmute itself into gold. Eating and drinking from cups and dishes made of that gold would prolong the emperor's life and enable him to meet the immortals. Then, after performing the major imperial ceremonies to heaven and earth, Emperor Wu would obtain immortality. The emperor is said to have followed Li Shaojun's suggestion.[5] This account shows that alchemy existed in China by the second century BCE, but it does not describe an actual method for making an elixir. Li Shaojun's elixir, moreover, was not meant to be ingested but only to be used for making vessels. The earliest mention of elixir ingestion is found in the *Yantie lun* ('Discussions on Salt and Iron'), a work dating from *c.* 60 BCE. On the other hand, as we shall presently see, the ritual aspects involved in Li Shaojun's procedure continued to play a major role in later times.

Details of the first clearly identifiable tradition of Waidan emerge about three centuries after Li Shaojun. Named after the heaven that granted its revelation, the Great Clarity (Taiqing) tradition originated in Jiangnan, the region south of the lower Yangzi River that was also crucial for the history of Taoism during the Six Dynasties (third to sixth centuries). Like most Waidan texts, the Great Clarity works are anonymous and undated, describing their doctrines and methods as transmitted, for instance, to the Yellow Emperor by the Mysterious Woman (Xuannü), one of his teachers in the esoteric arts. Later, around the year 200, an anonymous 'divine man' (*shenren*) revealed them to another *fangshi*, named Zuo Ci.[6]

In the Great Clarity textual corpus, the compounding of elixirs is the central act of a long ritual process that includes various stages, each of which is marked by the performance of rites and ceremonies. The alchemical practice consists of this entire process, and not only of the work at the furnace. It begins with the transmission of texts, methods and oral instructions. The disciple makes a covenant with his master by offering gold, silver, silk or cotton as tokens and pledges not to disclose the teachings he is about to receive. Then he retires to a mountain or an isolated place and performs the purification practices, by making ablutions and observing the precepts. He builds the Chamber of the Elixir (*danwu* or *danshi*, i.e. the alchemical laboratory) and places the furnace at its centre, protected by talismans and other objects – in particular, a sword and a mirror – deemed to ward off demonic beings. To begin compounding the elixir, the alchemist chooses an auspicious day based on traditional calendrical methods. When all preliminary spatial and temporal conditions have been complied with, and when the purification practices are completed, he can begin compounding the elixir. Before lighting the fire, he offers food and drink to several deities, asking them to watch over the process and ensure its successful conclusion. From that moment, the alchemist's attention focuses on the crucible, and he makes the elixir assisted by one or more helpers whose main tasks are pounding the ingredients and watching over the intensity of the fire (several elixirs require a long time for coalescing, for instance 100 days). When the elixir is ready, he offers different quantities of it to several deities. Finally, he again pays homage to the gods and ingests the elixir at the dawn of an auspicious day.[7]

With regard to the actual compounding, the main features of the process may be summarised as follows. The ingredients are placed in a crucible, which is covered over by a second, overturned crucible. Through the heat of the fire, the ingredients

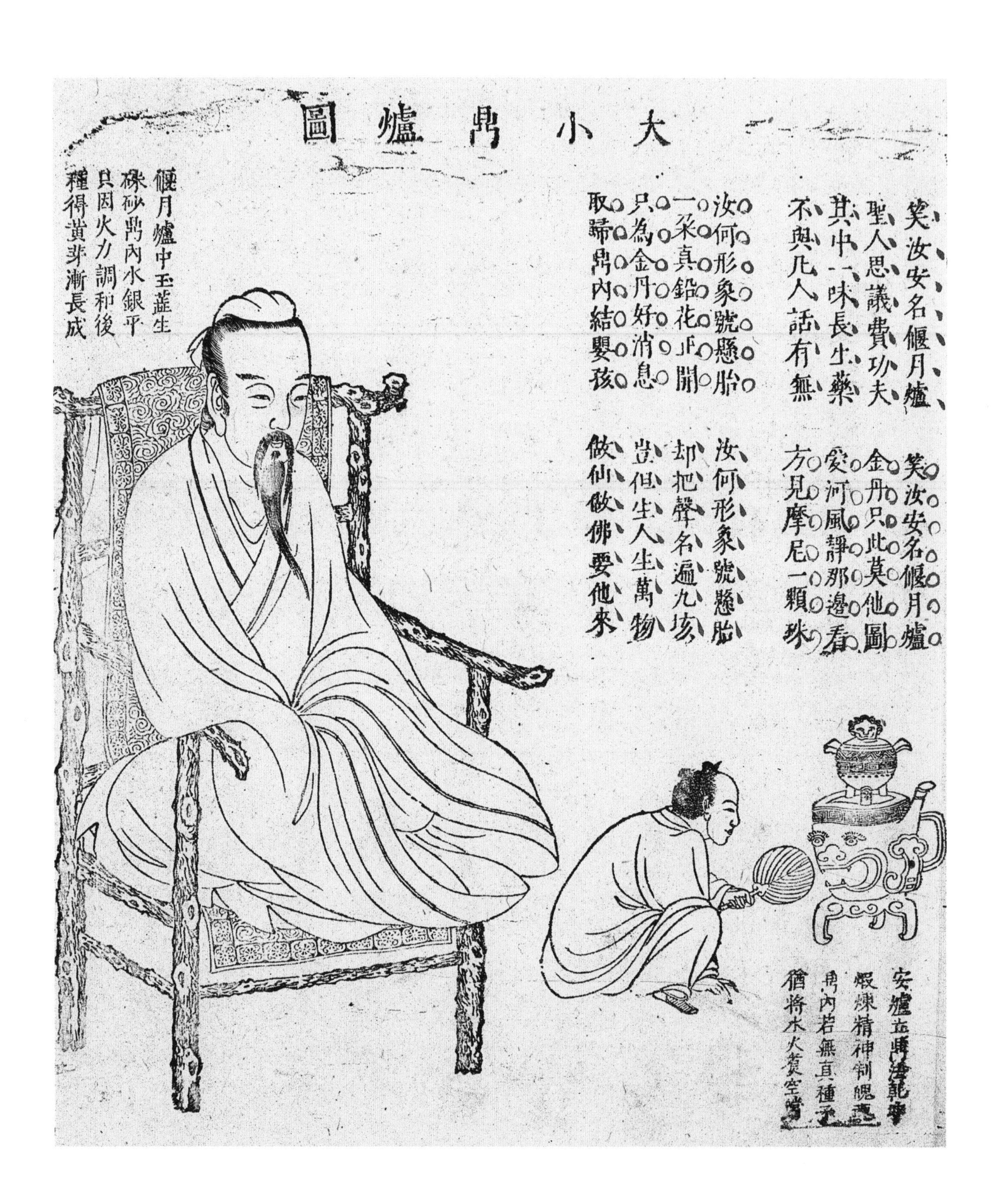
大小鼎爐圖
笑汝安名偃月爐
聖人思議費功夫
其中一味長生藥
不與凡人話有無
汝何形象號懸胎
一朵真鉛花正開
只為金丹好消息
取歸鼎內結嬰孩
笑汝安名偃月爐
金丹只此莫他圖
愛河風靜那邊看
方見摩尼一顆珠
汝何形象號懸胎
却把聲名遍九垓
豈但生人生萬物
做仙做佛要他來
偃月爐中玉蕊生
硃砂鼎內水銀平
只因火力調和後
種得黃芽漸長成
安爐立鼎法乾坤
煅煉精神制魄魂
鼎內若無真種子
猶將水火煑空鐺

transmute themselves and release their pure essences. At the end of the requisite number of days, the crucible is left to cool before being opened, and the elixir is found coagulated under the upper part of the vessel. It is carefully collected, usually by means of a chicken feather, and other substances (e.g. honey) are added to it. Sometimes it is returned to the crucible and heated again; otherwise it is stored to be ingested later.[8]

Mercury, realgar, orpiment, malachite, magnetite and arsenolite are the ingredients most frequently used in the Great Clarity methods. From both a technical and a ritual point of view, however, the main role in the alchemical process is not played by the ingredients but by the crucible. The vessel should be hermetically sealed in order to recreate within it the primal, inchoate (*hundun*) state of the cosmos before the emergence of the 'ten thousand things'. This is an essential task, as shown by several works stating that failure in the alchemical practice is due to errors made in preparing the crucible. For this purpose, a mud made of seven ingredients – one of which is named Divine Mud (*shenni*) – is spread on its outer and inner surfaces and where the two halves join. In several methods, a lead–mercury compound, representing the conjunction of heaven and earth, or Yin and Yang, is also spread on the vessel or is placed inside it with the other ingredients.[9]

In the crucible, the ingredients 'revert' (*huan*) to their original state. A seventh-century commentary on a Great Clarity text equates this refined matter with the 'essence' (*jing*) that, as the *Daode jing* says, is hidden within the Dao and gives birth to the world: 'Vague and indistinct! Within it there is something. Dim and obscure! Within it there is an essence.' The elixir is therefore a tangible sign of the seed through which the Dao generates the cosmos. With further work at the furnace, this pure *prima materia* can be transmuted into alchemical gold.

Ingesting an elixir is said to confer transcendence, immortality, admission into the ranks of the celestial bureaucracy and the ability to summon benevolent gods. Additionally, the elixir grants healing from illnesses and protection from demons, evil spirits, weapons, wild animals and even thieves. To provide these supplementary benefits, the elixir does not need to be ingested: it may simply be kept in one's hand or carried on one's belt as a powerful talisman to avert evil influences.

Alchemy and cosmology

The Great Clarity tradition shows that Chinese alchemy originated as a ritual practice performed to communicate with benevolent divinities and to expel dangerous spirits. The emphasis placed on ritual is closely related to another major feature: none of the Great Clarity texts describes the alchemical process using the language, emblems or images of Chinese cosmology. This system, which took shape in the third and second centuries BCE, does not pertain to any particular intellectual or religious legacy but to Chinese civilisation as a whole and lends itself to different applications, including the way of government. Several Taoist traditions have used it to explicate the relationship between the Dao and the cosmos, and to devise practices meant to enable practitioners to 'return to the Dao' (*huandao*, *fandao*).

At the basis of Chinese cosmology are several sets of emblems, three of which are especially important. The first is made of Yin and Yang, or the passive and active

Woodcut illustration of the 'Great and Small Cauldron and Furnace' from *Xingming guizhi* ('Principles of the Conjoined Cultivation of Inner Nature and Existence') by Yi Zhenren, a Daoist text on internal alchemy published in 1615.

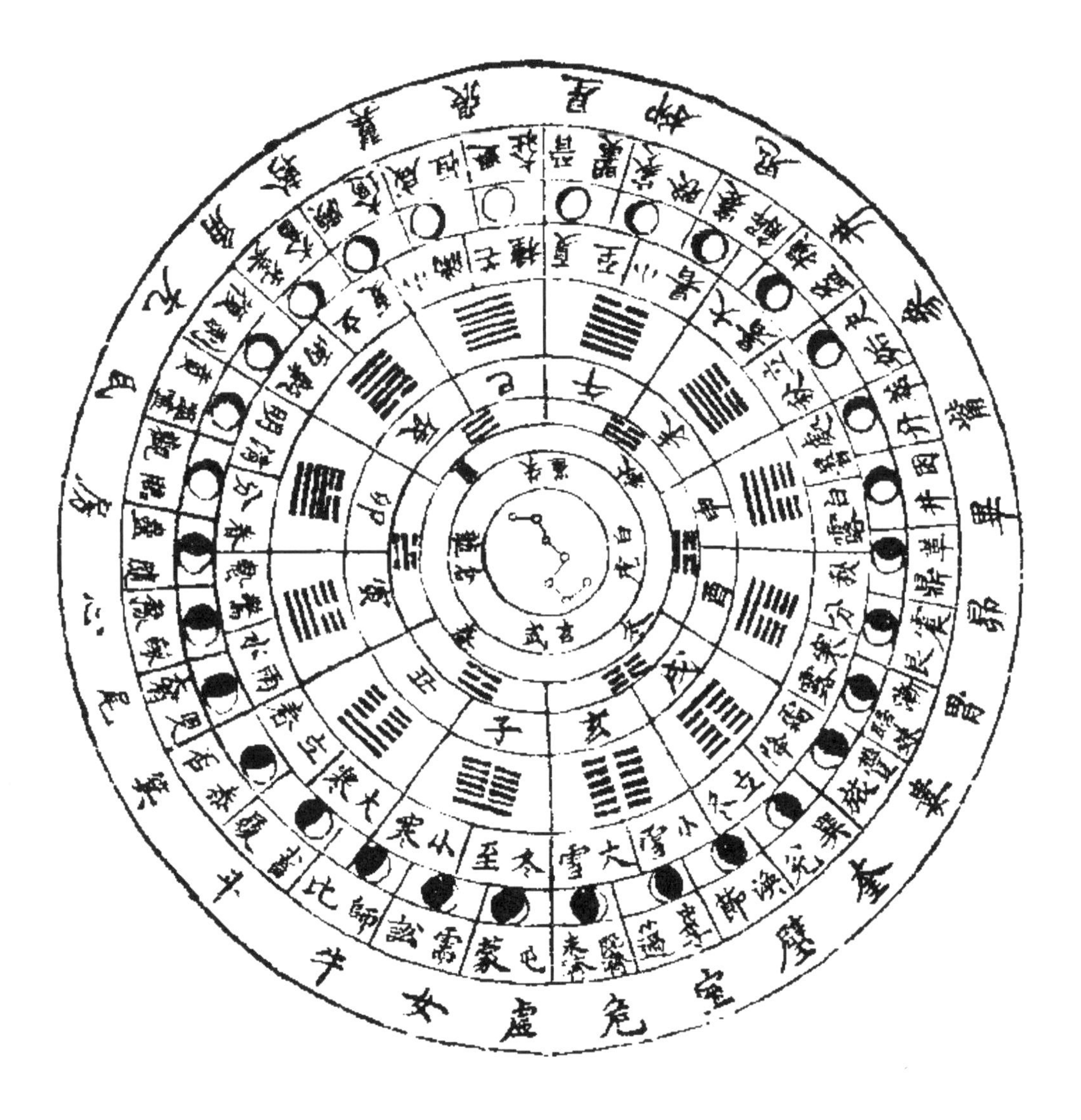

Chart of the Fire phases (*huohou*), showing the main cosmological emblems used in alchemy and their correspondences. From *Yiwai biezhuan* by Yu Yan, 1284.

principles. The second set consists of the five agents (*wuxing*, also rendered as 'five phases' or, somewhat incorrectly, 'five elements'). In the Taoist view, the five agents – Water, Wood, Soil, Metal and Fire – are different modes taken on by the One Breath (*yiqi*) of the Dao as it manifests itself and operates within the cosmos. The third set is formed by the eight trigrams and the 64 hexagrams of the *Yijing*, or *Book of Changes*. While this work is well known as the main Chinese divination manual, its emblems are used in alchemy, and in Taoism as a whole, to represent features of the cosmos; for instance, as we shall see, different states of the Yin and Yang principles and different stages in their continuous alternation.[10]

The textual source that provided the Chinese alchemical tradition with a new model of doctrine and practice – and in so doing, changed forever the history of Chinese alchemy – is the *Cantong qi* (translated as *Seal of the Unity of the Three*, *Kinship of the Three* and in several other ways). Almost entirely written in verse, this work is traditionally dated to the second century CE but was actually written (possibly based on one or more earlier texts) between the mid-fifth and the mid-seventh centuries. Under allusive language, complex terminology and a non-linear exposition, it hides the description of a doctrine that inspired a large number of other works belonging to Waidan and especially to Neidan.[11]

The *Cantong qi* and its tradition draw from the cosmological system several sets of emblems. The main set is formed by four of the eight trigrams of the *Book of Changes*, namely Qian ☰, Kun ☷, Kan ☵ and Li ☲ (the solid lines are Yang, and the broken lines are Yin). As they are used in alchemy, Qian, Kun, Kan and Li are formless principles that illustrate the relationship between the pre-celestial and the post-celestial domains (*xiantian* and *houtian*), i.e. the domains before and after the generation of the cosmos. The trigrams are images that represent those principles. Qian, made of three Yang lines, is the active ('creative') principle, and Kun, made of three Yin lines, is the passive ('receptive') principle; they are the True Yang (*zhenyang*) and the True Yin (*zhenyin*) of the pre-celestial domain. Their continuous conjunction gives birth to the cosmos and to every entity and phenomenon that exists or occurs within it. In a process that does not take place once and for all at the beginning of time but is continuous and everlasting, Qian entrusts his creative power to Kun, and Kun brings creation to accomplishment. In the corresponding trigrams, the Yang of Qian moves into Kun, and, in response, the Yin of Kun moves into Qian: Qian ☰ becomes Li ☲, and Kun ☷ becomes Kan ☵. Therefore, Kan and Li represent the Yin and Yang, respectively, of the post-celestial domain but harbour the pre-celestial Yang and Yin, as represented by their inner lines.

On the basis of these principles, the only form of alchemical practice sanctioned by the *Cantong qi* and by the works that follow its system is that which enables the conjunction of Qian and Kun, or True Yang and True Yin. According to the *Cantong qi*, only True Lead and True Mercury are 'of the same kind' (*tonglei*) as Qian and Kun. Within the post-celestial domain, in which we live, Yin and Yang are never found in their pure state: the Yang and Yin natural substances that contain those authentic principles are cinnabar and 'black lead' (native lead). The alchemist's task is therefore to extract True Lead from black lead and True Mercury from cinnabar and to join them together. This is accomplished in Waidan by means of the corresponding minerals and metals, and in Neidan by operating on various components of the human body, both physical and immaterial – for instance, the 'breath of the kidneys', which represents Yang within Yin (True Lead), and the 'liquor of the heart', which represents Yin within Yang (True Mercury).

Several facets of the Waidan traditions inspired by the *Cantong qi* show that, during the Tang period (*c.* seventh to ninth centuries), the alchemical methods were intended to mirror features of the cosmological system. Alchemists in this period often maintained that the work at the furnace reproduced the process by which nature transmutes minerals and metals into gold within the earth's womb. In their view, an elixir prepared in the alchemical laboratory has the same properties as the Natural Reverted Elixir (*ziran huandan*), which nature refines in a cosmic cycle of 4,320 years. This number corresponds to the total sum of the 12 'double hours' (*shi*) in the 360 days that form one year according to the lunar calendar. Through alchemical work, therefore, a process that requires an entire cosmic cycle to occur can be accomplished in a relatively short time.[12]

A similar intent inspires the method for heating the elixir, known as Fire phases (*huohou*). With another example of the use of emblems of the *Book of Changes*, the intensity of the fire is modelled on the 12 hexagrams used to represent the growth and decrease of Yin and Yang through a complete time cycle:

䷗ ䷒ ䷊ ䷡ ䷪ ䷀ ䷫ ䷠ ䷋ ䷓ ䷖ ䷁

The cycle goes from the rise of the Yang principle (the solid line at the bottom of the first hexagram) to its highest point of development (the sixth hexagram, formed by six solid lines), followed by its decline and the reversion to pure Yin (the last hexagram, six broken lines); then the sequence begins again. The 12-stage heating process therefore replicates the cyclical nature of time.[13]

From a large number of methods, two emerged in the Tang period as typical of the tradition as a whole. The first, which exists in several variants, is based on the refining of mercury (Yin) from cinnabar (Yang). Mercury is then added to sulphur (Yang) and refined again. The repetition of this process yields a substance deemed to be entirely Yang. The second method, which derives from the doctrines of the *Cantong qi*, is based on the conjunction of refined lead (True Yang, obtained from Yin black lead) and refined mercury (True Yin, obtained from Yang cinnabar). Again, the result is an elixir entirely devoid of Yin components and embodying the luminous qualities of Pure Yang (*chunyang*), the ontologic and cosmogonic stage prior to the division of the One into Yin and Yang.

The second method was ultimately successful. Under the influence of the *Cantong qi*, from the Tang period onwards lead and mercury became the main substances in Waidan, not only as ingredients of elixirs but especially as emblems of cosmological principles (True Yang and True Yin). Alchemical and cosmological images were deemed to be equivalent, with the result that the entire basic repertoire of terms and emblems of the cosmological system entered the vocabulary and imagery of alchemy. In this way, Waidan gradually turned into a symbolic system; the elixir and its ingredients became emblems that represent the different stages through which the Dao generates the cosmos and at the same time make it possible to trace those stages backwards, starting from the last to return to the first.

This, however, ultimately led to the decline of Waidan and paved the way for the birth of Neidan. The figurative language of alchemy is not only suitable to represent doctrinal principles but also lends itself to describing multiple forms of practice, providing that they are inspired by those principles. Using the same concepts and the same terminology, alchemical operations can therefore take place within a different microcosm: not the crucible or the laboratory but the human body, or rather the human being as a whole. From the Song period onwards (*c.* mid-tenth to mid-thirteenth centuries), Waidan gradually lost its core function, and virtually all of its soteriological content was transferred to Neidan.

Internal Alchemy (Neidan)

Neidan could easily be construed as a transposition of the Waidan practices to an inner plane, but this view would be reductive. Obviously, Neidan draws from Waidan several basic terms that refer to alchemical operations, instruments, ingredients and, most importantly, the idea of the elixir itself. The origins of Neidan, however, are more complex. Scholars mainly concerned with Waidan have suggested that the shift to Neidan was prompted by an increase in cases of elixir poisoning, which caused the death of at least two and possibly as many as four Tang emperors. However, scholars familiar with the history of Taoism have shown that Neidan evolved from the early traditions of Taoist meditation, which already incorporated imagery drawn from Waidan.

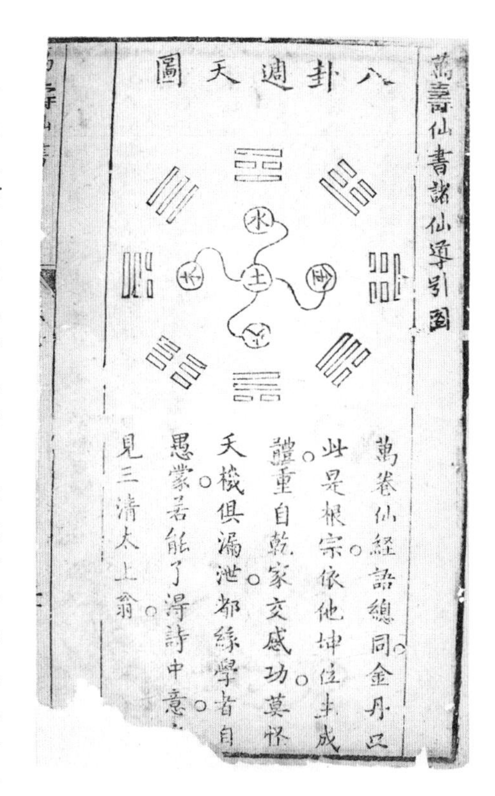

Chart of the eight trigrams showing the five agents at the centre. From *Wanshou xianshu (The Immortals' Book of Longevity)* by Wu Weizhen, who lived under the Ming dynasty.

Those traditions, transmitted in the same region (Jiangnan) and at the same time (*c.* third to fourth centuries) as Great Clarity alchemy, deemed each person to be the host of a veritable pantheon of inner gods. The innermost deity, called Red Child (Chizi) or Zidan (Child-Cinnabar), resides in the stomach – one of the multiple centres of the body – and is said to represent one's own 'true self' (*zhenwu*). In this, he is the precursor of the alchemical 'embryo' that Neidan adepts, centuries later, would generate and nourish by means of their practices. Moreover, to ensure that this and the other gods stay in their residences (their departure would cause death), one should nourish them and their dwellings. Among the sources of nourishment is the practitioner's own salivary juices, which serve to 'irrigate' (*guan*) the internal organs in which the gods reside. Their names have clear alchemical connotations; they include Jade Liquor (*yuye*), Golden Nectar (*jinli*) and even Golden Liquor (*jinye*, also the name of one of the Great Clarity elixirs).[14]

These early traditions were re-codified from the mid-fourth century by the Shangqing (Highest Clarity) school of Taoism, regarded as the main antecedent of Neidan and already involving practices for the generation of an 'immortal embryo'.[15] Elements of those practices were later combined with the cosmology of the *Cantong qi* and its alchemical model based on lead and mercury. While this led to the virtual disappearance of the inner gods, it resulted in the birth of Neidan.

In addition to alchemical language derived from Waidan, emblems drawn from the *Book of Changes* and fragments of methods inherited from early Taoist meditation, Neidan incorporates several other components, including teachings from the *Daode jing*; doctrinal concepts from Buddhism, Confucianism and Neo-Confucianism; views of the human body from traditional medicine; and several physiological practices (especially breathing). On the one hand, Neidan leaves a remarkable freedom to masters and authors of texts who, centred on a fundamental way of seeing, may use any or all of those components and combine them in order to frame their discourses and methods. On the other hand, this partly accounts for the complex terminology of Neidan texts and the various levels at which they can be read and understood.

The multiple roots and backgrounds in turn indicate that Neidan is not a unitary and uniform tradition. Instead, it has evolved through a large number of lineages, each with its own discourse and practices. The first identifiable lineage is Zhong-Lü, created during the second half of the Tang period and named after two illustrious immortals, Zhongli Quan and Lü Dongbin. This was followed by the Southern Lineage (Nanzong), whose five main masters lived between the late tenth and the early thirteenth centuries; they include Bai Yuchan, one of the central figures in the history of Taoism and Chinese alchemy and probably the *ex post facto* creator of the lineage. Not long later, the Northern Lineage was created by Wang Chongyang (1113–70) and his seven main disciples, who lived between the twelfth and the first few decades of the thirteenth centuries. The Yuan dynasty saw several important attempts to merge teachings and practices of the Southern and Northern Lineages, in particular by Li Daochun and Mu Changzhao (both active in the late thirteenth century) and by Chen Zhixu (1290–*c.* 1368). During the Ming and Qing periods (mid-fourteenth to early twentieth centuries) the Wu-Liu Lineage and its branches owe their origins to Wu Shouyang (1574–1644) and Liu Huayang (1735–99), who, more than a century after Wu's death, professed to be his disciple. Major representatives of Neidan in these periods also include Lu Xixing 陸西星 (1520–1606), creator of the Eastern Branch (Dongpai); Li Xiyue 李西月 (1806–56), originator of the Western Branch (Xipai); and Huang Yuanji (mid-nineteenth century), associated with the Central Branch (Zhongpai). Several other masters, including Liu Yiming (1734–1821) and Min Yide (1748–1836), were affiliated with one or another of the innumerable branches of the Longmen (Dragon Gate) Lineage, usually with little or no relation to the central monastic institution in Beijing.

The Red Child (Chizi) or Child-Cinnabar (Zidan), from the *Shangqing dadong zhenjing* ('True Book of the Great Cavern of the Highest Clarity'; *Daozang jiyao* ed.).

Especially in its forms that integrate elements derived from Buddhism and Neo-Confucianism, Neidan bases its discourse and practice on *xing* and *ming*, two cardinal concepts in its view of the human being. In the Neidan view, *xing* denotes the 'inner nature', innately awakened, which is related to spirit (*shen*) and pertains to the 'heart' or 'mind' (*xin*). Neidan texts often discourse on *xing* by using Buddhist terms and expressions, such as 'enlightened nature of true suchness' (*zhenru juexing*) and 'seeing one's own inner nature' (*jianxing*). *Ming*, a remarkably complex term, denotes one's individual existence, which is related to breath (*qi*) and pertains to the 'body' (*shen*). The term means in the first place the 'command' or 'mandate' conferred by heaven to each individual, but its senses also include 'destiny', 'life', 'lifespan' and, in the broadest sense, one's individual 'existence' as a whole. Different Neidan works define *xing* and *ming* as 'the root and foundation of self-cultivation', 'the essential for refining the elixir', 'the learning of the divine immortals' and 'the secret of the Golden Elixir'.[16]

These concepts have given rise to two main models of Neidan practice, formally linked to two of the lineages mentioned above. The first, associated with the Southern Lineage, gives initial priority to the cultivation of *ming* (existence), gradually shifting to the cultivation of *xing* (inner nature) as the practice progresses. The second, associated with the Northern Lineage, emphasises instead the cultivation of *xing* by means of 'clarity and quiescence' (*qingjing*). According to this model, cultivating *xing* includes cultivating *ming*. Several texts, nonetheless, mention the 'conjoined cultivation of *xing* and *ming*' (*xingming shuangxiu*); the difference between the two models essentially consists in which one is seen as the key for cultivating the other.

The Southern Lineage practice is based on the three main components of the cosmos and the human being, namely *jing* or 'essence', *qi* or 'breath', and *shen* or 'spirit', together called the Three Treasures (*sanbao*). The three components are related to a well-known passage of the *Daode jing*: 'The Dao generates the One, the One generates the Two, the Two generates the Three, the Three generates the ten thousand things.' According to one of several readings of this passage, this numerical sequence refers to the stages through which the Dao generates the cosmos, namely Dao → spirit → breath → essence. Through its own essence the Dao then gives birth to the post-celestial domain of the 'ten thousand things'. Here the three components take on different forms: with regard to the human being, *shen* refers to the mind (in particular, the 'cognitive spirit', *shishen*), *qi* appears as the breath of breathing, and the main materialisation of *jing* is semen in males and menstrual blood in females.

The practice, which exists in several variants but usually takes place in three main stages, intends to invert the process mentioned above and reintegrate each component into the one that precedes it in the cosmogonic sequence (i.e. essence → breath → spirit → Dao). The stages are known as 'Refining the Essence to Transmute it into Breath' (*lianjing huaqi*), 'Refining the Breath to Transmute it into Spirit' (*lianqi huashen*) and 'Refining the Spirit to Return to Emptiness' (*lianshen huanxu*), and are characterised by a gradual shift from intentional 'doing' (*youwei*) to spontaneous 'non-doing' (*wuwei*). The emblematic numbers 1, 2 and 3 also represent the progressive reduction of the components: 3 (essence, breath, spirit) → 2 (breath and spirit) → 1 (spirit) → 0 (emptiness). The practice focuses on the Cinnabar Fields (*dantian*), three immaterial centres of the person found in the areas of the abdomen, the heart and the brain. At the end of the first stage, the alchemical embryo is conceived in the lower Cinnabar Field. It is then nourished in or near the middle Cinnabar Field (the Crimson Palace, *jianggong*) for ten symbolic months, corresponding to the time needed for the gestation of the human embryo in the Chinese reckoning. Finally, the embryo is moved to the upper Cinnabar Field (the Muddy Pellet, *niwan*) and is delivered through the sinciput (one of whose names is Gate of Heaven, *tianmen*).[17]

Analogously to the human embryo, the initial conception of the alchemical embryo requires the conjunction of True Yin and True Yang. According to the so-called 'pure cultivation' branches (*qingxiu pai*) of Neidan, both components are found within the practitioner's person. In the 'Yin-Yang' branches (*yinyang pai*), however, the Yang male body is deemed to only contain True Yin, and True Yang should be collected from the Yin female body through sexual conjunction. Other practices are devised for women. Known as Nüdan (Women's Internal Alchemy), menstrual blood has a function analogous to that of the essence (semen) for males.[18]

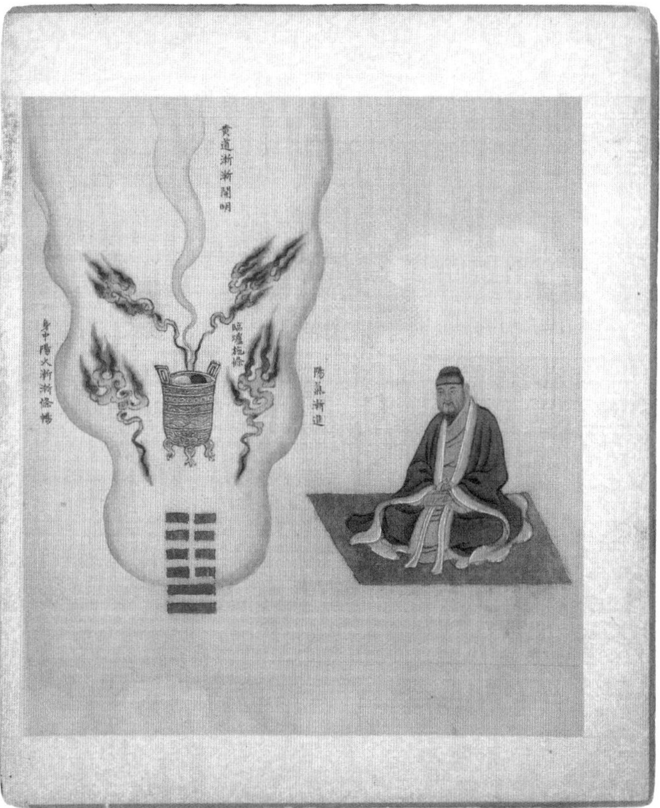

Pages from an early Qing dynasty album, depicting a girl riding on a dragon (true Yin within Yang) and a boy riding on a tiger (true Yang within Yin) discharging their energies into an alchemical crucible. 17th century.

The Beizong Lineage, as well as the branches of the Longmen Lineage that refer to it, are based on a different model. Since the ordinary mind, in the conditioned state, is the main agent that obscures one's own true nature, this model gives emphasis to such principles as 'making the mind empty' (*xuxin*), 'extinguishing the mind' (*miexin*) and 'being without thoughts' (*wunian*) in order to 'see one's own nature' (*jianxing*). The method to attain this state is to practice, mainly by means of meditation and breathing techniques, 'clarity and quiescence' (*qingjing*) of both mind and body, another expression drawn from the *Daode jing*.

In this way of seeing, the human being is envisaged differently compared to the Nanzong model. At its core is what several Neidan masters call the Mysterious Barrier (*xuanguan*) or the One Opening of the Mysterious Barrier (*xuanguan yiqiao*), a spaceless locus found at the symbolic centre of the body (or rather, according to one definition, 'neither within the body nor outside it'). Many of its synonyms – for instance, Gate of Yin and Yang, Altar of the Dragon and the Tiger, Opening of the Turtle and the Snake, Opening of Nature and Existence, and Barrier of Life and Death – allude to the conjunction of opposites. In one of his works, Liu Yiming describes it by saying: 'This is the opening that generates Heaven, Earth, and humans; this is the hometown of saints, buddhas and immortals. Here you arrange the furnace and set up the tripod; here you collect the ingredients and refine them; here you coagulate the elixir and deliver it. Being is here and Non-Being is here. The beginning and the end of all operations are here.' The entire alchemical work, therefore, takes place in this spaceless centre.[19]

Since inner nature is considered, as in Buddhism, to be fundamentally awakened, the elixir is deemed already to be possessed by every human being. The purpose of Neidan is to preserve the awakened state or to make it manifest again. Liu Yiming expresses this view by saying: 'Golden Elixir is another name for one's fundamental nature. There is no other Golden Elixir outside one's fundamental nature. All human beings have this Golden Elixir complete in themselves: it is entirely realized in everybody. It is neither more in a sage, nor less in an ordinary person. It is the seed of the immortals and the buddhas, the root of the worthies and the sages.'[20] Sharing this way of seeing, other Neidan masters state in their works: 'The Golden Elixir is in front of your eyes.'

The human body represented as a mountain, from the *Wuliang duren shangpin miaojing neiyi* ('Inner Meaning of the Wondrous Superior Book on Limitless Salvation'; Daozang ed.)

In both models of doctrine and practice summarised above, the achievement of the alchemical process results in the formation and delivery of the Yang Spirit (*yangshen*), often represented as a miniature replica of oneself issuing from the top of one's head. This 'body outside the body' (or 'person outside the person', *shen zhi wai shen*) is equated with the 'celestial mind' (*tianxin*), prior to the emergence of the ordinary mind, and with the 'dharma body' (*fashen*), the unmanifested body of a buddha, free from birth and death and therefore everlasting. This spiritual body, or embodied spirit, is the immortal body sought in Neidan.

OF ALCHEMY, ASTROLOGY AND MAGIC IN LATE ROMAN EGYPT

The Leiden and Stockholm Codices in Context

ANDREAS WINKLER

> To turn silver black like obsidian. Silver: 2 parts; lead: 4 parts; place on a fresh sherd (*ostrakon*), place unburned brimstone in triple quantity, put in the furnace, melt, then remove, hammer and do what you want, whether you want to make a hammered or cast figurine (*zōdion*). Then file and chisel. It does not rust.

Statuette of the sun god Helios, Egypt, 200-300 CE.

The instructions (L35) above stem from a collection of recipes focusing on metallurgy. Of a total of 99 recipes in the so-called Leiden Alchemical Codex (P.Leid. I 397), 88 concern various metal-related procedures. These involve gold, silver and other metals, such as tin, while the last 11 prescriptions deal with dyeing of fabrics.[1] The recipes of the first group concern the production of various metal alloys, debasing gold and colouring metals, as well as the manufacture of gold and silver inks. The later recipes deal mainly with production of alkanet-based dyes – a plant used to produce red colour – and purple. The collection was probably compiled in its current form in the fourth century CE in Egyptian Thebes by someone who around the same time assembled another treatise on similar matters: Papyrus Holmiensis.[2] The latter text, which today is kept by the Royal Library in Stockholm, focuses on slightly different processes. It contains 159 entries, of which only the first nine concern metals: silver and tin. For the latter, the aim is to make the tin appear as silver. The following 78 recipes concern the manufacture of (semi-)precious stones and the like. On page 14 of the codex, the scribe added decoration to indicate the start of a new section. The text that follows was written later, with a thinner stylus, and deals with various dyes and mordants. The areas covered by the two manuscripts correspond by and large to the content on which one of the 'founding fathers' of alchemy was writing. An author who wrote under the name Democritus, after the famous pre-Socratic atomist, but lived during the first century CE, Pseudo-Democritus is said to have treated exactly these four areas: the production of the two precious metals, as well as gemstones and dyes.[3]

ΑΠΟΠΛΥΝΟΝ ΘΑΛΑΣΣΗ ΑΛΗΛΙΑΝ ΚΑΙ ΨΥΞΟΝ ΚΑΙ ΕΝΙΠΠΟΥ
ΠΑΛΙΚΡΕΣ ΕΙΣ ΤΗΝ ΑΝΧΟΥΣΑΝ ΚΑΙ ΕΑ ΕΝΚΟΙΤΑΣΘΗΝΑΙ

ΑΛΛΟ

ΛΑΒΩΝ ΤΟ ΑΙΜΑ ΤΟ ΑΝΙΩΘΕΝ ΕΚ ΤΗΣ ΑΝΧΟΥΣΗΣ ΚΑΙ ΚΗ
ΚΕΙ ΤΑ ΟΛΙ ΦΑΚΙΝΗΝ ΜΙΑΝ ΚΑΥΣΑΣ ΟΠΤΑΝΙΩ ΚΑΙ ΤΡΙΨ
ΑΣ ΜΕΤΑ ΜΙΚΡΟΥ ΧΑΛΚΑΝΘΟΥ ΚΑΙ ΜΕΙΞΑΣ ΤΩ ΑΙΜΑΤΙ
ΖΕΣΑΣ ΒΑΠΤΑΙ ΠΟΡΦΥΡΑΝ

ΑΝΤΙ ΓΛΑΥΚΕΙΜΟΣ

ΑΝΤΙ ΤΟΥ ΓΛΕΥΚΙΣΑΙ ΛΑΒΩΝ ΣΚΩΡΙΑΝ ΣΙΔΗΡΙΟΥ ΚΟΨΟ
ΕΠΙΜΕΛΩΣ ΕΩΣ ΓΕΝΗΤΑΙ ΩΣ ΣΜΗΓΜΑ ΚΑΙ ΖΕΣΟΝ ΜΕ
ΤΑ ΟΞΟΥΣ ΕΩΣ ΣΚΛΗΡΟΝ ΓΕΝΗΤΑΙ ΚΑΙ ΧΑΛΑ ΤΟ ΕΡΙΟΝ ΠΡΟ
ΣΤΡΟΥΤΙΣΜΕΝΟΝ ΚΑΙ ΠΡΟΣΣΤΥΨΑΜΕΝΟΣ ΚΑΙ ΕΥΡΗΣΕΙΣ
ΑΥΤΟ ΠΟΡΦΥΡΙΖΟΝ ΚΑΙ ΟΥΤΩΣ ΒΑΠΤΕ ΧΡΩΜΕΝΟΣ ΑΣ
ΕΧΕΙΣ ΒΑΦΑΙΣ

ΔΙΟΣΚΟΥΡΙΔΟΥ ΕΚ ΤΟΥ ΠΕΡΙ ΥΛΗΣ

ΑΡΣΕΝΙΚΟΝ ΑΡΙΣΤΟΝ ΗΓΗΤΕΟΝ ΤΟ ΠΛΑΚΩΔΕΣ ΚΑΙ ΧΡ
ΥΣΙΖΟΝ ΤΗ ΧΡΟΑ ΤΑΣ ΤΕ ΠΛΑΚΑΣ ΛΕΠΙΖΟΜΕΝΟΝ ΑΣ
ΕΧΟΝ ΚΑΙ ΩΣΠΕΡ ΕΠΙΚΕΙΜΕΝΑΣ ΑΛΛΗΛΑΙΣ ΕΤΙ ΔΕ
ΑΜΙΓΕΣ ΕΤΕΡΑΣ ΦΥΣΕΩΣ ΕΣΤΙΝ ΔΕ ΑΥΤΟΥ Β ΕΙΔΗ ΤΟ
ΜΕΝ ΟΙΟΝ ΠΡΟΗΡΗΤΑΙ ΤΟ ΙΤΟ ΒΟΛΕΙΔΕΣ ΤΗ ΧΡΟΑ
ΣΑΝΤΑΡΑΚΙΖΟΝ ΔΕ ΥΠ[illegible]ΡΟΝ ΕΥΕΙΔΕ ΤΟ ΤΟΙΟΥ
ΤΟΝ ΟΥΠΑΤΑΙ ΔΕ ΟΥΤΟΣ ΕΠΟΣΤΡΑΚΟΝ ΚΕΝΟΣ ΕΠΕ
ΡΙΣΑΜΕΝΟΣ ΑΥΤΩ ΘΕΣ ΕΠΙ ΔΙΑΠΥΡΟΥΣ ΑΝΘΡΑΚΑΣ
ΚΑΙ ΣΤΡΕΦΕ ΠΥΚΝΟΣ ΩΣ ΔΑΝ ΠΥΡΩΘΗ ΚΑΙ ΜΕΤΑ
ΒΑΛΗ ΨΥΞΑΣ ΚΑΙ ΛΕΑΝΑΣ ΑΠΟΤΙΘΕΣΟ

ΣΑΝΔΑΡΑΚΗ

~~ΣΑΝΔΑΡΑΚΗ~~ ΠΡΟΚΡΙΤΕΟΝ ΤΗΝ ΚΑΤΑΚΟΡΗ ΚΑΙ
ΚΑΤΑΠΥΡΑΝ ΕΥΛΕΑΝΤΟΣ ΤΕ ΚΑΙ ΚΑΘΑΡΑΝ ΚΙΝΝΑ
ΒΑΡΙΖΟΥΣΑΝ ΤΗ ΧΡΟΑ ΕΤΙ ΔΕ ΘΕΙΩΔΗ ΤΗΝ ΑΠΟΦΟ
ΡΑΝ ΕΧΟΥΣΑΝ ΔΥΝΑΜΙΝ ΔΕ ΕΧΕΙ ΚΑΙ ΟΠΤΗΣΙΝ ΤΗΝ
ΑΥΤΗΝ ΑΡΣΕΝΙΚΩΝ

ΜΙΣΥ

ΜΙΣΥ ΠΑΡΑΛΗΜΠΤΕΟΝ ΚΥΠΥΡΙΟΝ ΧΡΥΣΟΦΑΝΕΣ
ΣΚΛΗΡΟΝ ΚΑΙ ΕΝ ΤΩ ΘΡΑΥΣΘΗΝΑΙ ΧΡΥΣΙΖΟΝ ΚΑΙ
ΑΠΟΣΤΙΛΒΟΝ ΑΣΤΕΡΟΕΙΔΩΣ

ΚΑΔΜΙΑ

ΑΡΙΣΤΗ Η ΚΥΠΡΙΑ ΕΠΙΚΑΛΟΥΜΕΝΗ ΔΕ ΒΟΤΡΥΙΤΙΣ
ΠΥΚΝΗ ΒΑΡΕΙΑ ΜΕΣΩΣ ΜΑΛΛΟΝ ΕΠΙ ΤΟ ΚΟΥΦΟΤΕΡΟ
ΡΕΠΟΥΣΑ ΕΧΟΥΣΑ ΤΗΝ ΕΠΙΦΑΝΕΙΑΝ ΒΟΤΡΥΩΔΗ
ΧΡΩΜΑΤΙ ΣΠΟΔΟΕΙΔΗΣ ΕΧΩΜΕΝΗ ΔΕ ΕΣΤΙΝ ΕΣΩ
ΘΕΝ ΜΕΝ ΚΥΑΝΙΖΟΥΣΑ ΕΞΩΘΕΝ ΔΕ ΔΙΑΦΥΣΙΣ ΕΧΟΥΣΑ
ΕΜΦΕΡΩΣ ΟΝΥΧΙ ΤΗ ΛΙΘΩ ΕΣΤΙΝ ΤΙΣ ΚΑΙ ΠΛΑΚΙΤΗ
ΛΕΓΟΜΕΝΗ ΩΣΠΕΡ ΕΙ ΖΩΝΑΣ ΕΧΟΥΣΑ ΤΑΣ ΔΙΑ

The two codices – perhaps even one work divided into two physical volumes – feature prominently in histories of early alchemy. Despite the resemblance of several of the recipes to those found in the works of Pseudo-Democritus, they are often referred to as recipe collections of an artisan – occasionally even described as a charlatan – who may have resorted to trickery to deceive his customers.[4] The text makes clear that certain procedures can fool even the most skilled craftsmen into thinking that the product is the real thing (L8, 37, 39, 40; H3, 48). Additionally, there is a procedure to check whether gold and silver is genuine or whether tin is pure (L31, 42, 43).[5] Although the two codices have been regarded as early witnesses to the practice of alchemy, opinions diverge.[6]

The rather prosaic appearance of the compositions and their straightforward approach to the subject matter prompted Lawrence Principe to pronounce that the texts

> provide a necessary background to the emergence of alchemy, but they are not themselves, strictly speaking, alchemical. Alchemy, like other scientific pursuits, is more than a collection of recipes. There must also exist some body of theory that provides an intellectual framework, that undergirds and explains practical work, and that guides pathways for the discovery of new knowledge.[7]

The definition takes account of practitioners such as Zosimos of Panopolis (fl. third to fourth century CE).[8] His preserved oeuvre can be divided into two main categories. One involves philosophical speculations, which delve into the nature of substances or allegorise the art he was practising. The other consists of practical recipes for producing alloys and tincturing metals. The recipes are generally devoid of any religious or philosophical reference, and thus resemble what we find in the two papyrus codices. The work attributed to Pseudo-Democritus also fits into this scheme, as it tinges the recipes with philosophical speculation on the ingredients and processes. The works also allude to secrecy and the necessity of being inaugurated into the art in order to perform it properly. While one of the recipes from the two codices mentions that it should not be revealed (H106), it has been explained as relating to secrets of the trade rather than referring to a mystery or the like.

A similar but perhaps more forgiving definition of alchemy states that it is

> a craft (*technē*) rooted in ancient artisanal traditions involving the coloration of metals. Alchemical literature, which began to appear in Roman Egypt, consists of recipes and notes about techniques and equipment; these often include religious and philosophical ideas about the transmutation of the qualities of matter, as well as legends about the origins and great masters of their craft.[9]

With such a definition, a collection of Demotic recipes for producing textile dyes kept by the priesthood of the crocodile god Soknopaios in Roman Soknopaiou Nesos – a village in the north-western corner of the Fayum oasis – would count for

Leiden Codex. This page includes instructions for dyeing wool purple.

ΑΝΛΗ

ΚΟΜΜΑΡΕΩΣ ΛΥΣΙΣ ΕΙΣ ΚΑΚΚΑΒΟΝ
ΥΔ ΠΗΓΑΙΟΝ ΜΕΙΞΕΣ ΤΑΣ ΔΥΟ ΤΟΥΤΟ
ΟΠΟΤΑΝ ΖΕΣΗ ΑΠΑΞ ΕΠΙΒΑΛΕ ΤΡΑΓΑ
ΚΑΝΘΗΣ ΤΟ ΤΡΙΤΟΝ ΚΑΙ ΤΟ ΚΟΜΜΑΡΙ ΚΕ
ΚΑΘΑΡΜΕΝΟΝ ΚΑΙ ΠΕΠΛΥΜΕΝΟΝ ΚΑ
ΛΩΣ ΛΕΙΩΣΑΣ ΕΝ ΧΥΤΡΑ ΕΠΙΒΑΛΕ ΤΟ ΤΡΙ
ΤΟΝ ΟΠΟΤΑΝ ΔΕ ΖΕΣΗ ΕΞΑΚΙΣ ΑΙΡΕ ΑΥ
ΤΟ ΑΠΟ ΤΟΥ ΠΥΡΟΣ ΕΣΤΩ ΔΕ ΜΑΛΑΚΩ ΠΥΡΙ
ΘΕΙΣ ΔΕ ΧΑΜΑΙ ΕΑΣΟΝ ΨΥΓΗΝΑΙ ΜΕΧ
ΡΙ ΗΜΕΡΩΝ ΤΙΝΩΝ ΚΑΙ ΟΥΤΩΣ ΧΡΩ ΕΞ ΑΥ
ΤΟΥ ΔΕ ΚΑΙ ΥΠΟΧΡΙΣΜΑ ΕΡΓΑΣΗ ΠΑΝΤΑ ΛΙ
ΘΟΝ ΕΥΧΡΗΣΤΗ

ΣΜΑΡΑΓΔΟΥ ΠΟΙΗΣΙΣ

ΧΡΥΣΟΚΟΛΛΗΣ ΟΛΚΗΣ ς — ΑΡΜΕΝΙΚΟΥ
ΟΛΚΗΣ ς — ΟΥΡΟΥ ΑΦΘΟΡΕΥ ΠΑΙΔΟΣ ΚΟ
ΤΥΛΑΣ ς ΧΟΛΗΣ ΤΑΥΡΙΑΣ ΤΟΥ ΥΓΡΟΥ ΤΟ
ΔΙΜΟΙΡΟΝ ΜΕΙΞΑΣ ΒΑΛΕ ΕΙΣ ΧΥΤΡΙΔΙΟΝ
ΟΜΟΥ ΚΑΙ ΒΑΛΕ ΤΟΥΣ ΛΙΘΟΥΣ ΠΑΝΤΑΣ ΙΝΑ
ΑΓΩΣΙΝ ΑΝΑ ς Β ΚΑΙ ς Δ ΚΑΙ ΠΩΜΑΣ
ΑΣ ΤΟ ΧΥΤΡΥΔΙΟΝ ΠΕΡΙΠΗΛΩΣΟΝ ΤΟ Π
ΩΜΑ ΚΑΙ ΚΑΕ ΕΝ ΕΛΑΙΝΟΙΣ ΞΥΛΟΙΣ ΕΛΑ
ΦΡΩ ΠΥΡΙ ΕΦ ΩΡΑΣ ΕΞ Ε ΕΠΤΩΝ ΕΑΝ ΔΕ ΓΕ
ΝΗΤΑΙ ΣΗΜΕΙΟΝ ΤΟΥΤΟΥ ΤΟ ΠΩΜΑ ΧΑ
ΡΟΝ ΜΗΚΕΤΙ ΚΕΕ ΑΛΛΑ ΚΑΤΑΨΥΞΑΣ Α
ΡΟΝ ΚΑΙ ΕΥΡΗΣΕΙΣ ΓΕΓΟΝΟΤΑΣ ΕΙΝ
ΑΣ ΟΙ ΛΙΘΟΙ ΚΡΥΣΤΑΛΛΙΝΟΙ ΠΑΣ ΑΔΕ ΚΡΥ
ΣΤΑΛΛΟΣ ΕΨΟΜΕΝΗ ΜΕΤΑΒΑΛΛΕΙ ΤΗΝ
ΤΗΝ ΧΡΟΑΝ

ΒΑΜΜΑ ΧΡΩΜΑΤΩΝ

ΒΑΜΜΑ ΔΕ ΕΧΟΥ ΤΡΙΑ ΒΑΜΜΑΤΑ ΓΕΙΝ ΤΑΙ ΚΡΙΜΝΟΥ
ΜΕΡΟΣ Β ΚΑΙ ΣΤΥΠΤΗΡΙΑΣ ΒΑΦΙΚΗΣ ΜΕΡΟΣ ΑΚΟ
ΟΝ ΚΑΙ ΦΥΡΑΣΟΝ ΜΕΘ ΥΔΑΤΟΣ ΚΑΙ ΧΡΩΜΑΤΑ ΕΡΙΑ
ΚΑΙ ΓΕΙΝΤΑΙ ΚΟΚΚΙΝΑ ΕΑΝ ΠΡΑΣΙΝΑ ΘΕΙΟΝ ΜΕ
ΘΥΔΑΤΟΣ ΤΡΙΨΑΣ ΕΠΙΒΑΛΕ ΕΑΝ ΔΕ ΜΗΛΙΝΑ ΝΙΤΡ
ΟΝ ΑΚΡΑΤΟΝ ΜΕΘ ΥΔΑΤΟΣ ΕΠΙΒΑΛΕ

this category.[10] The papyrus mainly contains prescriptions for dyeing, but at the outset it is stated that the recipes were composed by, or came from, the Memphite god Ptah, who was particularly connected to creation and crafts.

Page 14 of the Stockholm Codex.

Would the two codices fit such a description? In the following, I will try to situate the texts in their proper context and argue that they are indeed more than merely technical treatises belonging to an artisan, or a charlatan.

Let us begin with the 'great masters of their craft'. Some of the recipes in the codices are attributed to named authors: two procedures referring to dyeing techniques (H116, 141) are taken from a book written by a certain Africanus, perhaps to be associated with the famous historian and naturalist of the second and third centuries CE.[11] Another, for making silver (H2), belonged, according to the text, to the writings of Democritus. The recipe states that the attribution to the philosopher is taken from another author, a certain Anaxilaos.[12] This may be the neo-Pythagorean philosopher and magician from Larissa in Thessaly, who was expelled from Rome under Augustus. He may have been included due to the fact that he seems to have been an authority on natural history; Pliny the Elder (fl. 23/24–79 CE) quotes him extensively in his own work on the subject.

The last recipe with an attribution concerns Egyptian silver (L82) and is credited to Phimenas of Sais. This is the only person with a geographical designation, an Egyptian city in the Delta, but the name seems otherwise unattested. It might, however, be a variant form or misspelling of Pammenes.[13] A man by that name is said to have educated the Egyptian priests in alchemical lore, according to the Greek writings of Pseudo-Democritus. A Pammenes, who seems to have been an Egyptian priest, is cited by Aelian (fl. 175–235 CE) in his *De natura animalium* for the reported existence in Egypt of winged scorpions and double-headed snakes with feet.[14] The historian Tacitus (fl. first to second century CE) tells us about an astrologer of the same name who lived during the reign of Nero (r. 54–68 CE). Though there may be good reason to connect the three mentions of the name as referring to the same person,[15] there is no ancient source which could confirm such an identification. The attribution of the recipes to certain famous individuals shows that the scribe of the manuscripts, or the one compiling their sources, considered his work to be directly related to their activities.

But there is more. The last 96 lines of the Leiden codex contain excerpts from a treatise by the first-century CE medical writer Pedanius Dioscorides. The selected passages concern the materials used for the procedures described throughout the two codices. Our practitioner thus had a more profound interest in some of the substances used in the processes to which the two codices are devoted. Passages relating to arsenic, sandarach, *misy*, cadmium earth, chrysocolla, red ochre, alum, natron, cinnabar, and mercury are copied from the work of the famous physician and pharmacologist. This fact, furthermore, suggests that the copyist had contact with a vivid intellectual environment, which allowed access to technical literature.

While certain recipes, as mentioned, acknowledge that their products were not really what they appeared to be, a few texts give the impression that the practitioner believed that he had in fact produced the real thing (e.g. H1, 21, 35, 39):

To produce beryl: tie rock crystal with a string of hair and hang in a jar with urine of a female ass for 3 days, but do not let it touch the urine. The vessel should be closed. Then place the jar/vessel above a mild fire and you will find the finest (*aristos*) beryl. (H75)

Although appearing to avoid any overt reference to philosophical speculations, the text occasionally resorts to magical logic,[16] which indicates a common frame of reference between the genres. A recipe for cleaning pearls (H61) requires the milk of a white dog. Besides the potential chemical value of lactic acid, the recipe clearly relies on the ideas of sympathetic magic in the insistence on whiteness. While dogs – including white ones (*PGM* LXII) – feature prominently in magical recipes of all sorts, a procedure for bleaching pearls (H11) again involves dog milk, as well as donkey hair. There may be some chemical efficacy behind the use of these materials in the pearl procedure, but they once again suggest a connection with magic rationale, in particular since the mixture is said to cause leprosy in a man if it comes into contact with his skin. The remark is apparently based on the idea of like producing like – the disease is associated with white spots on the skin of those infected. An almost verbatim version of the recipe is also known from a Byzantine alchemical source.[17] Several magical texts contain spells not so different from the mixture described, which can induce illness, physical conditions or even death by ingestion of or contact with the concoction. The prescriptions to burn olive wood when producing emerald (H17, 41, 73, 83) may reflect a similar concept. Besides its use in magic and its religious connotations, the fruit produced by the trees from which the wood came was probably thought to transfer its green colour to the precious stone. Another ingredient tying recipes to the sphere of magic is the urine of an 'uncorrupted' (*aphthoros*) boy (L67, 74, 84; H13, 23, 43, 88, 149).[18] While urine surely was a useful chemical for several of these processes, various magical spells in contemporary collections from Egypt also involve a boy medium, and occasionally the child is described with the same Greek term as 'uncorrupted' (*PGM* I 86, II 55–6, V 376, VII 544).[19] The ingredient is also common in other alchemical recipes from later times. The idea of urine from uncorrupted boy may be taken as imbuing the product with the same purity and youthful vigour as was associated with the source of the prescribed substance.[20]

Other magical prescriptions – just like the texts under discussion – provide instructions for producing various inks. The recipes closely match the style of the ones found in the Leiden manuscript. Although not in itself a strong argument for connecting the two types of practices, the presence of such recipes in formularies shows a connection, particularly since some of the oldest recipes related to gold are found in magical compilations. Recipes for refining gold and for blistering the metal are inserted between a request for a dream oracle and a procedure for creating a ring functioning as a lucky charm in a bilingual (Demotic-Greek) compilation from Roman Thebes of the second century CE (*GEMF* 15).[21]

Given the connection between magical practices and those found in the two codices, one can ask who wrote the codices and for what purpose. While such a question may never get a satisfactory answer, the recipes provide some information. The passage quoted at the start of this essay may suggest that the person making use of the codices would be engaged in producing metal figurines. In fact, another recipe from the same codex refers to a similar task. It concerns the gilding of various objects, including small figurines and other types of images:

> Gilding: Gilding operating in the same way. Flaky orpiment, vitriol, golden realgar, mercury, gum, tragacanth, the inside of arum, in equal parts. Grind in goat's bile. Suitable for fired copper objects, silver objects, figurines (*eikones*), and tondos (*aspidia*).[22] Make sure the copper object has no burrs. (L73)

It has been suggested that the compiler of the two codices, although he possessed knowledge of the trade, was not a professional dyer, as he refers to such specialists as an external group.[23] Rather, he may have been involved in producing statues and other figurines. This activity would have been tightly connected to the Egyptian temple,[24] as it had been traditionally, but there is little evidence of still-thriving temples in the Theban region during the time the two codices were written; there was a stark decline of Egyptian indigenous temple activity from the end of the second century CE,[25] though still attested in the third century CE.[26] The possibility cannot be excluded, however, that the collection had been taken from such an environment after it had ceased operation, and that the texts were originally the tools of a 'sacred craftsman' (*hierotektōn*) engaged with some form of the 'sacred craft of Egypt' (*hiera technē Aigyptou*) – a term for the alchemical arts[27] – tied to the so-called House of Gold, the place in an Egyptian temple where divine statues were produced and ensouled. Numerous texts throughout Egyptian history inform us that the statues were made of various metals or were gilded and embossed with precious stones.[28] Among the temple servants attached to the House of Gold, we find both craftsmen and priests, who would have had knowledge of the areas subsumed under alchemy.

Other alchemists have been connected to such environments, including Zosimos. Although very little is known of his background, he has often been understood to be closely associated with a temple, either a priest connected to the House of Gold or a craftsman there,[29] though the difference between the two categories was not always sharply defined. While a description of the personnel tied to the House of Gold from the Dendera temple makes a distinction between the craftsmen producing the statue and the ritualists, who were initiated and thus priests,[30] such a division seems to have had little relevance in practice. For instance, a number of priests from the Karnak temple complex in Ptolemaic Thebes seem to have served both roles.[31] The same can be observed at other locations in the Roman period. From Tebtunis, a village in the south-eastern Fayum oasis, we have evidence for the combination of high-ranking priestly titles with those that indicate the physical production of statues.[32] At Oxyrhynchus, a certain Thonis, son of Phatres, grandson of Harthonis, who lived during the last quarter of the second century CE, styles himself 'priest (*hiereus*), lector priest (lit. 'feather carrier': *pteraphoros*), divine

ΗΛΙΕ ΒΕΡΒΕΛΩΧ ΧΘΩΘΩ ΜΙΑΧ ΣΑΝΔΟΥΜ
ΕΧΝΙΝ ΖΑΓΟΥΗΛ ΕΧΕ ΜΕ ΣΥΝΙΣΤΑΜΕΝΟΝ
ΚΟΙΝΑ ΚΑΙ ΤΟ ΤΕ ΕΓΧΡΙΟΥ ΚΑΙ ΑΥΤΟ ΠΙ ΤΗΣ

ΧΑΡΑΚΤΗΡΑ ΟΣ ΕΣΤΙΝ ΚΑΙ ΔΟΞΑΖΕΤΑΙ ΤΟ ΒΡΕΤ
ΕΠΙΚΑΛΟΥΜΑΙ ΣΕ ΤΟΝ ΕΝ ΤΩ ΘΕΩ ΜΕΓΑΝ ΚΥΡΙ-
ΟΝ ΜΕΓΑΣΘΕΝΗ ΙΑΩ ΟΥΩ ΙΩ ΑΙΩ ΟΥΩ
ΟΩΝ ΔΙΑΦΥΛΑΞΟΝ ΜΕ ΑΠΟ ΠΑΝΤΟΣ ΦΟΒΟΥ
ΑΠΟ ΠΑΝΤΟΣ ΚΙΝΔΥΝΟΥ ΤΟΥ ΕΝΕΣΤΩΤΟΣ
ΜΟΙ ΕΝ ΤΗ ΣΗΜΕΡΟΝ ΗΜΕΡΑ ΕΝ ΤΗ ΑΡΤΙ ΩΡΑ
ΤΑΥΤΑ ΕΙΠΩΝ ΑΣΦΑΛΙΖΟΝ ΤΟ ΦΥΛΛΟΝ ΚΑΙ
ΕΧΕ ΜΕΤΑ ΣΕΑΥΤΟΥ ΤΟ ΠΕΤΑΛΟΝ ΕΑΝ ΔΕ
ΣΤΑΧΕΙΡΩΝ ΕΠΙ ΤΗ ΧΕΙΡΙ

Η ΔΕ ΕΠΙΛΥΤΙΚΗ ΜΕΤΑ ΤΟ
ΟΛΙΣΘΟΥ ΓΡΑΦΟΜΕΝΗ
ΠΗΘ ΦΘΑ ΦΩΖΑ

ΜΩΥΣΕΩΣ ΑΠΟΚΡΥΦΟΣ Ζ΄ Η ΠΡΟΣ ΣΕΛΗΝΙΑΚΗ

ΟΙΝΕΛ ΒΙΟΥ ΧΝΟΥΒΟΥΣ ΑΧΟΡΟΜΒΟΥΣ
ΟΥΡΑΟΙ ΟΥΗ ΡΑΙ ΑΦΩΡΟΚΙ ΔΝΟΧ
ΒΑΡΙΝΘ ΜΑΝΙΚΟΥΡΦ ΑΕΙ ΑΕΙ ΗΑΕΙ
ΣΙΕ ΕΙΗ ΤΕΘΟΟΥΡ ΟΥΡΟΥΤΕΡΜΑΣΑΡΟΥ
ΡΧΟΥ ΤΑΙΣ ΕΧΡΗΖΗ ΕΧΡΙΖΗ ΜΑΡΜΑ
ΟΥΡΦ ΓΥΝΑΙΚΟΜΟΡΦΕ ΘΕΑ ΔΕΣΠΟΤΙΣ
ΠΟΙΗΣΟΝ ΤΟ ΑΠΡΙΣΜΑ ΑΝΥΞΙΣ ΛΑΒΩΝ
ΟΜΦΑΛΟΝ ΚΡΟΚΟΔΕΙΛΟΥ ΑΡΣΕΝΟΣ ΠΟΤΑΜΟ-
ΓΕΙΤΟΝΟΣ ΛΕΓΕΤΑΙ ΚΑΡΔΙΑΝ ΚΑΙ ΩΟΝ ΚΥΝΑΦΡΟ-
ΚΥΝΟΚΕΦΑΛΟΥ ΖΜΥΡΝΑΝ ΛΕΓΕΤΑΙ ΚΡΙ-
ΝΙΝΟΝ ΑΝΘΟΝ ΕΜΒΑΛΕ ΕΙΣ ΑΓΓΟΝ ΚΑΙ ΕΠΙ-
ΝΟΝ ΚΑΙ ΟΤΑΝ ΘΕΛΗΣ ΑΝΥΞΙ ΠΡΟΣ ΑΓΕ ΤΗ
ΘΥΡΑ ΤΟΝ ΟΜΦΑΛΟΝ ΛΕΓΕ ΤΟΝ ΝΗ
ΘΑΑΜ ΘΟΝΑΧ ΘΕΧΕΝΒΑΟΡ ΘΕΑΓΟΝ
ΠΕΝΤΑΘΕΣΧΙ ΒΩΘΗ ΕΝ ΤΗ ΒΑΝ ΤΗΝ
ΔΥΝΑΜΙΝ ΕΧΟΥΣΑΝ ΕΝ ΟΙ ΙΝΕΥΘΑΔΟΝ
ΑΡΗΝΟΙΕΙ ΗΙ ΟΤΙ ΛΕΓΩ ΣΟΙ ΣΑΥΑΜΒΑΧ
ΜΕΡΑ ΧΕΟΖΑΦ ΩΣΕΑΛΑ ΒΡΑΙ
ΒΗΛ ΠΟΥ ΟΤΟΥΤΟΥΑΙ ΟΙΡΗ ΡΕΤ ΑΡΝΟΧ
ΕΑΝ ΕΙ ΔΗ ΦΩΝΗΣΑΙ

ΜΩΥΣΕΩΣ

ΜΩΥΣΕΩΣ ΑΠΟΚΡΥΦΟΣ Η ΔΩΝΗ

craftsman (*hierotektōn*) of Thoeris, of Isis, of Sarapis, of the temple of the divine Augustus Caesar, and of the associated gods, and sealer of the sacred calves (*moschosphragistēs*)' (*P.Mich.* XVIII 788). Such titles underline that the same person could hold both priestly positions and those of a sacred craftsman of the same gods at the same time. The last known attestation of the title *hierotektōn* is found on a mummy label (*C. Étiq. Mom.* 44)[33] – an identification tag used by funerary workers so they could more easily distinguish the cadavers that they had been preparing. The small text dates to the third century CE; its origin is not known, but it at least shows that the institution survived in some form until that period. Which, if any, of those roles Zosimos had is uncertain, but there were still some traditional religious elements present at Panopolis throughout the third and up until the fourth century CE.[34]

It is possible to make further connections between the two codices and what can be described as an extension of an Egyptian temple environment. The papyri are part of what is known as the Theban Magical Library.[35] The term denotes a large collection of magical and related manuscripts in Demotic and Greek, Old Coptic and Greek, or Greek written between the second and the fourth centuries CE. The earliest texts can undoubtedly be connected to the sphere of the indigenous temple, but the exact nature of the collection, which is the work of several hands, is unknown, as is the precise find spot(s) of the manuscripts. That they were found together is highly probable. It has recently been suggested that the magical and alchemical cache was deposited in a tomb on the Theban West Bank, the so-called Tomb of Thoth (TT11). The tomb, which originally dates to the New Kingdom (*c.* 1550–1069 BCE), was repurposed in the Ptolemaic period (305–30 BCE), to serve as a burial place for votive animal mummies.[36] Whether the people behind the animal cult practised in the tomb were related to the ones who later deposited the magical texts there is not known. Nevertheless, the person or people who compiled and collected these texts apparently had an interest in all the genres represented, but whether they could still read the Demotic at the time the library was completed seems improbable.[37]

The person who compiled the two codices also copied two other manuscripts in the library, both of which are magical.[38] The first is a short spell (*PGM* Va)[39] – it is only three lines long – invoking the sun god Helios. The spell serves to make the god appear to the user, so he can put himself into direct communication with the divine sphere and deliver requests to the god. This text was written on what has been described as a loose leaf, which was inserted into the Stockholm codex. Besides invoking Helios, whose name is followed by a few magic words, so-called *voces magicae*, the text asks for its speaker to be 'united' with the deity. The spell continues with the remark 'to add the usual' and has a closing instruction to 'anoint yourself'. The purpose of the 'direct vision' is not stated, but 'add the usual' serves as a placeholder for the specific request that the practitioner has for the deity – perhaps in this case to assist in the processes described in the two codices under discussion.

The last text written by our practitioner, or at least in part,[40] is an instruction for

Stockholm Codex. Three-line spell, *PGM* Va, to the sun god Helios, from the Theban Magical Library.

PGM XIII, an instruction for initiation.

initiation (*PGM* XIII).[41] It exists in three versions,[42] with magical spells in between. These include instructions for how to keep a fire burning, send dreams, open doors, become invisible, make people attractive or unattractive, resurrect a deceased person and cross the Nile on the back of a crocodile. The spells work mainly by invoking a god's secret name to be accompanied by other actions and recitations. The name consists of the seven Greek vowels, which have to be uttered at the proper time following precise ritual steps, which include an invocation of the supreme deity and praises of him. The text then evolves into a description of how the god created the world. The work, which is called either the 'Unique' (*monas*) or the 'Eighth (Book) of Moses', mixes Egyptian, Hellenic, Gnostic and Jewish religious elements, and in one of the spells Christ is invoked – but the author can hardly be considered Christian, or Jewish. While several doctrinal features seem to point to the Egyptian temple, the text is not a direct product of such an environment but rather a development thereof, given the influence of elements that would be foreign to a proper Egyptian temple cult. As mentioned above, by the time the text was written, most traditional cults at Thebes seem to have ceased. Though the connection to Egyptian religious elements in the recipes is strong, the text also quotes the works of a number of sages who were broadly known in the ancient world, but none is necessarily associated with an Egyptian temple. Two of these citations deserve to be mentioned. One refers to a certain Erotylos, who is also quoted by Zosimos of Panopolis,[43] but he remains otherwise obscure. The papyrus also refers to a fifth book of the *Ptolemaica*.[44] This is most likely a pseudepigraphic work attributed to the Alexandrian polymath Claudius Ptolemy (fl. *c.* 100–178 CE), best known for his writings on astral science and geography. The work credited to him, however, appears to have little to do with the historical Ptolemy. The text concerns the birth of the spirit (*pneuma*) of fire and darkness. That is, one spirit is shared by the two opposites. Particularly relevant for the history of alchemy, it is also entitled 'All Is One' (*hen kai to pan*), a succinct summary of what is believed to be one of the basic tenets of alchemical thinking: one thing can be turned into another, given that all substances consist of the same matter.

The main goal of the magical text is to become initiated – the process is referred to as the *monad* – and to invoke the greatest god, who created the world and accordingly controls it. When the god reveals himself, he is to tell the magician his fate, if that is what is desired, but can also accomplish specific requests. We are later told by the text that the god can alter the stars in unfavourable positions in the practitioner's horoscope so as to improve his future.[45]

Without delving too deeply into how ancient horoscopes worked, it is relevant that the god was supposed to tell the magician about 'your star (*astron*), what the disposition of your daimon (*daimōn*) is, and your ascendant (*hōroskopos*), and where you will live and die'. The star could refers to which planet rules the astrological chart, but, given that the term astron is used, it could also designate an asterism, perhaps a zodiac sign. The daimon belongs to the doctrine of the astrological lots, whose calculations were based on the position of the ascendant, the zodiac sign rising at the moment of birth. The location of the practitioner's place of living and death are merely predictions that would be generated from the three astrological parameters (as well as other planetary positions).

Knowledge of astrology is also fundamental to the ritual described in the text as a whole. Besides correlating the days of the planetary week with the planets according to their geocentric distance[46] and associating flowers and types of incense with the planets,[47] the process is to be initiated 41 days prior to a new moon in Aries. The text also recommends specific configurations of planets as particularly favourable for producing amulets. A golden lamella inscribed with magical words is to be made

> when it [the moon] is rising and in conjunction (*sunaptō*) with a benefic star (*agathopoios*), either Jupiter or Venus, and when no malefic one (*kakopoios*), Saturn or Mars, is in aspect (*epimartureō*). It is best done when one of the three benefic stars[48] is in its own house, while the Moon is entering a conjunction, is in aspect, or opposition and the star is also rising. (*PGM* XIII 1027–35)

While the passage may appear bewildering, it is clear that the practitioner had to have insight into astrological doctrines and possess quite advanced astronomical knowledge, or at least be able to use, and have access to, astronomical tables. Such were usually reserved for professional astrologers, who could also engage in magical practices.

The simplest of the procedures, to establish the time of the new moon in Aries, requires at least two tables: one detailing daily positions of the celestial body and another indicating on which date the new monthly cycle begins. In fact, presupposing that the practitioner was able to follow the instructions, we have to assume that he possessed the same level of astral knowledge as a fairly erudite astrologer. Thus, the revelation of the horoscope would not have been strictly necessary, since with the astrological knowledge described the practitioner would have been able to cast it himself, but the divine communication may have had other goals. The fact that the all-powerful god needed to know the horoscope in order to change it may provide one reason for the higher being to reveal it, but perhaps the practitioner simply favoured divine revelation for such activities, rather than his own skills.[49] Could the first spell also be viewed in a similar light?

The astrological focus of our practitioner is not unparalleled. Olivier Dufault notes that the four areas of alchemical activity are also grouped together in certain astrological writings, such as the *Mathesis* (e.g. IV 14.20) of Firmicus Maternus (fl. fourth century CE). The astrologer writes that a certain configuration of the stars produces people who, among other things, will be experts in working with metal and precious stones, both polishing and tincturing them. These are tasks that can be connected with the recipes found in the two codices and thus are akin to alchemy. Other configurations yield makers of statues and images; temple musicians, which hints at certain sacerdotal activities; and knowers of heavenly things or celestial secrets – that is, the astral arts.[50] The connection between astrology and the production of statues becomes explicit in the writings of the Egyptian

Overleaf:
The funerary temple in Deir-el-Bahari, originally constructed during the 18th dynasty, Thebes, Egypt.

astrologer Hephaistion of Thebes (fl. fourth to fifth century CE). He dedicated part of a chapter of his *Apotelesmatika* (III 7.13–17) to the most crucial astral timings when producing statues of various deities, both Hellenic and Egyptian, for shrines.[51] That is, a statue-maker was not only supposed to know how to produce a figure, but he was also expected to be able to calculate the appropriate stellar positions for the undertaking. Astrologically informed manufacture was not an idiosyncratic invention of Hephaistion. The practice can also, for instance, be traced in the Graeco-Egyptian magical papyri (e.g. *PGM* IV 2390, 3145, V 245, 380, XIII 30–39),[52] though the statues seem destined for use outside the temple cult. If figurines were not produced for worship, it is however possible that they served more private purposes, such as magical ritual.[53]

Furthermore, astral timing – that is, to seek out a propitious moment – for tincturing metals is a topic raised by Zosimos of Panopolis. In one of his writings, 'On the Letter Omega',[54] the alchemist engages in an acerbic polemic against those who rely on such timing as well as on *daimons*, even astral *daimons*, in their pursuits instead of the proper craft.[55] In his view, they are subject to fate, something from which a proper practitioner is liberated. Given the focus on divine revelation in the magical text copied by our practitioner, it is possible that he was in fact one of those engaged in what Zosimos was excoriating. The Panopolitan alchemist also tended to describe the ones he polemised against as priests.[56]

Given the lack of evidence for a functioning official temple cult at Thebes in the given period, it is unlikely that our practitioner would have been a priest in a functioning sanctuary devoted to a pagan god. Nevertheless, it is tempting to connect him to such an environment, or at least to later continuations of some of its practices. There may be contemporary parallels from the Theban region that suggest a potential context for the two codices.

An association (*plēthos*) of ironworkers (*siderourgoi*) from Hermonthis, about 20 kilometres south of Thebes, where the traditional Egyptian religion can be traced until the mid-fourth century CE, visited the Theban west bank in the late third and early fourth centuries with some regularity. There, at the practically defunct temple complex of Deir el-Bahri, they left inscribed expressions of worship (*proskynemata*) and engaged in religious activities involving beer drinking and the sacrifice of donkeys.[57] Though the temple in which these acts were performed had a cult of two healing deities, Imouthes and Amenouthes,[58] earlier in the Graeco-Roman period there is no evidence of such sacrifices in their honour. It is thus unclear as to what deity or deities the rituals of the ironworkers were made, or for what purpose. The relevance of their trade here is that craftsmen – people who would not normally be considered priests or religious specialists – were engaging in sacerdotal activities and must therefore have had access to some degree of religious knowledge. Trade guilds could engage in both types of activities.[59] Given that the person who wrote the Leiden and Stockholm codices has been described as an artisan, and clearly also engaged in astrology, magic and theological speculation, it is tempting to draw a parallel between his activities and the ironworkers from Hermonthis. Both our practitioner(s) and members of the guild possessed insight into spheres of knowledge that were not purely technical. As the magical text was the effort of two scribes, the two individuals may also have been part of a larger group, perhaps a professional association that engaged in at least some of the practices embodied by the manuscripts.

How then should we understand the treatises of the Leiden and Stockholm papyri? Were they mere collections of recipes? Recipes are certainly what we have in front of us, and it is possible that whoever copied them could have made use of them, or at least the information they contained, for artisanal or 'charlatanic' purposes. Yet as the manuscripts show that the compiler knew of wider uses for some of the substances prescribed but at the same time display hardly any trace of use, and thus have often been referred to as 'library copies',[60] such an interpretation appears to reveal only part of their *raison d'être*.

Given the existence of the loose leaf and the longer magical codex, the scribe cannot have been a mere charlatan, even a charlatan interested in the substances he was using on a wider scale, given the presence of the extract from the medicinal treatise preserved in the Leiden papyrus. Although direct evidence is lacking, it is tempting to connect the 'direct vision' spell with the application of the recipes, at least those found in the Stockholm codex. The presence of the magical codex casts further light on the two texts under discussion, however. Besides extensive involvement with magic and astrology, the work also shows acquaintance with personalities associated with alchemy and discloses a familiarity with the essential doctrines of alchemy: all is one. The magical text further suggests that the compiler may have resorted to divine revelation or enlightenment through such a procedure rather than relying solely on his own skills, at least when it comes to astrology. In my view, it would thus be possible to transfer the same attitude to the metallurgical, dyeing and gemstone recipes. Though he could create metals, dyes, metallic inks and gemstones through technical processes, the 'creation' may have been aided by the divinity that the practitioner could conjure up through his spells.

The two codices are collections of recipes that occasionally resort to magical ingredients, but one could perhaps extend the interpretation one step further. The texts are not merely recipe collections. They should instead be interpreted as a skeletal version or partial reflection of practice, a practice that involved belief in or at least awareness of the theory that all is one and accepted that the results of the alchemical procedure could also depend on the workings of higher powers.[61] The fact that the alchemical recipes belong to separate manuscripts does not mean that they should not be analysed in tandem, also with the magical text. One could see the collection as forming a parallel to the preserved writings of, for instance, Zosimos of Panopolis. While the magical texts are less direct than what we find preserved in the oeuvre of the famous alchemist, they embody a similar attitude.

ALCHEMICAL PRACTICE IN ROMAN EGYPT

TOBIAS CHURTON

In August 1832, Sweden's Royal Academy of Letters, History and Antiquities wrote to thank Macedonian-born antiquities dealer Ioannis Anastasiou (1765–1860) for his gift of an alchemical codex. Discovered in either Thebes or Memphis, and dated from the mid- to late third century CE, the 'Stockholm papyrus' contained 154 recipes, of which nine concerned metals and 70 dealt with improving and imitating precious stones, while the rest were devoted to dyeing and mordanting cloth – activities shared with the 'Leiden papyrus' purchased from Anastasiou four years earlier by the Dutch government for Leiden's Rijksmuseum van Oudheden. Written in Greek and late Egyptian Demotic, these priceless papyri constitute the world's oldest physical remains of alchemical recipes from Roman Egypt.

Coloured glass-paste hieroglyphs decorating the sarcophagus of Zedthotefankh. From the necropolis of Hermopolis, Tuna el-Gebel, 7th–4th century BCE.

Scholar and diplomat Marcellin Berthelot published his study of the papyri in 1888, asserting that Leiden papyri labelled 'V', 'W' and 'X' showed 'precisely how the alchemical hopes and doctrines on the transmutation of precious metals were born out of the practices of the Egyptian goldsmiths to imitate and falsify them'.[1] That is to say, practitioners of metallurgical chemistry in Roman Egypt were interested in *dyeing* metals and stones, *not* transmuting them into something else. By the sixth century CE, misunderstanding of late antique recipes promoted the fallacy that the art centred on transmutation of ordinary metals into gold and silver.

Leiden's Papyrus V included a process to refine gold (*iosis chrusou*), a preparation for the colouring of gold, and a mystical ink recipe made of green vitriol, gum, oak apple, a blend of seven perfumes and seven flowers, and 'misy': apparently mixed ores of oxidised pyrite and copper and iron sulphates, possibly for writing magic formulae on nitre (potassium nitrate or saltpetre).

Papyrus X contained some 111 recipes, of which no. 8 revealed intentional deception: 'this will be asem [see below] of the first quality, which will deceive even the artisans'.[2]

We also find methods for purification of lead and tin; colouration, augmentation, falsification, testing and polishing of gold; making solder for working gold; doubling or 'diplosis' (increasing the volume of gold by adding another metal that didn't change its appearance); preparing liquid gold; whitening copper; making copper appear like gold; purification, colouring, testing and gilding silver; fixation and falsification of alkanet (a herb whose roots make a red dye); and seven recipes for making greatly valued purple dyes. Quantitative recipes for alloy manufacture are the earliest known, and according to chemist Earle Caley: 'the papyrus is of the highest historical importance chemically in showing the real starting point of the alchemical ideas of the transmutation of metals'.[3] Minerals referenced include arsenic (our orpiment); sandarac (our realgar); misy; cadmia (impure zinc oxide, mixed with copper oxide, or even lead oxide, antimony oxide, arsenic acid, etc.); gold or chrysocolla solder (meaning both an alloy of gold and silver or lead, or malachite and various congeners); sinope rubric (vermilion, or minium, or blood); alum (our alum and various other astringent bodies); natron (nitrum of the ancients, our soda carbonate, sometimes also soda sulphate); cinnabar (our minium and mercury sulphide); and mercury.

Leiden Papyrus V contains instructions for a magic ring incorporating an ouroboros, aimed at attracting favours.

Other important early texts include third-century quotations mistakenly attributed to pre-Socratic philosopher Democritus (*c.* 460–370 BCE). Matteo Martelli dates these fragments of Pseudo-Democritus's now lost *Four Books* to *c.* 54–68 CE, making them the oldest alchemical *texts* (not artefacts) from Roman Egypt.[4] Summary texts from the *Four Books* exist in *Physika kai Mystika* ('Natural and Secret Questions') and *Peri asēmou poiēseō* ('On Making Silver'), both treatises in the

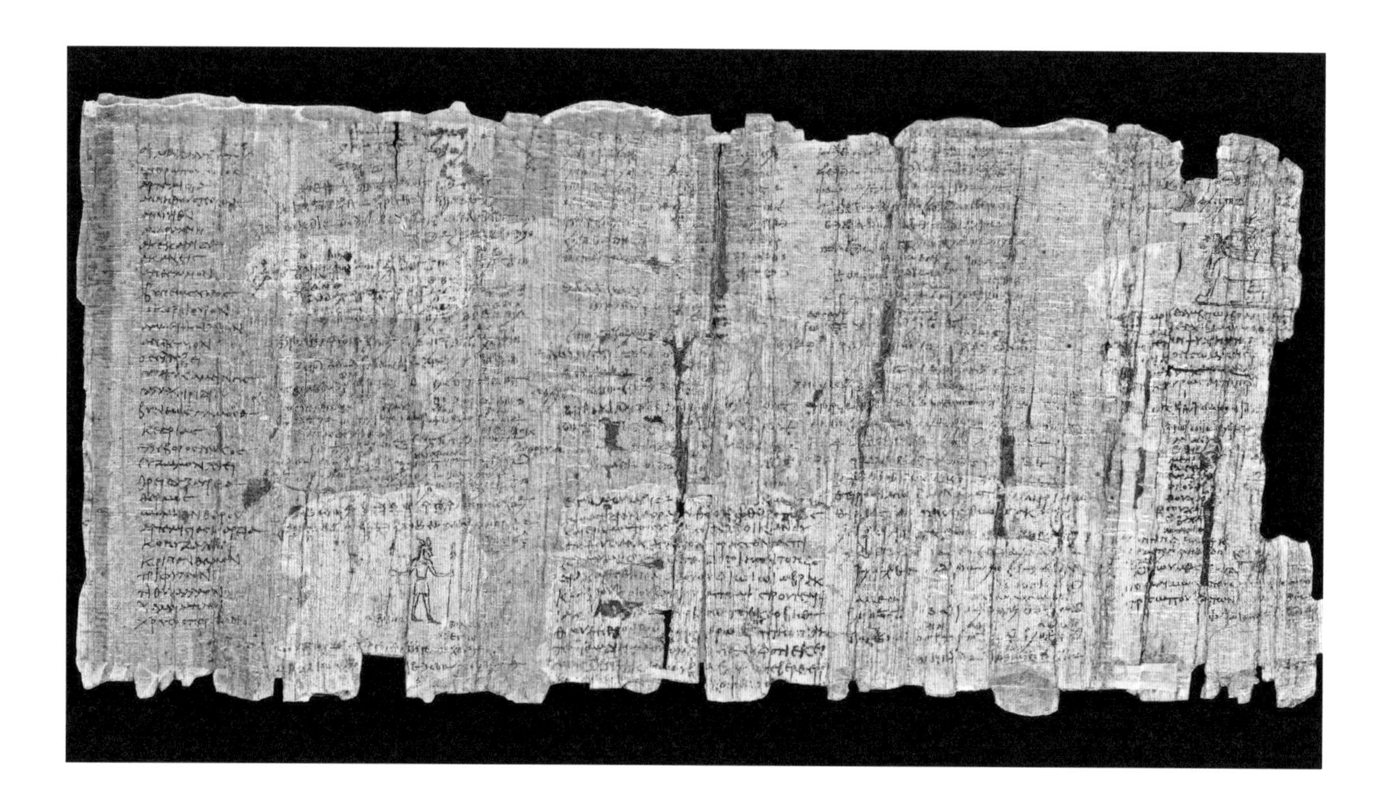

manuscript collection *Marcianus graecus* 299 (tenth–eleventh century; originally Byzantine, now in Venice).[5] Paris holds two significant Greek manuscript collections: the thirteenth-century *Parisinus graecus* 2325[6] and fifteenth-century *Parisinus graecus* 2327.[7] Cambridge[8] and London[9] hold more complete Syriac and *garshuni* manuscripts of Pseudo-Democritus and of alchemist Zosimos of Panopolis (*c.* 300 CE).[10]

Despite Pseudo-Democritus attributing his sagacity to his alleged teacher Ostanes of Persia, the texts cover the same topics of silver and gold (colour) making, artificial stones and purple wool dyeing as the Leiden and Stockholm papyri, suggesting these were the trade's principal concerns.

Recipes in the Stockholm and Leiden papyri show the art's practical nature rather than its symbolism, though instructions for a magic ring attracting glory, power and wealth in Leiden's Papyrus V has a stone depicting a snake biting its own tail – the ouroboros, still regarded as alchemy's comprehensive symbol, showing nature as a unified cyclic system of birth and rebirth, expressing the mystical 'One', while reinforcing a revelation attributed to Ostanes, and oft repeated in Pseudo-Democritus, that 'nature delights in nature, nature conquers nature, nature masters nature'. Papyrus X's recipe no. 90 displays astrological symbols for the sun and moon, representing gold and silver respectively.

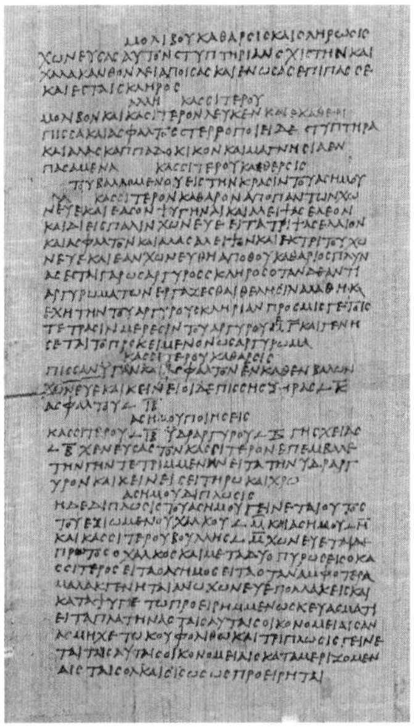

Leiden Papyrus X contains recipe no. 5 with instructions (lines 25–29) for making asem.

Papyrus X also deals with the vital constituent 'asem', meaning silver, alloys of silver and gold (including electrum), or a jeweller's alloy resembling these, appearance being more significant than composition. Perhaps later practitioners saw asem-making as transmutational, for, depending on treatments, asem could provide what was called 'pure silver' or 'pure gold'. Recipe no. 5 gives one asem-making method: 'Tin, 12 drachmas; mercury, 4 drachmas; earth of Chios [a clay], 2 drachmas. To the melted tin, add the crushed earth, then the mercury, stir with an iron, and put in use.'[11]

Recipe no. 89 is significant: the 'invention [preparation] of sulphur water'. The Greek *hydōr theion* can mean either 'sulphur' or 'divine' water. Texts play heavily on this nominal ambiguity. Take a handful of lime and another of fine powdered sulphur place them in a vessel containing strong vinegar or infant's urine. Heat from below until the supernatant liquid appears like blood. Carefully decant the liquid to separate it from the deposit – in fact, it's a preparation for calcium polysulphide.

'Sulphur' water's effect on metals impressed practitioners, producing coloured precipitates of black, yellow and red among other shades, and metal salts and oxides. Polysulphides can dissolve most metal sulphides, colouring metal surfaces with distinctive tints. They can even dissolve gold. All important metal colouring included applied mercury (also used in the alloys), which appeared silvery. Gilding methods included a blend of lead and powdered gold attached to an object before burning off the base metal. Gold amalgam for silver gilding was clearly explained.

Stockholm papyrus recipes include variants for making and doubling silver; tin purification; pearl making and whitening; preparing amethyst, lychnis (ruby red, according to Pliny's *Natural History*), chrysolite, chrysoprase, lapis lazuli, beryl, emerald, verdigris (for emerald), 'green stone' and 'sunstone'; emerald softening;

softening, purification and preservation of crystal; making ruby (from crystal); corrosion of crystal, stone and 'sunstone'; the boiling and opening of stones; crystal bleaching; dissolving comarum (plant); cleaning wool with soap weed; mordanting (to 'fix' dyes to fabric) for Sardian and Sicilian purple, genuine purple (note the word 'purple' could also mean red); dissolving alkanet; cold-dyeing of purple, dark yellow, gold colour ('Take safflower blossom and oxeye, crush them together and lay them in water. Put the wool in and sprinkle with water. Lift the wool out, expose it to the air, and use it'); dyeing in rose colour, scarlet; orchil (red or violet dye from lichens) dyes for Phoenician colour; dissolving orchil and alkanet; and dyeing of 'madder' and what the papyrus heralds as Tyrian 'Guaranteed Superior' purple.

References to crystal generally mean rock crystal or quartz. Surfaces were treated by heating and dipping them in oil, wax or solutions of alum, native soda, common salt and calcium sulphide in varying mixture to make the stones rough, porous and receptive to vegetable dyes, and those of celandine, cedar oil, pitch and numerous resins. The recipes in this and the Leiden papyrus provide our earliest specific dyeing instructions: another boon to science.

Wool fragment dyed with costly Tyrian purple. From Karanis, Egypt. Roman period, 2nd century.

One of numerous gold-making recipes in Pseudo-Democritus instructs the operator to extinguish mercury by alloying it with another metal, or by uniting it with sulphur or with arsenic sulphide, or by putting it beside certain earthy materials. A resultant paste is spread over copper to whiten it. Add electrum or gold powder, and one obtains a coloured metal in gold. A variant method has the copper bleached with arsenic compounds or decomposed cinnabar. As Berthelot observed, this is apparently a process of silver-plating copper before a superficial gilding.[12]

The Codex Marcianus claims Ostanes author of a precious stones recipe using animal gall and copper rust for emeralds and 'yakinthos' (herb) and woad for making blue stones.[13]

The concentration on dyeing perhaps explains why Roman Egypt's alchemy was dominated by women's names: Mariam (Mary), Cleopatra and Theosebeia, among others. Theosebeia's famous mentor Zosimos (*c.* 300 CE) lived some 100 kilometres north of Thebes, at Panopolis (Greek: Chemmis), which city, according to Strabo, was noted for textiles,[14] including dyed wools, suggesting 'alchemists' probably operated at the high end of this and related industry, developing technology that before Rome's conquest had been exclusive to priests.

Purple-robed woman in a painted mummy portrait. From Fayum, Egypt. Trajanic period, 98–117 CE.

The Ptolemaic name Cleopatra ('her father's glory') appears in an anonymous first- or second-century CE text, 'Dialogue of Cleopatra and the Philosophers', in the *Book of Komarios*, where symbolism overlays chemical processes. Ostanes says to Cleopatra, 'In thee is concealed a strange and terrible mystery. Enlighten us, casting your light upon the elements.'[15] Cleopatra sees analogues between plants and distillation; unions twixt bride and bridegroom as nature rejoices in nature; nourishing of ingredients in fire like a mother's womb nourishing the coming child; an apparent consigning to Hades of ingredients that later mount from the flames like the child from its mother. The symbolic-analogical approach would fascinate medieval alchemical enthusiasts, who saw in it an eternal recounting of faith's mystery and key to the spiritual working of the universe.

A Codex Marcianus treatise on asem cites Cleopatra's invention of a vase or crucible: 'After obtaining the asem, if you want to purify it, throw into the crucible of Cleopatra's glass and you will have pure asem.'[16] Inventiveness also distinguishes the most famous female 'alchemist'. Mary the Jewess is credited by Zosimos with the airtight ('hermetically sealed') copper-topped *kērotakis* for vapour collection, the *tribikos* (a three-armed alembic for collecting distillates)[17] and the bain-marie: a double boiler vessel to limit one liquid's temperature to the boiling point of another. Mary also mastered several distillation procedures. Again, the setting seems more a world of elite product industry, especially textiles and decorative or private sacred wares, with lay guild associations, than one of male priests in temples.

Zosimos's treatise *On the Measure of Yellowing* testifies to Mary being 'first' of 'ancient' practitioners:

> Mary (places) in the first line molybdochalchos ['black lead'] and manufacturing (processes). The burning operation (is) what all the ancients advocate Mary, the first, says: 'The copper burned with sulphur, treated with natron oil, and resumed after having several times undergone the same treatment, becomes an excellent gold and without shade. This is what God says: Know all that, according to experience, by burning copper (first), sulphur produces no effect. But when you burn (first) the sulphur, then not only does it make the copper without stain, but again it brings it closer to gold.' Mary, in the description below the figure, proclaims it a second time, and says: 'This was graciously revealed to me by God, namely that copper is first burnt with sulphur, then with the body of magnesia; and one blows until the parts that are left have escaped into the shadows: (then) the copper becomes without shadow.'[18]

According to Zosimos's *On the Apparatus and the Furnaces*,[19] 'Mary the Prophetess', guided by God, also devised constructions for manufacturing divine (sulphur) water and different furnaces, taking advantage of the first century BCE's revolutionary innovation of glass-blowing, whose first dateable evidence (50–40 BCE) included open-ended tubes, discovered in Jerusalem in 1971. Judaean products were possibly acquired by Alexandria's Jewish community for industry, and, arguably, a golden age of guild-supported industrial progress.

Mary famously likened the 'body' (solid) and 'incorporeal' (condensed or vaporous) materials to male and female: 'Join the male with the female and you will find what is sought.' This union is effected by fire, suggestive of passion: 'Do not touch with your hands for it is an igneous preparation.'[20] Thus we see what appeared enigmatic, or mystical, was sound industrial sense.

Three primary factors spurred industrial growth in the late first century BCE: Egypt's absorption into the Roman Empire after 30 BCE (curtailing sacerdotal power), the invention of glass-blowing (creating practical receptacles and tubes, as well as precious stones, and a vast market for decorative objects), and furnace development for the heat-adjustable melting of principal metals over 1,000°C. Glass had properties of chemical neutrality (mercury did not corrode glass as it did other container materials) as well as high temperature endurance.[21] Gem fabrication and colouring already existed. When Zosimos affectionately described Theosebeia as 'purple-robed', one might reasonably speculate she made purple dyes for textiles and was wealthy enough to sport them herself. Insights into sellable objects appear in Book VI of the Syriac Zosimos extracts:

Artisans working gold. From the tomb of Rekhmire, Sheikh Abd el-Qurna, Thebes. *C.* 1504–1425 BCE.

Manufacture of black metal blades, or Corinthian alloy – One uses it for the work of images, or statues that one wants to make black. One also operates on statues, or trees, or birds, or fish, or animals, or on objects that you want.

Copper of Cyprus, a mina [unit of weight: around 100 drachmas]; silver, eight drachmas, that is to say an ounce; gold, eight drachmas. Melt, and after melting, sprinkle with sulphur, twelve drachmas; untreated ammonia salt, twelve drachmas. Take and put in a cleaned vase, placing below the ammonia salt. Then sprinkle over with ammonia salt what was thrown. Allow to cool. Then take, heat and immerse this preparation in vinegar, two half-heminas [*hēmina*: a measure of half a sextary; sextary = approximately an English pint]; lively black vitriol, eight drachmas: all for a copper mina [or wealth of copper]. If you want to operate on more or less, take the preparation in proportion, and let it cool in the ingredient.

Take, roll the metal, but do not roll it to the length of more than two fingers. Then heat, and every time you heat, plunge into the ingredient and remove the filth, so that it gives a radiance.

This copper will retain its blackness when it is ground and reduced to powder; when it is melted, it will also remain saturated with its black colour.[22]

Zosimos explained the making of dyed images of men and women, while expressing views on priestly attitudes,[23] objecting to priestly use of secrets to make 'idols' and castigating their blaming outsiders for using recipes. Zosimos condemned bragging about secrets. Everyone, he says, knows the secrets are in the books of Hermes, but he insists only the spiritually worthy should practise. He strongly objected to a priest, Neilos, to whom Theosebeia had given ear. Neilos spoke of secret oaths and placating daimons; Zosimos dismissed him as an ignoramus, while declaring a Memphis temple furnace unfit for purpose. This also suggests, importantly, that Hermetic tracts represented a perceived liberation of sacerdotal knowledge for use by a cultured *demos*.

Zosimos scholar Shannon L. Grimes attributes 'the best explanation for the emergence of alchemical texts in Roman Egypt' to 'the rise of trade guilds in this era, which disrupted traditional temple economies and created new networks for the exchange of materials, ideas, and techniques'.[24] While insisting the 'sacred art' originated exclusively, even secretly, under priests with metallurgical skills making temple paraphernalia, Grimes sees craft secrets circulating more freely thanks to guilds. However, loss of priestly power to organise Egyptian society, formerly protected by the Ptolemies, followed Octavian's absorption of Egypt after 30 BCE and the Roman imposition of governor Gaius Cornelius Gallus. Christianity's long advance accelerated democratisation. It's unlikely Jewish women like Mary had recourse to pagan priests. Assessing the fragmentary evidence, the setting is almost certainly commercial industry and, hypothetically, its protection of trade secrets through forms of initiation, exclusivity and, quite likely, appropriate ritual. Indeed, such a notion of independent settings would complement a possible *sitz im leben* of the second–third century Hermetic philosophical corpus as well (Zosimos often reminds Theosebeia – whose name means 'God-fearer', suggesting conversion to Judaism – of what they owe to the Hermetic 'race').[25]

A collection of practical alchemical recipes. This medieval manuscript fragment from the Cairo Genizah testifies to the continuity of Jewish artisan guilds working in ancient and medieval Egypt.

Jewish philosopher Philo of Alexandria observed Jewish artisans in guilds for goldsmiths, silversmiths, weavers, coppersmiths and blacksmiths.[26] In the second century, Tanna rabbi Judah, visiting the city, noted five guilds occupying their own seats in the synagogue basilica.[27] Other than training apprentices, the guilds provided social services for members, including help for families in adversity. 'Alchemy' would have found a welcome home amid these and kindred structures. Zosimos strongly favoured Jewish culture. The ingenuity brought to the chemical processes of the 'holy art' makes it plain that transmutation to authentic gold was *not* Graeco-Egyptian alchemy's interest. Besides, as Roman historian Pliny observed: 'gold, for which all mankind has so mad a passion, comes scarcely tenth in the list of valuables, while silver, with which we purchase gold, is almost as low as twentieth'.[28] Our knowledge is all practical, not theoretical. No recipe supports the still widely repeated view that alchemy based itself on the existence of a kind of proto-metal or substrate common to all metals that, if acted upon by a secret ingredient, enabled protean transformation. While some practitioners doubtless saw the art as integral to a universe of daimonic causation and sanction, Zosimos exerted himself to keep the practice on the right side of the good angels, in tune with the highest mind (*nous*) and dependent on correct empirical knowledge of natural processes.

It is likely the Greek designate for the art, *chēmeia*, is derived from Akkadian, Egyptian (*kam* or *chem*) and Hebrew (*cham*) words denoting cognate ideas of 'roasting', 'black', and 'heat'. *Chēmeia* (origin of 'chemistry') would then mean *the art of heat*, involving flame, burning, roasting, blackening and sometimes evaporating to bring forth colour, value and appearance of life. Rising as vapour toward a startling transformation is arguably a visual analogue for transmigration of soul. Hot spirits rise, or fall, by attraction.

Mary used spiritual analogues for chemical processes, naming 'fixed and not fleeting things' *bodies*, saying 'the body of the magnesia is the secret thing that comes from lead, from the summer stone and from the copper'.[29] The 'body' comes from lead *to* magnesia, which is not otherwise a 'body', for the treatise informs us that magnesia, pyrite, mercury and chrysocolla (a blue-green crystal with high copper content) and their likes are 'incorporeals'. The *bodies* are copper, iron, tin and lead; they don't evaporate in the fire. However, when a body is mixed with an incorporeal, according to Mary, the bodies become incorporeal, and the incorporeals body. In a corresponding axiom, she declares: 'If two don't become one; that is to say if the volatile (materials) don't combine with the fixed materials, nothing will happen of what is expected.'[30] It is vital that the bodies become incorporeal by becoming one with the incorporeals, so that, 'reduced in spirit', they can rise as sublimated vapour and become bodies again.[31] This of course denotes the distillation process which so impressed Zosimos, and for which blown glass was such a boon.

The First Book of Zosimos the Theban's Final Account opens by reminding Theosebeia that all the realm of Egypt depends on two arts: suitable dyes, and minerals – confirmation that we are looking not into a sacerdotal setting but an industrial one wherein *technai* could exploit the speedy volatilisation of some minerals. Mercury's low boiling point helped to extract the metal from cinnabar; Pliny described cinnabar being put in an iron shell in flat earthenware pans before being 'covered with a convex lid smeared on with clay', whereafter 'a fire is lit under the pans and kept constantly burning by means of bellows, and so the surface moisture (with the colour of silver and fluidity of water) which forms on the lid is wiped off it. This moisture is also easily divided into drops and rains down freely with slippery fluidity.'[32] The process would have required two vessels, a lower one (the *lōpas*) and an upper *ambix* (Arabic: 'al-anbiq'; our 'alembic') which was widest at the middle with the upper part tapered off like a cone to a narrower aperture, from which condensed mercury could be collected.

Matteo Martelli and a team at Bologna University put three ancient types of hot extraction of mercury from cinnabar to an efficiency test: simple heating of cinnabar; heating it in a sealed vessel with iron (a procedure of Dioscorides's); and heating it with 'nitron oil', according to Pseudo-Democritus.[33] The overall conclusion showed that details in ancient recipes were highly significant and sometimes revelatory, contradicting assumptions. For example, an iron shell was not simply a handy material for containment, but was determinative in chemical results; likewise copper in cold extraction – initially just the pestle's metal. Nitron oil aided extraction yield. Furthermore, the choice of nitron itself may have been as cultural a stimulus as a technical one. Nitron had usage in sacred settings, such as nitron balls used to clean mummies, and of course its association with purity in embalming the dead and as a health aid. Investigating *what* alchemists did became for the team an analytical exercise into how and why they did it.[34]

We may presume it was an art pursued for profit, though examining Roman Egypt's greatest alchemical legacy might suggest profit was means rather than end. The name Zosimos means 'survivor', and Zosimos's reputation persists after 1,700 years. When alchemy is said to require spiritual piety, whether by Byzantines or Renaissance sages, we hear the eager genius of Panopolis warning against false paths.

Was Zosimos Egyptian? His treatise *On Apparatus and Furnaces* refers to 'the first man is called Thoth among us', meaning Egyptian tradition, while among 'them' – the Jews – the first man is called Adam. Zosimos most respects 'Hebrew' monotheism and technology. Apparently influenced by a syncretic form of Christian gnosis, he is best known as a devotee to the spiritual and philosophical tractates attributed to Hermes Trismegistos. His thought is wedded to them, making him, in the words of Hermetic scholar A. J. Festugière, the 'father of religious alchemy': first to portray alchemy as a dual-purpose technique, purifying metals and purifying human souls.[35]

Despite Hermes's reputation for religious magic, Zosimos insists solely on natural (*physikos*) tinctures. Only the earth-body-bound pervert the art with daimonic supplication. Zosimos summons practitioners to identify with the divine dye that raises soul. The 'metal' (body) must be willing to suffer the process – a doctrine permeating his startling series of dream narratives: contrived teaching pieces, taken by Carl Jung for real dreams.[36]

Shannon Grimes favours a picture of Zosimos as Egyptian priest, overseer of craft work associated with temple worship.[37] However, when describing priest

Neilos's interests in Egyptian magic and temple idols, Zosimos is scornful. His *The Final Account* (or 'Quittance') contains a mighty polemic against daimonic deception through temple idolatry and 'unnatural tinctures' reliant on daimonic propitiation.[38] Referring to the Jewish 'Book of the Watchers' (part of *I Enoch*) as 'scripture', Zosimos traces the art's origin to sinful angels ('daimons') who left heaven with knowledge taught to unworthy humans. Its source, however, remains heavenly – hence the need for heavenly minds to work it; for daimons (*hoi kata topon ephoroi* = 'local overseers') lurk on earth to mislead.[39] Dealing with such daimons makes for slaves, he warns Theosebeia, declaring statue worship a bribe exacted from untrustworthy priests.[40] Zosimos's title, *The Final Account*, possibly suggests adherence to Enochic doctrine that the time when daimons would be finally judged and destroyed was nigh. Zosimos also castigates priests for hiding alchemical secrets in temples. Instead of realising the divine bounty of the whole of nature and natural processes, would-be adepts were compelled to resort to a magical system corrupting, enslaving and vain. Such prescience is arguably redolent of seventeenth-century scientific mentalities. Zosimos's revelation apparently came through Jewish prophecy, Christian liberation from daimonic causation, and from a vision of the divine creation in the works of Hermes.

The Ibis-headed god Thoth, known to the Greeks as Hermes Trismegistus, associated with divine knowledge and writing. From the tomb of Ramses V, Valley of the Kings, Luxor.

Zosimos's syncretic, Gnostic eclecticism suffused his 'alchemy', and would subsequently wield great influence on alchemy's development. In attempting to guide Theosebeia into producing the right dyes profitably, Zosimos revealed his beliefs, beliefs that he really saw reflected in chemical processes. The *practical* essence of this simultaneous two-dimension or two-world practice seems to come from devotion to the works of Mary, notably in her distinguishing 'bodies' from 'spirits', while applying these conceptions to developmental processes, particularly distillation. One can see what it was in Mary's writings that especially appealed to Zosimos; she learned from God – fate or destiny only dominating those bound to the corporeal.

Due to daimonic temptations, most human beings live under fate's dominion, and *that*, according to Zosimos, is the position of Theosebeia's wayward advisors, and it is one he determined to identify, isolate and free her from. Employing a Gnostic conception, Zosimos called for Theosebeia's spiritual liberation from fate and the earthly Adam. The result would, he believed, be a purified and truly holy art, performed without secrecy or guilt.

It is not uncommon today to read of alchemy being a historic transmitter of Gnostic ideas. According to Berthelot, 'the history of magic and Gnosticism is closely linked to that of the origins of alchemy: the current texts provide new evidence in this regard in support of what we already knew'.[41] There are for sure intriguing parallels of language between alchemical tracts from Roman Egypt and Gnostic texts from Egypt and Syria in late antiquity. Perhaps *not* authentically Zosimian, *Authentic Memoirs on Divine Water* are striking for their extravagant claims for the divine mercury:

> This is the divine and great mystery; the object we seek. This is the All. From him [comes] the All, and through him (exists) the All. Two natures, one sole essence [Greek: *duo phuseis, mia ousia*]; for one attracts one; and one dominates one. This is silver water, [according to Berthelot: philosopher's mercury and ordinary mercury] the hermaphrodite, which always flees, which is attracted to its own elements. It is the divine water, that everyone has ignored, whose nature is difficult to contemplate; for it is neither a metal, nor water always in motion, nor a (metallic) body; it is not dominated.[42]

Cross-cultural worship. Priests and worshippers paying homage to the Egyptian god Isis. Fresco from Pompeii, *c.* 60–80 CE.

The second sentence is reminiscent of logion 77, Gospel of Thomas: 'Jesus said, "*It is I who am the all. From me did the all come forth, and unto me did the all extend.* Split a piece of wood, and I am there. Lift up the stone, and you will find me there."'[43] (my italics). The Christ-Mercurius-Lapis (Stone) identification emerges as a significant motif in late medieval and Renaissance alchemy; it is perhaps lodged in early alchemical tradition, or inferred from it.

The memoirs continue: 'It is the All in all things; it has life and spirit and is destructive. He who understands this possesses gold and silver. The power has been hidden, but it is recovered in erotylos [Greek: *anachaitai de tōi erōtulōi*].' That there's something 'destructive' in the 'All' strikes a curious chord with Thomas Lambdin's translation of saying 70 from the Gospel of Thomas: 'If you bring forth what is within you, what you bring forth will save you. If you do not bring forth what is within you, what you do not bring forth will destroy you.' As for *erotylos* ('love stone'?), Democritus praised this gemstone in Pliny's *Natural History* for divination purposes. It was perhaps a stone from which mercury could be extracted, possibly by heating a sulphide ore, with air combining with the sulphur to form sulphur dioxide and mercury liberated at a temperature above its boiling point (cf.: 'Lift up the stone, and you will find me there').

Zosimos believed the original man (*Phōs* = 'light') was not subjected to daimonic fate and applied the principle to alchemical processes and to freeing Theosebeia from daimonic advice, using a doctrine employed by Sethian Gnostics:

> He [the Son of God] appeared to men deprived of all power, having become man (himself), subject to suffering and beatings. (However), having secretly stripped his own mortal character, he felt (in reality) no suffering; and he had seemed to trample on death, and to push it back, for the present and until the end of the world: all this in secret. Thus stripped of appearances, he advised his followers also secretly to exchange their mind (or spirit) with that of the Adam which they had in them, to beat him and to put him to death, this blind man being led to compete with the spiritual and luminous man: it's thus that they kill their own Adam.[44]

– a conception closely akin to the account of the Passion in the Coptic Second Treatise of the Great Seth (second or third century CE), where Jesus says he 'visited a bodily dwelling'. *That* summarised his relationship with the corporeal: *just passing*

through, 'For my [Jesus's] death, which they think happened, [happened] to them in their error and blindness, since they nailed their man unto their death.'[45]

Sayings 22 and 23 of the Gospel of Thomas are reminiscent of Mary the Prophetess's distinction of bodies and incorporeals as male and female: 'Jesus said to them, "When you make the two one, and when you make the inside like the outside and the outside like the inside, and the above like the below, and when you make the male and the female one and the same, so that the male not be male nor the female...then will you enter the kingdom."'[46] As we may recall from page 55, Mary declares: 'If two don't become one; that is to say if the volatile (materials) don't combine with the fixed materials, nothing will happen of what is expected.'[47] This arguably suggests an analogy between the practical sphere of operation and the blue-sky Gnostic message. Indeed, making 'the inside like the outside' is the literal meaning of 'transmutation' in the treatise *Synesius and Dioscorus*, where Dioscorus asks Synesius: 'What transformation is he [Democritus] talking about?' Synesius replies that Democritus speaks of the 'bodies' – that is, metals – and in answer to a subsequent question about how to 'turn the nature inside out', Synesius says it's necessary to 'transform their nature, for nature has been hidden within'.[48] The Greek verb translated as 'transform' ([*ek*-]*strephō*) means to 'turn inside out', exposing something formerly within by getting inside matter. The sense in this passage is of a *return* to a former, if secreted, state. *Mercury* then symbolises *pneuma* (spirit), being elusively liquid and metallic and secreted in bodies.

The apocryphal Gospel of Philip's Gnostic message even calls on the divine analogy of tinctures as salvific metaphor *and* divine identity: 'God is a dyer. As the good dyes, which are called 'true', dissolve with the things dyed in them, so it is with those whom God has dyed. Since his dyes are immortal, they become immortal by means of his colours.'[49] Later we learn that 'Levi' [possibly referring to Jewish priesthood] owned a *dye works*: 'The Lord went into the dye works of Levi. He took seventy-two different colours and threw them into the vat. He took them out all white. And he said, "Even so has the Son of Man come as a dyer."'[50]

Dyeing, to the Gnostic author, is transformation. Later in the text we find the unity theme recapitulated in a statement echoing the Hermetic 'As above, so below' type: '[The Lord] said, "I came to make [the things below] like the things [above, and the things] outside like those [inside. I came to unite] them in that place."'[51] Is Gnostic symbolism alchemical, or alchemical symbolism Gnostic? In the above instances, they seem curiously united.

An unanticipated side effect of such cross-symbolism was detrimental to the 'sacred and divine art' (Greek: *hē hiera kai theia technē*), for transformation too easily passed into miraculous transmutation, with gold representing the highest spiritual attainment. This scientifically retrograde emphasis may stem from a rarity of texts following an event recounted in the tenth-century Byzantine lexicon *Suda*, where we find *chēmeia* defined as

The preparation of silver and gold...due to the Egyptians' rebellious behaviour Diocletian [emperor, 284–305 CE] ... [sought] out the books written by the ancients concerning the alchemy of gold and silver, he burned them so that the Egyptians would no longer have wealth from such a technique, nor would their surfeit of money in the future embolden them against the Romans.[52]

What seems to have occurred is that after Zosimos's time, study of surviving texts fell largely into the hands of writers wishing to garner for themselves the respect accorded to Alexandrian Neoplatonic philosophers such as Synesius (late third/ early fourth century), Olympiodorus (late fifth/sixth century), and Stephanus of Alexandria (late sixth/early seventh century). Alchemical works attributed to Synesius and Olympiodorus were probably not written by the philosophers of those names. Stephanus of Alexandria, however, is almost certainly author of *On the Great and Sacred Art of Making Gold*, a commentary on earlier texts. His works occupy a large part of the Codex Marcianus. A process of Neoplatonising the mystical potential of the tradition, combined with Christian and Aristotelian philosophical interpretation of the texts, was likely to have been responsible for the generation of, and fixation on, the 'philosophers'-stone-to-transform-lead-into-gold' idea. Transformation mutated into transmutation as the textual tradition lost contact with dyeing industry.

In fact, as a term for the transmutational agent, the 'philosophers' stone' does not appear until the seventh century, while our word 'alchemy' comes from the Arabic for *chēmeia*: 'al-kīmīyā". Medieval Latin translations of post-seventh-century Arabic 'alchemy' texts conflated the errors and disseminated a false idea of metallurgical chemistry in Roman Egypt from those times to our own.

ALCHEMICAL EVOLUTION IN MEDIEVAL CULTURES

FROM ALEXANDRIA TO BYZANTIUM

The Systematisation of Alchemical Tradition

CRISTINA VIANO

Byzantine Egypt and the period of the commentators[1]

The Byzantine period of Egypt begins at the death of emperor Theodosius I in 395 CE, when the province of Aegyptus came under the Eastern Roman Empire. It ends under the reign of Heraclius, with the Arab conquest in 640 CE. Byzantine Egypt experienced a period of peace, which extends from the fifth to the beginning of the seventh century, during which Alexandria is at the centre of intense intellectual and spiritual activity. Philosophical and scientific debates continue to flourish, and lively doctrinal disputes arise around the tenets of Christianity, which intersect with the doctrines of Gnosticism and Hermetism.

In this bustling atmosphere, Greek alchemy experiences a crucial moment in its development: in that period the doctrines and operations, and the conceptual tools for thinking, that will be the basis for all subsequent periods are developed and defined.

This period is characterised by a generation of 'commentators' tied to the Neoplatonic milieu, like Synesius (fourth century CE), Olympiodorus (sixth century CE) and Stephanus (seventh century CE). Their writings, designed primarily to clarify the thinking of the great figures of previous generations, including Democritus and Zosimos, represent the most advanced stage of ancient alchemical theory.

We witness a genuine process of defining and systematising alchemical doctrine through the intellectual tools of philosophy available to these authors. This process, already begun by previous commentators, now finds its full realisation. From this perspective, through the systematic search for causes, *historia* of the recipes is integrated through *theōria*. Indeed, these authors, seeking to develop the links between theory and practice, between nature and *technē*, between the doctrine of transmutation, philosophical theories of matter on one hand and technical processes on the other, laid the basis for a reflection on the possibility and on the nature of alchemy as an autonomous knowledge.

Hermes Trismegistus teaching Ptolemy the World System. Byzantine silver plate with relief from the Eastern Mediterranean, *c.* 500–600.

Previous vignette: The Alchemist, from 'The Working World' cycle after Giotto. Fresco by Nicolo and Stefano da Ferrara Miretto, c. 1450.

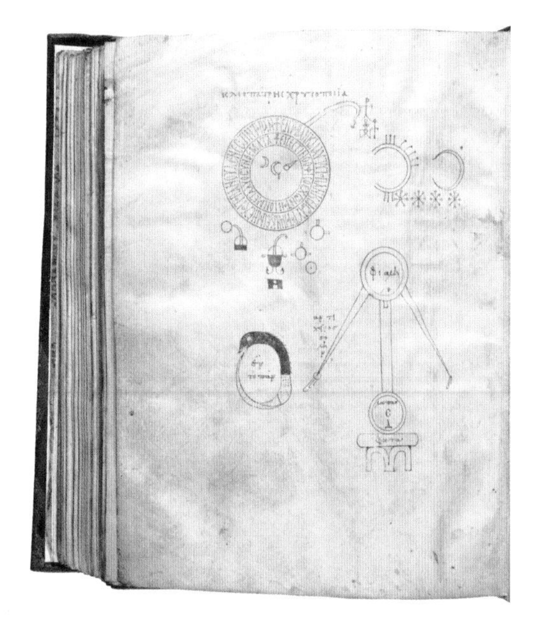
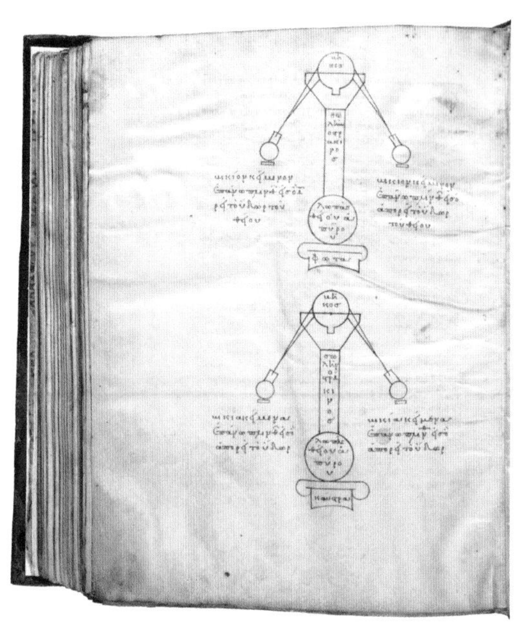

From manuscript Marcianus 299: Cleopatra's Chrysopoeia, a diagram showing alchemical tools and one of the earliest depictions of the ouroboros (left); alembics (right). 9th–10th century.

It was also during that period, around the seventh century, that the corpus of alchemical texts began to be assembled in its very particular form of an anthology, essentially of extracts, as found in a large number of manuscripts. Of these, the oldest and most beautiful is the Marcianus Graecus 299 (M) (tenth to eleventh century), brought back from Byzantium by Cardinal Bessarion in the fifteenth century and currently kept at the Library of St Mark in Venice.

Finally, it is in the fifth century that authors external to alchemy begin explicitly to speak of alchemy as a contemporary practice to produce gold from other metals. Proclus (fifth century CE) compares astronomers who make astronomical tables to 'those who claim to produce gold by the mixture of certain species (of metals)'.[2] Aeneas of Gaza (fifth to sixth centuries CE), Christian philosopher and orator, pupil of the Neoplatonist Hierocles, talks about the possibility of improving the material of bodies by changing their form and offers the example of those who produce gold by melting together and dyeing silver and tin.[3] Afterwards, in the Byzantine and the Islamic worlds, the reflections of those interested in the transformation of matter will multiply, and the possibility of alchemy will be a real object of philosophical debate.

Here it is proposed to develop a picture of the most characteristic aspects of the alchemy of that period, starting with the specific contributions of its most representative protagonists. I seek to answer two closely related questions, which are essential for identifying and understanding this complex and paradoxical knowledge, which will not even get a proper name until a relatively late period. Indeed, the Greek term *chēmeia* is found in Stephanus in the seventh century, and the Latin term *alchimia*, an Arabic derivation, appears only in the Western world in the twelfth century.

The first question is essentially internal to the texts: how did the alchemical authors view their knowledge? We seek to understand, through the methodological reflections of the authors, how they defined, and what epistemological status they attributed to, their field.

The second question is external and concerns our epistemological approach to this knowledge: how should we study the alchemical texts? Can one sketch the rules of a proper approach that can at once take account of the multiple facets and also of the unique specificity of this cultural phenomenon we call Graeco-Alexandrian alchemy?

The protagonists and the question of pseudepigraphy

Greek alchemical literature is usually divided into three periods. The first is located between the first and third centuries CE. It includes the chemical recipes of the *Physika kai Mystika* attributed to a 'Democritus' (first to second centuries CE) and the anonymous papyri of Leiden and Stockholm (third century CE). These recipes focus on the imitation of gold, silver, precious stones and purple. The model of production of gold seems to be that of an imitation through colouring, which acts on the external properties of bodies. This notion of imitation is the crux of the old conception of the art and, as we shall see, contains in embryo the idea of transmutation. At this stage we also see reported a series of short quotes or treatises of the mythical 'old authors' such as Hermes, Agathodaimon, Isis, Cleopatra, Mary the Jewess, Ostanes, Pammenes and Pibechius (between the first and third centuries CE).

The second period is that of alchemical authors properly so called: Zosimos, Pelagios and Iamblichus (third to fourth century CE). Zosimos appears as the greatest figure of Graeco-Egyptian alchemy. Coming from Panopolis in Egypt, he perhaps lived in Alexandria around 300 CE. From his work, we have fragments gathered in four groups: the *Authentic Memoirs*, the *Chapters to Eusebia*, the *Chapters to Theodore* and the *Final Account*, with two excerpts from the *Book of Sophē*. Among the most famous pieces are the three *Visions*, which are part of the *Authentic Memoirs*; these describe dreams that unveiled to Zosimos the properties of metals. Metal processing operations are accompanied by a ritualisation of the symbols of death and of resurrection, and of purifying the mind of matter. Indeed, the concept of metals is often paralleled in Zosimos with the concept, inspired by Gnostic and Hermetic thought, of the double nature of humans, composed of body and spirit.

Finally, the third and final period is precisely the one that interests us: that of the commentators. The most important are Synesius (fourth century CE), Olympiodorus (sixth century) and Stephanus (seventh century). Close to Stephanus are four poems transmitted under the names of Heliodorus, Theophrastus, Hierotheus, and Archelaus. Later, perhaps between the sixth and eighth centuries, two anonymous commentators, commonly called the Christian Philosopher and the Anonymous Philosopher, lead directly to the period of the most extensive compilation of the main manuscript of the collection, the Marcianus Graecus 299. Indeed, it is assumed that this anthology was compiled in Byzantium in the seventh century, during the reign of Emperor Heraclius, to whom the index of the manuscript attributes three alchemical works. The compiler was probably a certain Theodore, who wrote the verse preface to the manuscript which is found at the beginning of this and who was probably a pupil of Stephanus. Thereafter, the alchemical tradition in Byzantium continues with Michael Psellus (eleventh century), Nikephoros Blemmydes (thirteenth century) and Cosmas (fifteenth century). Among the alchemical works in Greek, one must also mention the *Anonymous of Zuretti*, a Byzantine treatise produced in Calabria in 1378 and compiled from Latin sources.

The issue of identification of the commentators Olympiodorus and Stephanus with their namesakes the Neoplatonic commentators was raised very early by historians of alchemy and has made much ink flow. In fact, pseudepigraphy is a frequent phenomenon. In the alchemical literature, we can find Plato, Aristotle, Democritus and Theophrastus mentioned among the alchemical authors. From a chronological point of view, however, Olympiodorus and Stephanus constitute the borderline between these obviously false attributions and authentic attributions to known characters, such as Psellus.

In the corpus of Greek alchemists these two authors are defined as 'the masters famous everywhere and worldwide, the new exegetes of Plato and Aristotle'.[4] And there is good reason to attribute the writings of Olympiodorus and Stephanus, at least in their original versions, to their Neoplatonist namesakes. Indeed, the latest studies are turning more and more towards the hypothesis of identity, but for

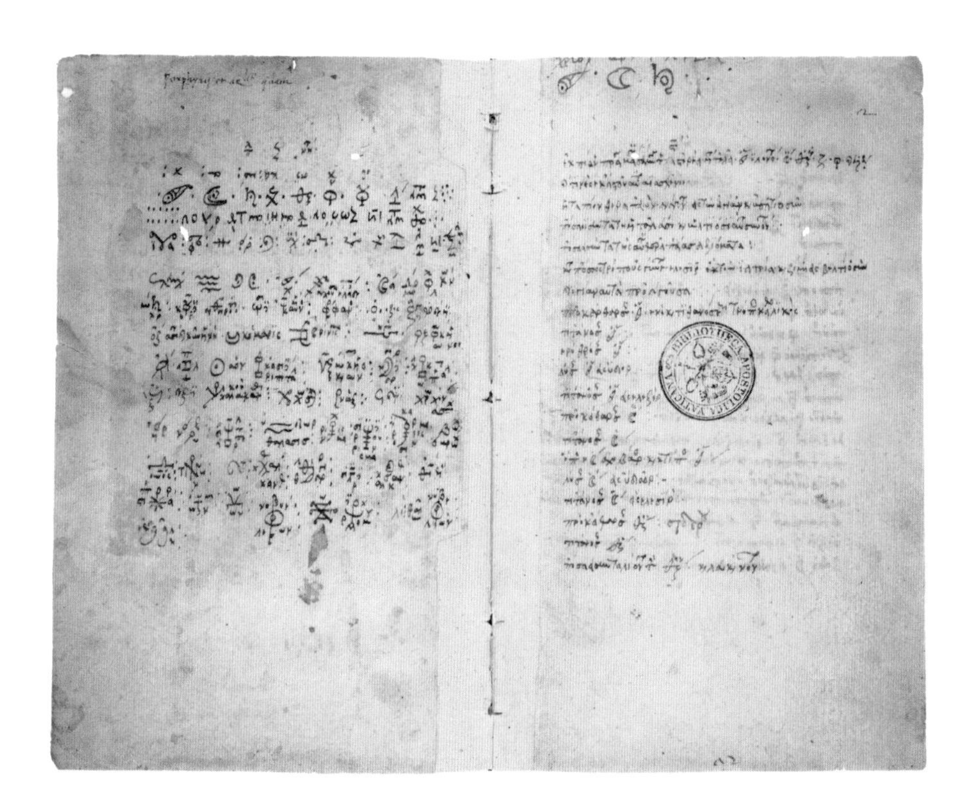

Opening pages of the *Anonymous of Zuretti*, a Byzantine treatise on alchemy. Calabria, 1378.

Emperor Heraclius Bearing the Holy Cross. Oil on panel by Cristóvão de Figueiredo, 1522–30.

Olympiodorus, because of the especially composite and discontinuous form of his work, the question of attribution is more complex and delicate than in the case of Stephanus, who offers a more homogeneous collection of treatises. As we shall see, the commentary of Olympiodorus the alchemist is an exemplary product of the alchemical literature.

Synesius

Synesius is the author of a commentary on the *Physika kai Mystika* of 'Democritus', in the form of a dialogue entitled *Synesius to Dioscorus, Commentary on the Book of Democritus*.[5] Synesius is unknown to Zosimos but cited by Olympiodorus, who inserts long sections of Synesius into his commentary *On the Kat'energeian of Zosimus*. Dioscorus was, as indicated by Synesius, a priest of Serapis in Alexandria. Synesius has been identified with the homonymous Christian bishop of Cyrene, a Neoplatonist and student of Hypatia, but the dedication to Dioscorus, pagan priest, makes this argument difficult to sustain. In addition, this dedication shows that the work of Synesius pre-dates the destruction of the Alexandrian Serapeion (391 CE).

The exegesis of Synesius relies on both practical explanations (for example, 'the dissolution of metallic bodies' means bringing metals to the liquid state)[6] and on general principles (for example, the statement of the principle that liquids derive from solids, relative to colouring principles provided by dissolution, called 'flowers').[7]

As in most of the texts of that period, the object of the research is identified with agents of transformations of matter.[8] The cause of the transformation is an active principle, called 'divine water' (or 'sulphur water'), mercury, *chrysocolla* or raw sulphur and acts by dissolution. Mercury is at once the dyeing agent and the prime metallic matter, understood as the common substrate of the transformations and the principle of liquidity.[9]

One can detect, in the explanations of the general principles of the transformation of metals, the strong influence of Aristotelian terminology. First, the object of the research is identified as an efficient cause. Then, the fabrication of metals is conceived as a mixture (*mixis*), especially among liquids (which according to Aristotle is the optimal condition);[10] the preliminary condition is that of dissolution, which in Aristotle represents the culmination of the separation of compounds, thus of mixtures.[11] The transformation is conceived as a change of specific quality, generally through colour. Mercury is compared to the material worked by the artisan,[12] who can change only the form. The distinction between potential and activity is applied to the colouring activity of mercury: 'in activity it remains white, in potential it becomes yellow'.[13]

Along with other authors of this period, Synesius presents a natural conception of alchemy: it is always nature that, ultimately, is the true principal agent of the operations. The task of the artisan is to create the conditions so that the active properties, buried in the substances, become operative and act on the substances themselves by virtue of their affinity.

Olympiodorus

Olympiodorus is one of the most interesting authors of the alchemical corpus. The question of attributing the commentary *On the Kat'energeian* (*On Action* or *According to the Action*) *of Zosimus* to his namesake touches on two issues vital to the understanding of Graeco-Alexandrian alchemy: the constitution of treatises in the corpus and the interest of Neoplatonist exegesis on Aristotle in alchemy. For this reason, it is worthwhile to devote more detailed analysis to him.

The work of Neoplatonic commentator Olympiodorus is a rich source of information on cultural conditions and educational methods of Alexandria in the sixth century. In particular, his commentary to the *Meteorologica* is an extremely interesting work for the history of science. Olympiodorus completes and fixes the Aristotelian classification of meteorological and chemical phenomena, a tremendous task of systematising notions sometimes barely sketched by Aristotle, like that of 'chemical analysis' (*diagnosis*) of homogeneous bodies in Book IV.[14] Olympiodorus's commentary on Book IV of the *Meteorologica*, the first 'chemical' treatise of antiquity, is the most widely used, not only by Arabic and Renaissance authors but also by Greek and medieval alchemists.

It is therefore not surprising that there has survived under the name of Olympiodorus one of the most 'philosophical' writings of the corpus of Greek alchemists, which presents itself as the commentary on a (lost) treatise of Zosimos and on the sayings of other ancient alchemists.[15] In the principal manuscript of the corpus, the Marcianus Graecus 299 (M), the treatise has the title 'Olympiodorus, philosopher of Alexandria, *On action* (or *According to the action*) of Zosimos [and] everything that was said by Hermes and the philosophers'.

The author deliberately presents his commentary as a work at once exegetical and doxographical. He explicitly claims that Greek philosophy, including pre-Socratic philosophy, is the epistemological basis of transmutation. Indeed, near the middle of the commentary,[16] Olympiodorus sets out the opinions of nine pre-Socratic philosophers (Melissus, Parmenides, Thales, Diogenes, Heraclitus, Hippasus, Xenophanes, Anaximenes and Anaximander) on the sole principle of things and then establishes a comparison between these theses and those of the principal masters of the alchemical art (Zosimos, Chymes, Agathodaimon and Hermes) on the efficient principle of transmutation, designated as 'divine water' (*theion hudōr*).

Like most texts of the corpus of Greek alchemists, the commentary of Olympiodorus presents a composite and seemingly unstructured nature. It has neither preface nor conclusion: it begins and ends abruptly.

One can divide the text into two sections. Only the first[17] presents a coherent structure: the author begins by commenting on a saying of Zosimos on the operation to extract gold flakes from ore, through 'maceration' (*taricheia*) and 'washing' (*plusis*).[18] These follow the typical schema of Olympiodorus the Neoplatonic commentator: first the lemma, the phrase of Zosimos to explicate, then a general explanation (*theōria*) and after that the detailed exegesis of terms (*lexis*). Then he introduces gold 'soldering' (*chrysocolla*),[19] which consists of collecting the gold particles obtained into a homogeneous body. These two specific operations, separation and reunion, are here interpreted as allegories of the transmutation of metals. The three types of dyeing of the ancient alchemists come next:[20] one that dissipates, one that dissipates slowly and one that does not dissipate. The third attributes to metals an indelible nature. This means, in operative terms, to fix the colour of a metal permanently.

The second section – the more substantial part of the text[21] – consists of a suite of unstructured *excerpta* and digressions, accompanied by notes on the main alchemical operations. In particular, the above-mentioned comparison between the principles of pre-Socratic philosophers and those of the principal masters of the alchemical art is found in this section.[22]

Despite this apparent disorder, we can detect a rational and coherent design in the way the treatise unfolds, which reveals itself in two factors. The first is the logical sequence that ties together the alchemical operations, the principles and fundamental substances. This sequence demonstrates a development through the presentation of the constituent elements of alchemy, from the fundamental operations (grinding, melting, dyeing) to its active and material principles, ending with epistemological considerations on this discipline as a *technê*.

The second factor is the use of phrases that one might call 'connective and associative'. The author speaks in the first person and marks the transitions between different sections as well as the aim, method and internal organisation of his text. His work thus turns out to be an *epitomê*, a summary with protreptic intent, providing a selection of testimonies, with commentary, taken from the writings of the ancient alchemists and also of philosophers properly speaking, on the basic elements of the art (operations, ingredients and also its history). It seems to be addressed to a young man of high rank with the aim of providing him with 'a comprehensive view of the complete art'.[23]

This suggests that the text is based on a work by Olympiodorus, now lost, composed in a more structured way. The text we possess would be composed of at least two layers: Olympiodorus's commentary on Zosimos's *On Action* and the arrangement of a compiler. The compiler copied Olympiodorus up to a point and then added a sequence of remarks on the main alchemical operations, accompanied by *excerpta* from Zosimos and other alchemical authors, organising the whole work by the criteria mentioned above.

Presumably, the opening sections as well as a sizeable part of the doxography on the pre-Socratics come directly from the commentary by Olympiodorus the Neoplatonist. It is also quite plausible that Zosimos's *On Action* itself was already a doxographical work as far as the opinions of the alchemists are concerned, and that Olympiodorus's commentary added a doxography on the pre-Socratics, which is structured according to the characteristic scheme of Neoplatonic commentaries.

The issue of pseudepigraphy among Greek alchemists thus re-joins that of the place of alchemy with respect to the official philosophical knowledge of its period. We will return below to why Olympiodorus the commentator might have been interested in alchemy.

Stephanus

Stephanus is the author of nine *Praxeis* (lessons) on the divine and sacred art, as well as a letter to Theodore.[24] Lesson 9 is addressed to the Emperor Heraclius and therefore can be dated in the years of his rule (610–41 CE). Some astronomical data in his work would moreover enable us to date it to exactly 617 CE.

We saw that in the corpus of Greek alchemists, Stephanus is mentioned, with Olympiodorus, as among 'the masters famous everywhere and worldwide, new exegetes of Plato and Aristotle'. Indeed, the Emperor Heraclius appointed him 'worldwide professor', i.e. professor of the imperial school of Constantinople. The current scholarly trend is to consider this Stephanus of Alexandria as identical to the Neoplatonic commentator on Plato and Aristotle, author of a commentary on the *De interpretatione* and one on the third book of the *De anima*, and to Stephanus of Athens, commentator on Hippocrates.

In his alchemical work, Stephanus comments in a very rhetorical style on the ancient alchemists and connects alchemy to medicine, astrology, mathematics and music. He declares alchemy compatible with Christianity and defines it as 'mystical' knowledge, woven into in a cosmology based on the principles of unity and of universal sympathy. Alchemical transformations are considered natural and enter a close relation of analogies and correspondences between the microcosm and the macrocosm, the human body and the four elements, the heavenly bodies and earthly bodies.

The *Praxeis* of Stephanus are very interesting philosophically, both from the point of view of method as well as in terms of their contents. Indeed, on the one hand, Stephanus plans to build a new system through the critical comparison of theories and admission of their difference. This form of *status quaestionis* of existing theories is one of the most 'scientific' aspects, in the modern sense, of his work. On the other hand, he creates a synthesis of Aristotelian, Platonic and Neoplatonic doctrines to build his alchemical doctrine.

In particular, he presents a model of matter and the transformations of metals that is one of the most original in the corpus of Greek alchemists, since it appears to be based both on the theory of surfaces in Plato's *Timaeus* and on the theory of exhalations in Aristotle's *Meteorologica*. The vaporous exhalation ('dyeing spirit', *pneuma*, 'cloud') responsible for composing and colouring metals is likened to the planar surface. An abstract geometric principle is thus identified with something physical and elemental (*pneuma*, humid exhalation, made of water and air), but subtle and rarefied, at the limit of the body.

The work of Stephanus was well known in the Arab world. According to the Arab-Latin tradition, it was one of his students, the monk Morienus (or Marianos), who propagated alchemy in the Arab world between 675 and 700 CE by initiating the Ummayad prince Khalid ibn-Yazid. Moreover, in the Byzantine era, the spread of alchemy in the Near East and the frequent exchanges of technical knowledge between the two cultures are testified by various sources.

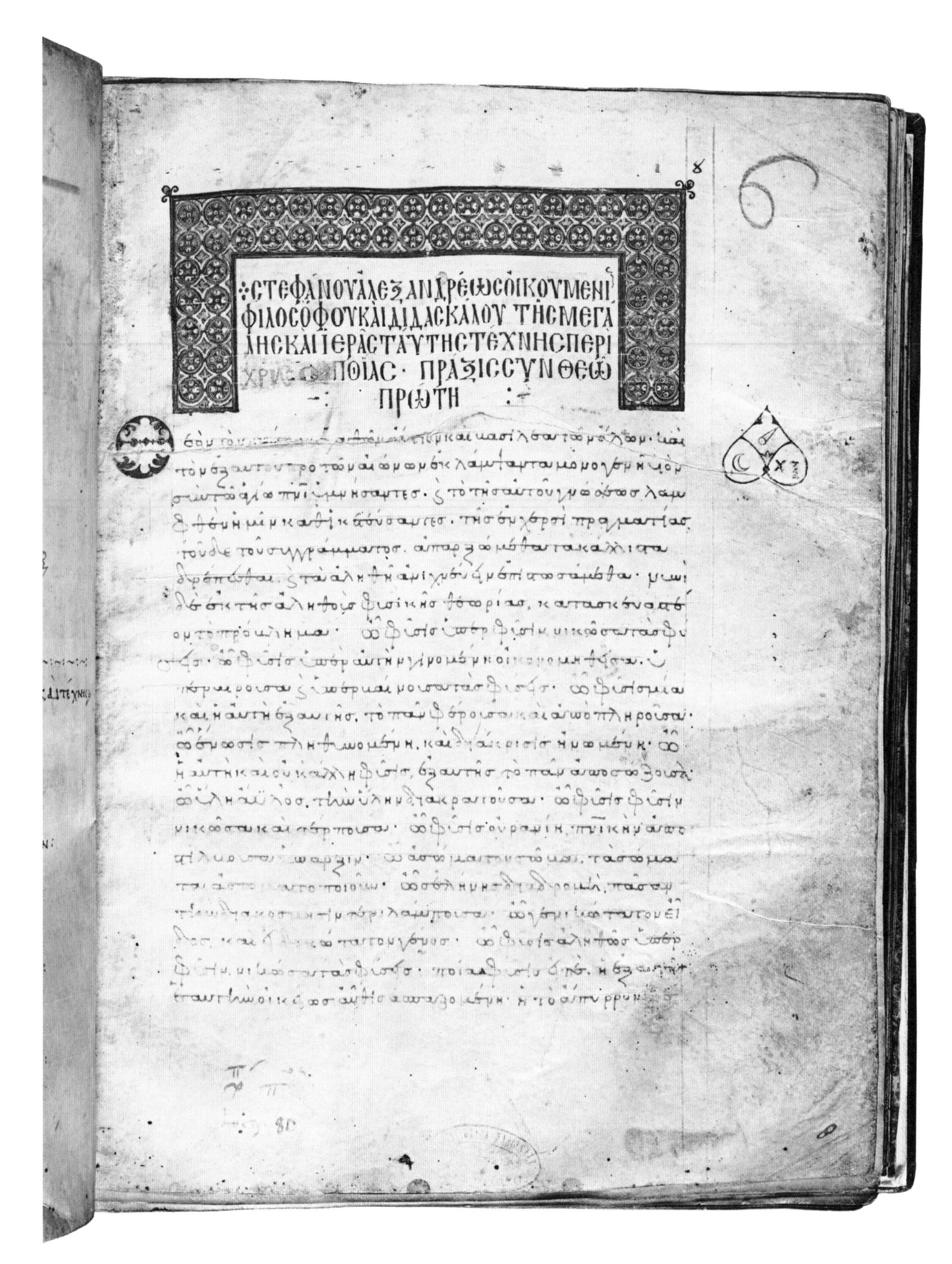
:ΣΤΕΦΑΝΟΥ ΑΛΕΞΑΝΔΡΕΩΣ ΟΙΚΟΥΜΕΝΙ(ΚΟΥ)
ΦΙΛΟΣΟΦΟΥ ΚΑΙ ΔΙΔΑΣΚΑΛΟΥ ΤΗΣ ΜΕΓΑ
ΛΗΣ ΚΑΙ ΙΕΡΑΣ ΤΑΥΤΗΣ ΤΕΧΝΗΣ ΠΕΡΙ
ΧΡΥΣΟΠΟΙΑΣ· ΠΡΑΞΙΣ ΣΥΝ ΘΕΩ
ΠΡΩΤΗ

The Christian Philosopher, the Anonymous Philosopher and the four alchemical poems

We thus arrive – with Stephanus and two anonymous commentators commonly called the Christian Philosopher[25] and the Anonymous Philosopher[26] – at the period when the first collection of Greek alchemists was constituted.

As with other commentators, these two anonymous works are presented as compilations, with commentaries, based on ancient writers (Hermes, Zosimos, Democritus) and about specific topics or questions. These compilations, especially that of the Christian, follow the general system adopted by Byzantines of the eighth and tenth centuries, which was to draw excerpts and summaries from ancient authors, such as those by Photius and Constantine Porphyrogenitus, a method that has preserved fragments but also contributed to the dismemberment of the texts.

As for 'divine water', the active principle of transmutation, the Christian insists upon the apparent disagreement among the ancient alchemists as to its designations, and especially as to the meaning of its unity.[27] As Zosimos would already have done, the commentator wants to show the basic consensus among the authors about the specific unity of this principle.[28] In particular, he shows that Democritus speaks of the unique species in general, and that Zosimos speaks of its multiple material species,[29] and he concludes that ultimately all multiplicity is reduced to unity. The substances are classified according to the four parts of an egg: shell, membrane, albumen, yolk.[30]

Some considerations bear on the method. The distinctions of materials and treatments show the influence of the descriptions of states of physical bodies (liquids, solids, composite nature) and transformative processes (cooking, melting, decomposition by fire or liquid) in Book IV of Aristotle's *Meteorologica*. The treatments are compared to planar geometric figures,[31] a comparison that recalls the concept of metals by Stephanus and Plato's *Timaeus*. Finally, the Christian applies the dialectical method of Plato, which divides and unites by species and genera, to the explanation of the operations, with the aim of clarity.[32]

The compilations of the Christian show a direct application of the conceptual tools of philosophy, especially of Aristotle and of Plato, to alchemical exegesis. One also notes some features of classical exegesis by the commentators, such as the search for agreement among opinions and the effort to derive the multiplicity of principles from a single one.

The Anonymous Philosopher presents a doxography on the 'prime ministers' of aurifaction. He mentions Hermes, John the Archpriest, Democritus, Zosimos and then 'the famous worldwide philosophers, commentators of Plato and of Aristotle, who used dialectical principles, Olympiodorus and Stephanus': they deepened aurifaction, they composed vast commentaries and they bound by oath the composition of the mystery.[33]

In particular, the 'Anonymous' examines the mixture of substances by liquid means, without the assistance of the fire, of which Olympiodorus also speaks.[34] There is still, as we saw with Synesius, influence from the Aristotelian theory of mixture, the basic composition of all natural bodies.[35] As for methodology, the

The first page of the alchemical lectures of Stephanus of Alexandria. *On the Great and Sacred Art of Making Gold.*

'Anonymous' commentator makes an original analogy between the general and specific instruments of music and the general and specific parts of the alchemical science.[36] It is a completely new musical theoretical system in Byzantine culture connected to the search for a mathematical language to decipher alchemical operations.

Finally, the four iambic poems on the divine art, placed under the names of Heliodorus, Theophrastus, Archelaus and Hierotheus (seventh and eighth centuries CE). These poems, highly mystical in inspiration, contain litanies about gold and have parallels with Stephanus in style and in content. Some scholars think the names probably refer to a single character, namely Heliodorus, who said he sent his poems to the emperor Theodosius, probably Theodosius III (716–17 CE).

Signs of ancient gold mines in Egypt.
A Ptolemaic period mill in Kompasi (Daghbag), Egypt (top).

An ore-crushing area at the Samut North site, in the Eastern desert of Egypt (middle).

A view of the two mills of Samut North (bottom).

The alchemists and their knowledge – Transmutation and its principles

Although these authors have their individual characteristics, from their writings we can reconstruct the lines of a fairly homogeneous theory of transmutation.

The idea of transmutation is based on the concept that all metals are constituted of the same material. The production of gold results from the synthesis of a common and receptive prime metallic material, onto which are incorporated the 'qualities' – i.e. substances – which are responsible for the colouration or transmutation into gold, according to the principles of sympathy.

Among these substances, 'divine water' (*theion hudōr*) or 'sulphur water' (*hudōr tou theiou*) plays a fundamental role. This water is frequently indicated as the goal of research and the principal agent of transmutation. It is an active principle derived from the metallic material itself, endowed with a double power, generative and destructive, which acts on the material itself. The common metallic material is not a substrate inseparable from the form, unknowable and indeterminate in itself, but a tangible body with an independent existence. It can be black lead or mercury.

Similarly, the active principle is identified with dyeing agents, which in practice are volatile substances such as mercury vapour. The distinction that the alchemists made, starting with Zosimos, between two components in metals, the one non-volatile (*sōma*) and the other volatile (*pneuma*), was surely inspired by observing the colouring action of some vapours on solid metals, such as mercury and arsenic vapours, which give a silvery colour to copper.

Often, transformation into gold is described as a deep dyeing. From this perspective, the colouring agent and the coloured body become a single thing through transmutation.

The discipline and its method

Let us now turn to some reflections of the alchemists themselves on the nature and method of their knowledge.

Olympiodorus sometimes defines the discipline as *technē*, sometimes as philosophy. The inextricable link between the two is expressed early in his doxographic statement, where he says that the ancient alchemists were properly philosophers

and addressed themselves to philosophers, that they introduced philosophy into *technē* and that their writings were doctrines and not works.[37]

Stephanus, too, speaks of 'philosophy', which he identified with the imitation of god.

He terms this discipline *chēmeia* and distinguishes it as 'mythical' (*muthikē*, fabulous) and 'mystical' (*mustikē*, symbolic, allegorical, but also for insiders): 'The mythical is reduced to a mass of empty statements, whereas the mystical chemistry methodically deals with the creation of the world by the Word, so that the man inspired by God and born of him is instructed by a proper effort and by divine and mystical statements.'[38]

For the alchemist, to know and to make, or better, to remake, are inseparable moments of a single act: it is through analysis, the reconstruction of the unity and accuracy of the process, that the work of the craftsman reproduces the organisation of the world. Note that the analysis is not just about the distinction of the components but also about the theories that concern the compositions.

The Christian Philosopher characterises the knowledge in question as both 'divine science' (*theia epistēmē*) and 'valuable and excellent philosophy' (*entimos kai aristē philosophia*).[39] We saw that he applies to the operations the dialectical method that divides and unites by species and genera.

The Anonymous Philosopher, for his part, compares alchemy to music in order to show the affinity of the structure of these two disciplines, characterised by the development of multiple practical applications rigorously regulated by a single principle.[40]

The epistemological status of this discipline is that of a reflection at once on the theory and practice, on the natural world and on the rational method of the *technē*. Theory and practice are always dialectically and indissolubly linked. Stephanus speaks of 'theoretical practice' (*theōrētikē praxis*) and 'practical theory' (*praktikē theōria*).[41]

We can now summarise some characteristics of the alchemical literature of the commentators.

First, most of these treatises are excerpts and summaries of other lost works, but they nevertheless have an order and purpose. The exegetical intention is often declared and focuses especially on the deliberate obscurity of the authors. This obscurity has a double explanation: first it is a strategy for defending the doctrine against those who do not deserve it, and second, it has the pedagogical and protreptic function to exercise the intelligence of adepts and push their minds towards the ultimate principles.

Next, the exegesis of the Alexandrian alchemical commentators touches on both the practice of operations and theoretical and methodological principles, frequently expressed through well-known concepts of Aristotelian natural philosophy (e.g. notions of mixing, of change of species, of potential/actuality, of matter/form), or of Platonic natural philosophy (such as elementary surfaces which form bodies).

Finally, all the Greek alchemical commentators, having identified the basic purpose of research with the principle responsible for transmutation, generally identified with the 'divine water' (the 'philosophers' stone' of the Middle Ages), which represents, in Aristotelian terms, a form of efficient and effective causality. Alchemists consider this goal, like the art and method concerned with it, unique. On this point, one can observe, especially in doxographies, research on the agreement among opinions, both of alchemical authors as well as of philosophers, such as the pre-Socratics, Plato and Aristotle. However, the agreement among the doctrines of Aristotle and Plato on a single object of research is also a common topos of Neoplatonic exegesis.

Philosophy and alchemy: The case of Olympiodorus

If one can easily spot among Greek alchemists the influence of the philosophy of their period, testimonies about alchemy are very rare in the writings of contemporary philosophers. Thus, even if one can perceive many similarities between the alchemical commentary of Olympiodorus and the commentary on the *Meteorologica* as well as with other texts of Olympiodorus the Neoplatonist, in contrast, in the commentary on the *Meteorologica* there is no explicit connection with the art of transmutation.

One may thus wonder what interest a Platonist philosopher like Olympiodorus could have in alchemy.

To get out of this impasse, one might reflect on the fact that what we now mean by 'alchemy' would not be perceived in the same way in the period of Olympiodorus and Stephanus. When we talk about alchemy, we immediately think of transmutation, of knowledge defined and characterised by a precisely determined goal, the transformation of lead into gold, etc. Indeed, while this may be true for the alchemy of the Middle Ages, both Western and Arabic, the boundaries of this knowledge would have seemed much more fluid in the Graeco-Alexandrian world.

First, the proper name of this knowledge, *al-chemia*, is an Arabic term (*al-kīmīyā'*) consisting of the article *al* and a Greek word of disputed etymology, *chēmeia* or *chumeia*. Stephanus used it on one occasion, and the Anonymous Philosopher is the first to use *chumeutês* to indicate an alchemist.[42]

The Byzantine lexicon *Souda* (tenth century) defines *chēmeia* as the art of making money and gold (*X*–280). But the term *chēmeia* is rare in the corpus of Greek alchemists. As we have seen, the authors speak instead of the 'divine art', of the 'great science', of 'philosophy'. Its scope is not only the production of gold and other precious metals, or the path of self-transformation, but the primary recipes also concern the colouring of stones and fabrics, that is, the production of pigments. Hence the use of entire repertory of organic and inorganic substances and of processes that affect matter and its transformation. The revolutionary concept – revolutionary in the Greek world – of transmutation is absent from the first 'technical' treatises but appears in the more philosophical authors such as Zosimos, and then in the commentators.

And even among those authors who speak of transmutation there are also substances and clearly identifiable procedures, which are in no way mysterious or metaphysical. This is the case with the descriptions of the distillation devices of

Zosimos, whose *ambix* (a term that will, via the Arabic *al-anbīq*, give us the well-known 'alembic').

It is therefore understandable that Olympiodorus, the commentator on the *Meteorologica*, could very well be interested in these texts we group in the category of Greek alchemy, to fill out his commentary and update the Aristotelian data, especially those about craft skills from Book IV. For example, Olympiodorus mentions glass artisans,[43] while Aristotle never mentions artisanal glass. Olympiodorus describes techniques of purifying and refining metal, effecting a separation of metal from its impurities, primarily of an earthy nature, or of one metal from another, as in the case of silver and gold.

So, it is not absurd to suppose that he might have wanted to go further and comment on a work by one of the most prominent authors of this science under construction, namely the *Kat'energeian* of Zosimos of Panopolis, which probably was already itself a doxographic and protreptic work on the foundations of alchemy.

That's why Olympiodorus represents an emblematic case of Alexandrian alchemy and constitutes a fundamental step in the epistemological identification of this fluid knowledge and the transition from the chemistry of the *Meteorologica* to alchemy. This transition will in turn be theorised and formalised in the Middle Ages by authors such as Albert the Great, Avicenna and Averroes.

Towards a multidisciplinary approach

Now we come to the second question: how should we study Byzantine alchemy? What is the approach most consistent with its specific nature?

In fact, this question is crucial for all periods of the history of alchemy. But that of the commentators is privileged because it contains an explicit epistemological reflection on an already established tradition. From this, one can envisage an interdisciplinary approach, which can account, in a fruitful way, for the composite nature of the writings and for the wealth of content that the tradition conveys.

As noted, the nature of the Graeco-Alexandrian alchemical knowledge is undeniably twofold: theoretical and practical. It comprises texts and recipes that concern at once mystical, physical and cosmological ideas, and the production of concrete and historically identifiable objects, such as the working and colouring of metals, fabrics and precious stones. It is not just the ideal goal, dreamed of but never attained, of aurifaction – that is to say the production of gold out of other metals. The earliest texts are probably artisans' notebooks, published in the milieu of the goldsmiths of the Egyptian pharaohs. That is why we can consider Greek alchemy a domain shared between the history of philosophy and of religion, between philology and the history of science and technology. It is a composite subject that therefore demands the sharing of many competences, not only theoretical and historical but also practical and technical, in direct contact with matter, such as archaeology, metallurgy and chemistry, which studies the materials and their transformations by artistic processes.

On this point, I would like to cite a recent and emblematic example of the fertility of an interdisciplinary collaboration around a common object of study.

In the first lines of his commentary on the *Kat'energeian* of Zosimos, Olympiodorus speaks about maceration (*taricheia*), the paradigmatic operation of processing gold ore, which involves several stages. These technical stages are described in detail by the geographer Agatharchides, tutor to Ptolemy III (second century BCE), who left a vivid account of the activities of the gold mines in the Eastern Desert.[44] In reality, this testimony is not entirely outside the corpus since we find an abstract in the alchemical manuscript Marcianus 229 (ff. 138–141).

The precise descriptions of Agatharchides on the four fundamental operations of ore processing – crushing, grinding, washing (or levigating) and refining – confirm that the passage from Olympiodorus referred to real procedures. But there is also another very recent and concrete testimony on the procedure for extracting and washing the gold ore, which represents for us an element of crucial importance for reconstructing the operations of the Greek alchemists. These are the results of excavations in 2013, at the gold mining sites of the Ptolemaic period (late fourth to mid-third century BCE) at Samut, by the French mission in the Eastern Desert.

The great clarity of the surface remains revealed facilities illustrating different stages of the work: first the mechanical phase of the sorting; crushing block of gold-bearing quartz; transformation into 'flour' (powdered ore) by mills; then the washing phase, in washing basins, for separating the metal particles to melt; and finally the metallurgical phase of refining on site, shown by the presence of an oven.

By putting together the pieces of this puzzle, we can advance a hypothetical reconstruction of what Olympiodorus tells us in his commentary and we can show that Olympiodorus, or Zosimos, refer to very real operations.

This example illustrates very well the fecundity and the necessity of applying a multidisciplinary approach to the Greek alchemical texts.

Indeed, the appeal to other disciplines and evidence, whether literary, archaeological or even chemical, allows us to interpret the alchemical texts. At the same time, alchemical texts shed light on historical and archaeological investigations.

In the current state of this field, it is essential to continue research on a multidisciplinary front and enhance the systematic and positive side of alchemy. This is a legitimate undertaking, because the ancient authors themselves often opposed natural and rational research to a deceptive practice subject to the laws of chance and the will of demons.

HERMES TRISMEGISTUS'S EMERALD TABLET

PETER J. FORSHAW

> In the place where I lived there was a stone statue raised on a wooden column; on the column one could read these words: I am Hermes to whom knowledge has been given; I made this marvellous work in public, but then I hid it by the secrets of my art, so that it could not be discovered except by a man as learned as myself. On the statue's chest one could also read these words written in ancient language: If anyone desires to know the secret of the creation of beings, and in which manner nature has been formed, let him look under my feet.[1]

Many came and looked under the statue's feet but discovered nothing, until the narrator, Balīnās, understood that he had to dig beneath them. He discovered an underground passage, so dark that he couldn't see, and when he tried lighting a burning torch it was blown out by winds that blew incessantly. In despair he curled up to sleep and had a dream in which an old man told him to put his light in a transparent vase, so that it wouldn't be extinguished by the wind.

Following this advice, he entered a chamber and discovered an old man seated on a golden throne and who was holding a tablet of emerald in his hand, on which was written: 'This is the art of nature.' In front of the old man was a book on which could be read: 'This is the secret of creation of beings and the knowledge of the causes of all things.' Balīnās eagerly took the book and learned what was written in this *Secret of the Creation of Beings*, understood how nature had been formed and acquired knowledge of the causes of all things, becoming famous, a master of the art of talismans and of wondrous things.[2]

Detail of *Tabula Smaragdina* ('Emerald Tablet'), a copperplate engraving by Matthaeus Merian, 1618.

The passage above is the earliest account of the discovery of the famed *Tabula Smaragdina* or *Emerald Tablet* of the legendary forefather of the alchemists, the Egyptian Hermes Trismegistus. It is from a ninth-century Arabic manuscript, the *Kitāb Sirr al-khalīqa* ('Book of the Secret of Creation'), attributed to Balīnās, better known in the West as the famous thaumaturge Apollonius of Tyana (*c.* 15–100 CE).[3] Other accounts describe the discovery of a plaque of precious emerald in Hermes's tomb in the Vale of Hebron by a woman called Zora, Zara or Sara;[4] or of Alexander the Great, during a visit to the Oracle of Ammon, discovering the tomb of Hermes containing a *Tabula Zaradi*.[5]

Balīnās's *Kitāb Sirr al-khalīqa* was translated as the *Liber de secretis naturae* ('Book of the Secrets of Nature') in the first half of the twelfth century by the Spanish scholar Hugo of Santalla. It is a confluence of different traditions: Greek and Syriac, Islamic and pre-Islamic, philosophical and alchemical. It opens with a discussion of the nature and names of God, followed by 'a description of the birth of the cosmos and the formation of minerals and stones, vegetable and animal life, and finally man'. The last part of the book introduces Hermes's *Emerald Tablet*.[6]

'Thrice Great' Hermes is famed for both his philosophical and technical *Hermetica*.[7] The former, the *Corpus Hermeticum*, is a collection of 17 religio-philosophical Greek texts, believed to originate in antiquity, compiled in the Middle Ages and translated into Latin in the fifteenth-century Renaissance by the Italian Hermetic philosophers Marsilio Ficino (1433–99) and Ludovico Lazzarelli (1447–1500).[8] These tracts, written as dialogues, usually include Hermes in conversation with his son Tat or with pupils, including Asclepius and Hammon, and concern the relationship between God, man and the cosmos, and a way of heavenly ascent leading to spiritual rebirth and deification through divinely revealed knowledge. Such is the case in the first tract, *Pimander*, in which Hermes is enlightened by Pimander, the divine mind, from whom he learns the nature of things and comes to know God. The work contains an elaborate account of creation, which must surely have resonated with alchemical readers. Only one philosophical Hermetic text was available in Latin during the Middle Ages, the *Asclepius* or *Perfect Discourse*, which contains the notorious passages about humans making gods, as well as the 'Hermetic Apocalypse', describing the plight of the Egyptian people after the gods withdraw from earth to heaven.[9]

Besides these theosophical texts, there are also technical texts, with a focus on the operative practices of works of medicine, astrology, magic and alchemy attributed to Hermes. The *Tabula Smaragdina* (*Emerald Tablet*) is the most representative work of the alchemical technical Hermetica and the one that we shall focus on in this chapter.

The earliest versions of the *Emerald Tablet* appear in Arabic: in the *Kitāb Sirr al-khalīqa* ('Book of the Secret of Creation'), ascribed to Balīnās;[10] in the eighth-century Pseudo-Aristotelian *Sirr al-asrār* ('Secret of Secrets'), and the eighth- or ninth-century *Kitāb Ustuqus al-Uss al-Thānī* (*Second Book of the Elements of Foundation*) attributed to the Persian alchemist Jābir ibn Ḥayyān (d. *c.* 806/16).[11]

More than one Latin translation of the *Emerald Tablet* survives. The earliest was that of Hugo of Santalla, although it was not really picked up by later alchemists. Another version, in the Italian priest Philip of Tripoli's thirteenth-century translation of the *Sirr al-asrār* as the *Secretum secretorum* ('Secret of Secrets', *c.* 1230–40), is presented as a lecture by Aristotle to his pupil Alexander the Great (in which

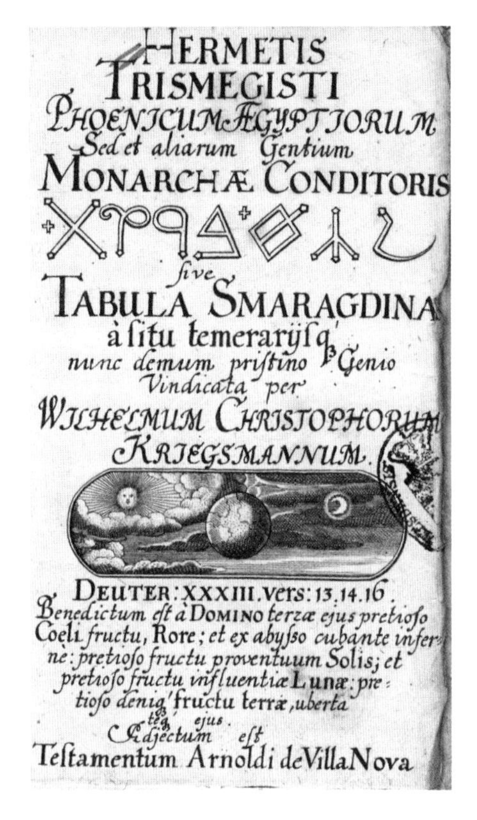
HERMETIS
TRISMEGISTI
PHOENICUM ÆGYPTIORUM
Sed et aliarum Gentium
MONARCHÆ CONDITORIS
sive
TABULA SMARAGDINA
à situ temerariisq'
nunc demum pristino Genio
Vindicata per
WILHELMUM CHRISTOPHORUM
KRIEGSMANNUM.
DEUTER: XXXIII. vers: 13.14.16.
Benedictum est à DOMINO terræ ejus pretioso Coeli fructu, Rore; et ex abysso cubante inferne: pretioso fructu proventuum Solis; et pretioso fructu influentiæ Lunæ: pretioso denique fructu terræ, ubertatisq' ejus.
Adjectum est
Testamentum Arnoldi de VillaNova

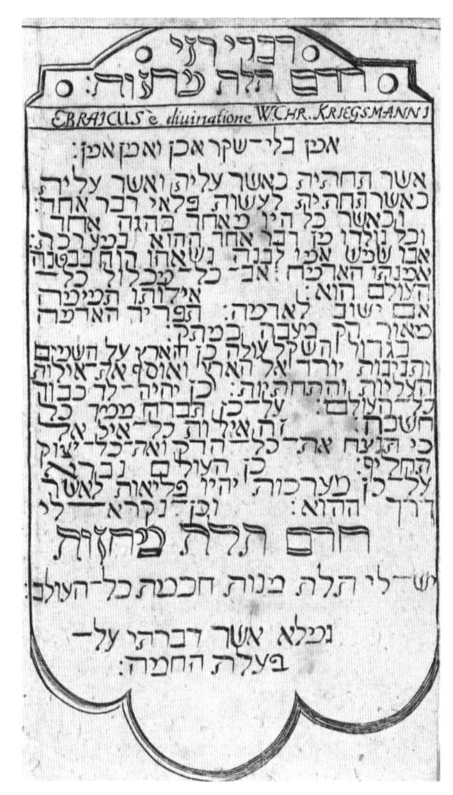

Title page and the *Emerald Tablet* from Wilhelm Christoph Kriegsmann, *Hermetis Trismegisti Phoenicum Aegyptiorum*, Leipzig, 1657.

Overleaf:
The second circular figure, of Adam Androgyne, from *Amphitheatrum Sapientiae Aeternae* by Heinrich Khunrath, 1595.

Hermes's name has been deformed into Hermogenes).[12] It was a third version, however, that became popular with alchemists, from a variant translation of the *Kitāb sirr al-halīqa* by the twelfth-century Italian mathematician and astronomer Plato of Tivoli, which was included in works such as *Liber Hermetis de alchimia* ('Hermes's Book of Alchemy'), the *Liber dabessi* or *Liber Rebis* ('Book of the Two-Thing' [Hermaphrodite]).[13] Later enthusiasts, such as the German pietist and Hermetic philosopher Wilhelm Christoph Kriegsmann (1633–79), even provided Hebrew and Phoenician renderings of the *Emerald Tablet*, in his *Hermetis Trismegisti Phoenicum Aegyptiorum sed et aliarum gentium Monarchae Conditoris sive Tabula Smaragdina* ('The Emerald Tablet of Hermes Trismegistus Phoenix of the Egyptians but also founder of the monarchy of other nations', 1657).[14]

As the *Emerald Tablet* is so famous in the history of alchemy and is a surprisingly short work, which can be divided into approximately 14 statements that have fascinated myriad readers, including the English 'father of modern science' Isaac Newton (1642–1727) and the Swiss founder of depth psychology Carl Jung (1875–1961), it seems appropriate to provide it here, in translations of Latin and German versions present in one of the engravings from the German alchemist Heinrich Khunrath's *Amphitheatrum Sapientiae Aeternae* ('Amphitheatre of Eternal Wisdom', 1609).[15]

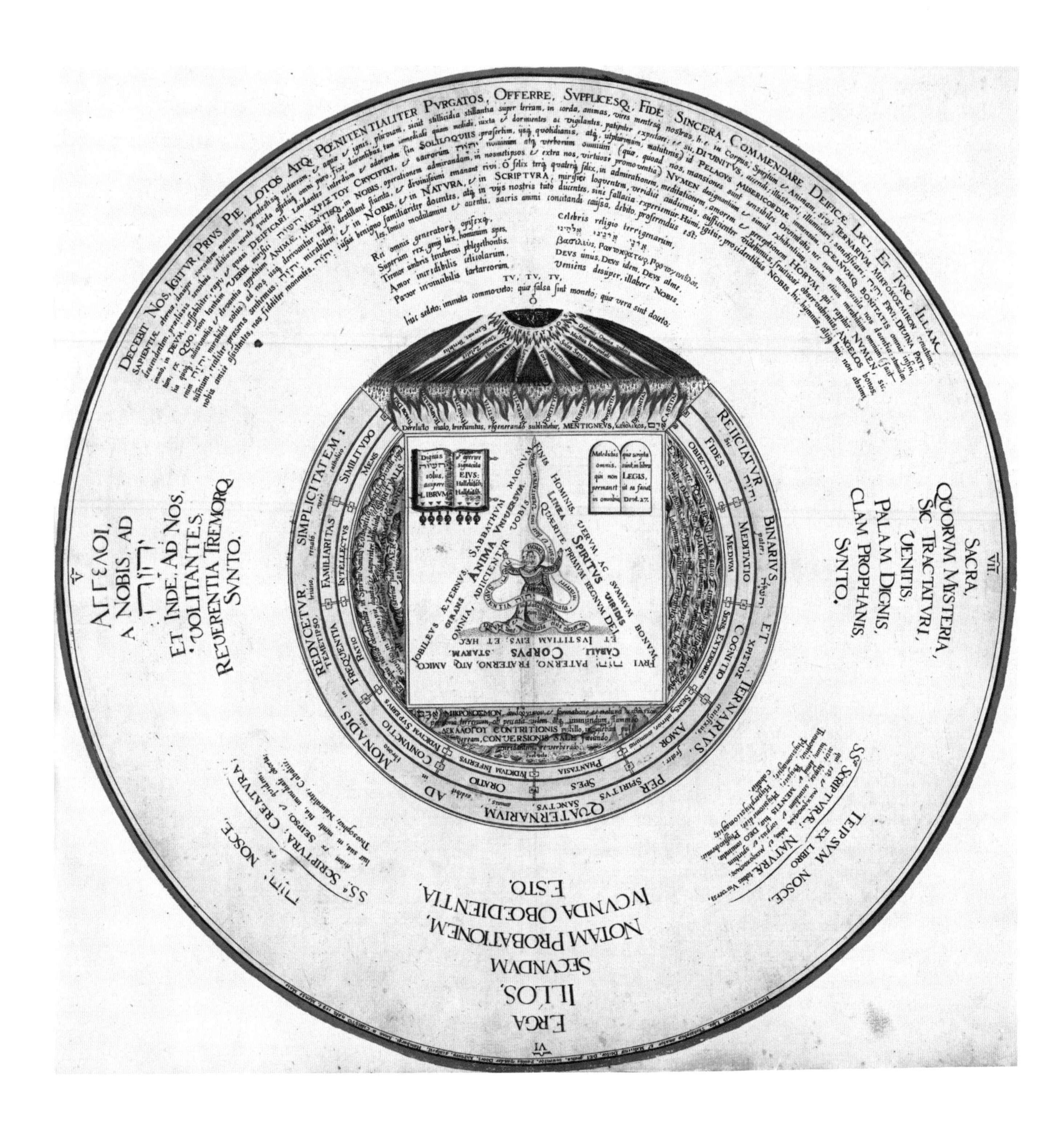

Sacra, Quorum Mysteria, Sic Tractaturi, Venitis, Palam Dignis, Clam Prophanis, Sunto.
Erga Illos, Secundum Notam Probationem, Iucunda Obedientia, Esto.

	From the Latin	*From the German*
1	True, without falsehood, certain and most true, what is below is like what is above; and what is above is like what is below; for performing the miracles of the one thing.[16]	Truly, without lies, certainly and most truly, that which is Below is like that which is Above; And that which is Above is like that which is Below; whereby one can achieve and perform the miracles or wondrous signs of one thing.[17]
2	And as all things were from the one, by the meditation of the one; so all things were born from this one thing by adaptation.	And just as All Things are created from one thing alone, by the will and command of the one who conceived it; so do All Things have their source and origin from this same One Thing, being brought together by providence and unification.[18]
3	Its father is the sun, its mother is the moon, the wind carried it in its belly; its nurse is the earth.	The Sun is its Father, and the Moon is its Mother; the Wind carried it in its belly. Its Nourisher or Wet-Nurse is the Earth.[19]
4	It is the father of all the perfection of the whole world.	This is the Father of all Perfection in all this World.[20]
5	Its power is Absolute.	Its Power is Perfect.[21]
6	If it is turned into earth, you will separate the earth from the fire, the subtle from the gross, smoothly and with great cleverness.	If it is transformed into Earth, then you should separate the earthly matter from the Fire, and the subtle from the fat or coarse, most gently, with great discretion and understanding.[22]
7	It ascends from earth into heaven, and again descends to earth, and receives the power of [those] things above and [those] below.	It rises from Earth to Heaven, and [descends] from Heaven to Earth again, and thus acquires the power of [the things] Above and [those] Below.[23]
8	In this way you will have the glory of the whole world!	Thus will you have all the glory of the whole world![24]
9	Therefore all obscurity shall flee from you.	Therefore cast from you all incomprehension and impotence.[25]
10	It is the strong fortitude of all fortitude; because it will overcome every subtle thing and penetrate every solid.	This is the Strong[est] strength of all strengths; for it can overcome all subtlety and penetrate any solid.[26]
11	In this way the world was created.	Thus was the world created.[27]
12	From it there will be wonderful adaptations, of which this is the method.	Hence occur rare combinations, and many Wonders are worked; [and] this is the way to work them.[28]
13	And so I am called Hermes Trismegistus, having the three parts of the Philosophy of the whole world.	And therefore am I called Hermes Trismegistus, having three parts of the Wisdom of the entire world.[29]
14	What I have said concerning the operation of the sun is completed.	Everything I have said concerning the Work of the Sun has been fulfilled.[30]

Over the centuries there have been many interpretations of this enigmatic text. Here let us focus on two commentaries, one medieval, from the fourteenth century, the other early modern, from the sixteenth century. As it would take too much space to discuss all variant readings of even this short text, we shall consider just a few of the most popular motifs.

The first commentary is one of the earliest and best known, that of Hortulanus, tentatively identified as the Englishman John of Garland.[31] In his *Liber super textum Hermetis* ('Book concerning the Text of Hermes)', Hortulanus makes it clear that he considers Hermes to be talking about the creation of the philosophers' stone, explaining the references in the first statement to 'above' and 'below' in this way: 'the stone is divided into two principall parts by Art: Into the superiour part, that ascendeth up, and into the inferiour part, which remaineth beneath fixe and cleare: and yet these two parts agree in vertue: and therefore hee sayeth, *That which is above, is like that which is beneath*.'[32]

This sense of the 'above' and 'below' is presented visually in a fourteenth-century manuscript of *The Book on the Silvery Water* [*Mercury*] *and the Starry Earth* [*Sulphur*] by the tenth-century Islamic alchemist Abu Abdullah Muhammed Ibn Umail al-Tamīmī (*c.* 900–60), where we see a figure, presumably Hermes, holding a tablet on which can be seen various images, including that of two birds, one winged above and one wingless below, taken to symbolise the volatile (winged) mercury being fixed by sulphur.[33] The fifteenth-century *Aurora Consurgens* ('Rising Dawn') repeats this motif, although at least one manuscript has the interesting variant of a bird above and a fish below.[34] In his *Symbola Aureae Mensae* ('Symbols of the Golden Table', 1617), the German physician Michael Maier (1568–1622) introduces Egyptian Hermes as the first alchemical representative of 12 nations, with an engraving in which we see a man dressed in Eastern costume, holding aloft an armillary sphere, indicative of his astrological knowledge of celestial influences above, while pointing downwards with his left hand to a terrestrial and alchemical sun and moon (gold and silver/sulphur and mercury), with the element of fire between them.

Returning to Hortulanus, we learn that 'The miracles of the one thing' refer to the philosophers' stone, used for gold-making, with the suggestion that the transmuted metal be used for healing, with the declaration that 'the golde ingendred by this Art, excelleth all naturall gold in all proprieties, both medicinall and others[35] In accordance with the words of Genesis 1:2 that 'the earth was without form, and void' we learn that the stone 'is borne, and come[s] out of one confused masse, containing in it the foure Elements[36] Considering the third statement in the *Emerald Tablet*, Hortulanus interprets the 'Sunne is his father' as philosophers' gold, while the Moon, as philosophers' silver, is its mother, a 'fitte and consonaunt receptacle for his seede and tincture[37] When these two conjoin, the stone is conceived in the Wind's belly, which Hortulanus interprets as the air, which he glosses as 'life', meaning the 'Soule', which 'quickneth the whole stone[38] This stone, as a child, requires nourishment, which it receives from the earth as its nurse. Through this the stone is turned into ferment, like the yeast that causes the dough to rise when making bread,[39] so that, after its constituent parts (earth and fire) have been separated (i.e.

'Hermes in the Treasure House holding his tablet', from *Aurora Consurgens*, in the alchemical miscellany *De lapidibus* by Pseudo-Albertus Magnus, and other writings.

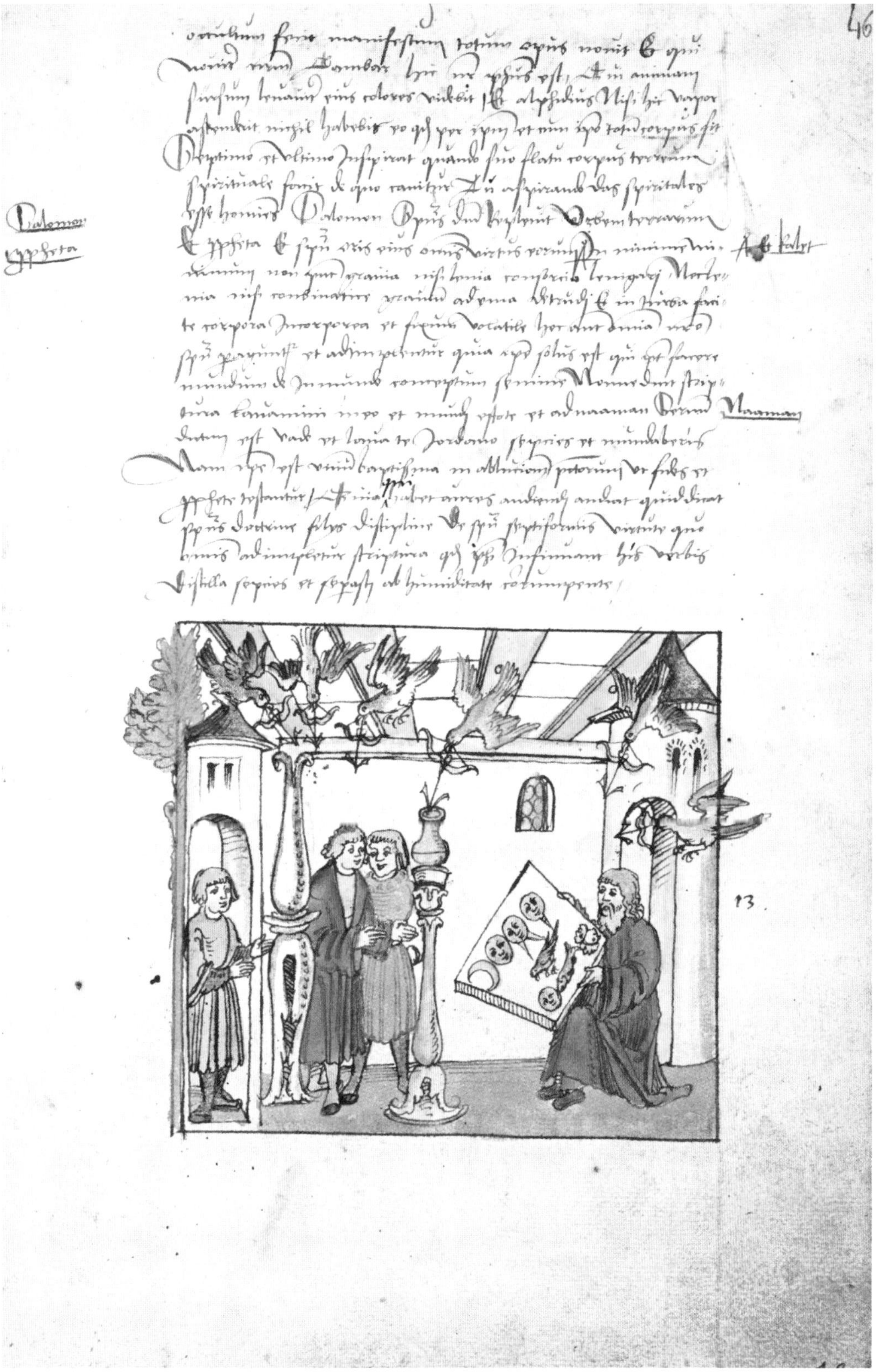

occultum fuit manifestum totum opus novit et qui
venit vinum Sambar hic in prophetis est. Et in animam
siccam lavavit eius colores vidit sicut alphidius. Nisi hic vapor
ascenderit nichil habebis eo quod ipse opus et sine ipso totum corpus fit
Septimo et ultimo inspirat quando suo flatu corpus terrenum
spirituale fecit et opere terminatur. Et in aspirando dat spiritales
esse homines. Salomon. Spiritus domini replevit orbem terrarum
et propheta. Et spiritus oris eius omnis virtus eorum. Et numerus do-
minium non erit gratia nisi tinctura corporis congregat. Morienus nisi continuatur granum ad terram. Utrum et in terra fac-
te corpora incorporea et fecerunt volatilia hoc autem omnia non
spiritu perficiuntur et adimplebuntur quia ipse solus est qui potest facere
mundum et inmundum corruptum. Scitur Nam dicit scrip-
tura lavamini in eo et mundi estote et ad naaman dicitur
dictum est vade et lava te in Jordano septies et mundaberis
Nam ipse est unum baptisma in ablutionem peccatorum ut fides et
prophete testantur. Et qui spiritum habet aures audiendi audiat quid dicat
spiritus doctrine filiis discipline de spiritu septiformis virtute quo
omnis adimpletur scriptura quod philosophus insinuant hiis verbis
Distilla septies et separasti ab humiditate corrumpente.

Salomon
propheta
[illegible] Kalyt
Naaman

13

dissolved) and purified, it can multiply whatever it is used to create. This takes place through the 'unfixed' part ascending 'from the earth into heaven' by sublimation and subtilisation, which then descends from heaven 'into the earth', to remain fixed and flowing.[40] All this is in imitation of God's creation of the world in the book of Genesis.[41] Hortulanus also alludes to biblical verses that were particularly well-known to alchemists, from John 12:24: 'unless the grain of wheat falling into the ground die, itself remaineth alone. But if it die, it bringeth forth much fruit.' He associates this notion of death with alchemical processes of mortification and putrefaction, and the related colour changes perceived in the alchemical vessel:

> For this cause is the Stone saide to be perfect, because it hath in it the nature of Minerals, Vegetables, and Animals: for the stone is three, and one having foure natures, to wit, the foure elements, & three colours, black, white and red. It is also called a graine of corne, which if it die not, remaineth without fruit: but if it doo die...when it is ioyned in coniunction, it bringeth forth much fruite.[42]

All these, according to Hortulanus, represent the transformations of the four elements in 'one onely thing, namely in the Philosophers Mercurie'.[43]

Our second commentator is the prolific promoter of Paracelsian alchemical philosophy, the Flemish writer Gerard Dorn (*c.* 1530–84). In contrast to Hortulanus, Dorn does not consider the *Emerald Tablet* to be about gold-making but rather about *spagyria* – that is, the chemical medicine advocated by Paracelsus, who had praised Hermes's work, while condemning current academic medicine, in *De tinctura physicorum* ('On the Tincture of the Natural Philosophers'): 'The ancient *Emerald Tablet* shows more art and experience in Philosophy, Medicine, Magic and the like than could ever be taught by you and your crowd of followers.'[44]

Dorn discusses his spagyric interpretation of the *Emerald Tablet* at some length in his *Clear and very brief exposition of the Truthful and Enigmatic Discourse of Trismegistus, left by Mercurius Hermes Trismegistus.*[45] He concedes that the *Tablet* could be understood in an allegorical sense as concerning the metamorphoses of the metals, but it is evident that he really considers it of benefit to students and patients of the medical art. The *Tablet*, he believes, does not simply concern metals but also vegetables and animals.[46] encouraging him to 'interpret it from a medical point of view, and not from a metallurgical but from a human point of view'.[47] He explicitly states that Hermes is 'the interpreter of spagyric wisdom'.[48]

Concerning the *Tablet*'s 'above' and 'below', Dorn resorts to the trope of alchemy as 'lower astronomy'.[49] found in works like the early medieval *De perfecto magisterio* ('On the Perfect Magistery') of Pseudo-Aristotle and in John Dee's *Monas hieroglyphica* (1564).[50] He writes of the 'generations of meteors...not in the sky, but in the elements of air and fire under the earth, which are not unlike the ethers, which are helped by celestial actions and impressions to the generations of minerals, stones, and of metals'.[51] All this has 'proof deduced from the physical truths of Genesis'.[52] with Dorn arguing that the very words 'What is below is like what is above' make it clear 'that Hermes Trismegistus was taught Genesis by the Hebrews, although he was an Egyptian', as had been argued by Augustine in *The City of God*.[53]

Engraving of Hermes Trismegistus in *Symbola Aureae Mensae Duodecim Nationum* ('Symbols of the Golden Table of the Twelve Nations') by Michael Maier, 1617.

One of the major products of alchemy, the quintessence, useful for preserving the body from decay, had been called *Coelum* or 'Heaven' since the days of John of Rupescissa (*c.* 1310–62).[54] and here Dorn states that the '[spagyric] heaven which is below is therefore the same as the [celestial] heaven which is above...to perform the miracles of one thing, that is to say universal medicine'.[55] This heaven constitutes 'the long and healthy life of natural things' and 'whichever physician, therefore, knows how to unite the incorruptible spagyric heaven of long and healthy life with sick bodies and the internal and languid heaven of short life, will restore a short and healthy life (God granting) by the miracle of one thing'.[56]

In support of this view, Dorn could have adduced Hortulanus's medical interpretation of the *Tablet*'s statement that 'all obscurity shall flee from you', as meaning 'all want and sicknesse, because the stone thus made, cureth everie disease'.[57] Instead, he draws inspiration from the German abbot Johannes Trithemius's (1462–1516) meditations on Pythagorean symbolic numbers in relation to magic and alchemy.[58] and advises that 'before you begin this work, you must reduce yourself to the simplicity of unity through the morally appropriated art of chemistry, by which you will first learn to drive away from yourself all the darkness of the mind, and afterwards, illuminated from above, you will perceive the light and the medicine that can drive away the darkness from your body and mind and ignorance'.[59] For Dorn, the penetrative powers of the philosophers' stone are less about penetrating alchemical matter in the vessel and more about its ability to overcome 'the subtle and spiritual disease in the human mind' – by which he means madness, mania, fury, stupidity and so forth.[60] – and drive away all the internal and external defects of the body.[61] He argues that the *Tablet*'s references to overcoming every subtle thing and penetrating every solid refer, respectively, to 'spiritual darkness that covers the mind' and the eradication of 'every gross and hard bodily defect'.[62]

Dorn's younger contemporary Andreas Libavius (1560–1616) expands on this in 'a hieroglyphic explication of the *Emerald Tablet* of Hermes'.[63] in the first volume of his *Rerum Chymicarum Epistolica Forma* ('On Chymical Matters in Epistolary Form', 1595), arguing that obscurity refers to the dispelling of ignorance, as when in Plato's

Republic the enlightened philosopher walks out of the dark cave of ignorance into the light of understanding.[64]

Although some writers, like Newton, working from a variant transmission of the *Emerald Tablet*, replaced the word 'meditation' with 'mediation.[65] in the phrase 'as all things were from the one, by the meditation of the one', Hortulanus and Dorn both favour 'meditation', the former writing that this means 'the cogitation and creation of one, that is the omnipotent God'.[66] the latter that 'all things were created mentally by one GOD, by the meditation or conception of one world through an idea'.[67] Dorn is critical, here, of the textual transmission, commenting that

> the author's sentence seems to me distorted, and mutilated by the omission of a single term, namely, *created*, by the negligence of the translators rather than of the printing, and I suggest that it should be restored in this way: *Just as all things were created by one, by meditation, &c*. With these words Hermes therefore hints at this universal medicine, to be made in imitation of the creation of the world.[68]

With this in mind, like Hortulanus, Dorn compares Hermes's work to the divine act of creation, as well as alluding to a verse from Wisdom 11:21 popular with alchemists:

> Not unlike the way in which God first created one world by meditation alone, he [Hermes] created one world in a similar way, from which indeed all things were born by adaptation. Under meditation he wants to denote the simplicity of unity in mental creation, and afterwards by adaptation to suggest arrangement or order under the appropriate number, measure, and weight of matter.[69]

While gold-making alchemists draw an analogy between the sun in the heavens and gold in the earth, spagyric alchemists are concerned with the relation between the sun and the human heart. Pondering the *Tablet*'s statement concerning the Sun, Moon, Wind and Earth, Dorn writes,

> Hermes speaks of the genealogy of the Spagyric offspring. His father (he says) is the sun, &c. What the sun and the moon are in the upper heaven is not for us to explain...As the source of the life of the human body, it is the centre of his heart, or rather that which conceals the secret in it, in which natural warmth thrives.[70]

This microcosmic, anatomical, spagyric interpretation is very different from the chrysopoetic, gold-making interpretations. In the thirteenth century, in *De Mineralibus* ('On Minerals'), Albertus Magnus (*c.* 1200–80) had compared 'the Wind carried it in its belly' to the distillation of rose water by steam distillation, in which the steam rises and passes through rose petals and absorbs their essence, then cools and condenses into essential oil and rose water.[71] Libavius also considered the *Emerald Tablet* a work of transmutational alchemy and stated that 'the Wind carried it in its belly' describes how philosophical mercury absorbs gold.[72] explicitly warning his reader away from the 'crazy Paracelsian physicians' and their mistaken iatrochemical interpretation of the text.[73] The already-mentioned *Aurora Consurgens* took a different approach and interpreted the Sun, Moon, Wind and Earth not as alchemical substances but instead as processes, respectively, of calcination, solution, distillation and coagulation.[74]

Dorn, however, sticks to his biological analogies, discussing the conception of the philosophers' stone with reference to the Genesis account of Adam and Eve, reminding his readers that 'man and woman were one person, were made two, afterwards joined together again in the flesh'.[75] He describes the generation of the foetus of the Sun and Moon and the alchemical processes of solution and fixation:

> The opinion of Father Hermes is this: The fire, under the centre of the bottom of the vessel, acting from the outside on the centre of matter, from this propels the spagyric foetus to the higher, as the words have it: It ascends from the earth to heaven, that is, from the lower part of the vessel (which is sometimes called the earth) to the higher, which is also sometimes termed heaven by the philosophers. This passage can be understood differently in this way. The foetus of the sun and the moon ascends, that is, its thick earthly substance is converted or resolved into heaven, that is, into the subtlest heavenly substance.[76]

Dorn tells his readers that 'this was once wrapped up in a riddle by the philosophers: Make the fixed (they say) volatile, and again the volatile fixed, and you will have the whole magistery.[77]

Much more could be said about the influence of Hermes Trismegistus on the history of alchemy, be that gold-making, iatrochemical or later spiritual alchemical interpretations. An attentive reader can detect his enduring presence in the texts and images of scores of alchemical books and manuscripts in a multiplicity of languages throughout the centuries, such that, for example, the German physician Daniel Sennert (1572–1637), writing in the seventeenth century, could cite Albertus Magnus from the thirteenth century, justifiably saying that 'Hermes is the root on which all philosophers are supported'.[78]

materia rem ex qua trahit
lutio fit: tertium
materia lapidis:
Geber:
phus:
Deus et
natura no
faciunt frusta:
RIOVRABETI

ALCHEMY IN THE MEDIEVAL ISLAMIC WORLD

SALAM RASSI

The 8th-century alchemist Jābir ibn Ḥayyān, possibly illustrated by Giovanni Bellini. From a manuscript in Biblioteca Medicea Laurensiana, Florence, Italy, 1460–75.

Alchemy (*al-kīmiyā'* in Arabic, from the Greek *chēmeia*) played an important role in the intellectual life of the medieval Islamic world. Flourishing from the eighth century into the modern era, the discipline was chiefly concerned with the investigation and transformation of matter. Its significance, however, extended far beyond the realms of science and technology. It permeated various social, literary and religious dimensions of medieval society, shaping and enriching the tapestry of intellectual pursuits of that time. Yet among the many disciplines that were widespread, alchemy is perhaps the least studied. One reason for this has been its contested and often marginalised status in colleges, universities and other learned institutions.[1] Alchemy's public image has also suffered in various ways, with many medieval writers regarding it as a gateway to charlatanry.[2] Meanwhile, until very recently, modern historians have overlooked its importance to the history of science.[3]

Fortunately, contemporary scholarship has progressed, particularly among historians of Christian Europe who have attempted to approach alchemy through a meticulous examination of its texts.[4] However, the study of Arabic alchemy remains in its infancy, despite its pivotal role in shaping the history of this discipline in the West. Although a number of pioneering studies appeared in the twentieth century,[5] several Arabic alchemical texts remain without critical editions, and fewer still have been translated into modern languages. To foster further exploration, this chapter offers an overview of some central themes. I will survey the intellectual doctrines of medieval Arabic alchemy, from its earliest beginnings to about the twelfth century of the Common Era. My focus will be on the theoretical and technical aspects of alchemy, rather than its social perception in Muslim lands.

Between translation and legend: The appearance of alchemy in the Islamic world

Roman Egypt is often regarded as the birthplace of alchemy, though earlier strands may have first appeared in ancient Mesopotamia.[6] According to some scholars, the practice emerged from metallurgical techniques used by Egyptian temple artisans to create polychromatic statues.[7] Others, meanwhile, have traced the origins to ancient discourses on magic.[8] In the first four centuries of the Christian era, Greek authors like Pseudo-Democritus and Zosimos of Panopolis combined ancient Egyptian cosmological theories with elements of Hermeticism, Gnosticism and Stoic physics.[9] Alchemy eventually moved beyond Egypt and influenced the intellectual culture of the Byzantine Empire in late antiquity, culminating in works by Olympiodorus, Christianos ('The Christian'), Stephanus of Alexandria and others.[10]

These developments laid the groundwork for the migration of alchemy into Islamic lands. The eighth century marked the beginning of the so-called Translation Movement, supported by the early Abbasid caliphs.[11] Greek texts were translated into Arabic during this period, sometimes through Syriac intermediaries. Much of this process has been well documented. We know, for example, that the Christian physician and philosopher Ḥunayn ibn Isḥāq (d. 873) played a prominent role in translating works by Aristotle, Galen and Hippocrates for both Muslim and Christian patrons.[12] Attested translations during Ḥunayn's lifetime cover various subjects such as logic, medicine, physics and metaphysics. However, there are fewer historical sources regarding the translation of Greek alchemy. The reason for this is unclear. It could be due to the secretive nature of alchemy, passed down as it was from practitioner to pupil. Another factor may be that alchemy was not a prominent part of any systematic curriculum.

Despite the limited information, it is evident that Greek alchemical writings became available to Arabic readers through direct translation, paraphrasing and epitomisation. One of the earliest Greek alchemical writers to be translated into Arabic was Zosimos. Arabic readers were familiar with his *25th* and *27th Epistles*, which derived from authentic Greek originals.[13] Some of Zosimos's works, like *The Sulphurs*, exist only in fragmentary form in Greek but are preserved in Arabic.[14] Other works such as the *Keys to the Art* and *Seven Epistles* cannot be linked to extant Greek originals but have been indirectly authenticated as works of Zosimos.[15] Others still fall in a grey area between authenticity and forgery. The *Tome of the Art* and *Tome of Images*, for instance, are dialogues addressed to the female alchemist Theosebeia, a prominent figure in Zosimos's genuine works, and provided valuable information to medieval Arabic writers about the Greek alchemical corpus, including truncated versions of the sayings of Ostanes, Pseudo-Democritus and Maria the Jewess.[16]

The *Fihrist* ('Catalogue') compiled by the Baghdad bookseller Ibn al-Nadīm (d. 995) is another source of information about the transmission of Greek alchemical authorities into Arabic. However, as with Zosimos's works, there is a blurred line between verifiable material and false attributions. Ibn al-Nadīm ascribes the origins of alchemy to the legendary sage Hermes Trismegistus, a central figure in a body of texts commonly referred to as Hermetica. These works appeared in Roman Egypt during the first three centuries of the Christian era and covered a wide range of topics, from cosmology to magic. The association of Hermes with alchemy emerged in the Roman period and continued among Byzantine and Arabic writers. In Ibn al-Nadīm's list, there are 13 books on alchemy associated with Hermes, among

Detail from *Kitāb al-Aqālīm al-sabʿa* ('Book of Seven Climes') by Abū al-Qāsim al-ʿIrāqī. 18th century.

صفة كما وجدت نقلت من مصحف زوسيموس واوتاسيا

خذ من حجر ما شئت وهو الكبريت الاحمر الذي لا يخلوا منه مكان والقي
من الكبريت الابيض مثله واسحقه فانه يذهب بصلابته والق مثلهم زيبقا مدبرا
واعملهم في النار ساعة ثم اعيد عليهم السحق والسقي الي ان يعجبك لونه
فالقي منه علي حجر يصير ذهبا ابريزا كاملا والحمد لله تعالى

اعلم ان المركب المشار اليه له اوزانا تخص به
ليكون معتدلا حتى لا تغلب حرارته علي
برودته ولا يبوسته علي رطوبته لان ما
اعتدلت طبايعه كان خالدا لا يتغير
ابدا ولم يعتدل طبايعه وقع في التغيير

ضوء النهار

سواد الليل

اعلم ان ضوء النهار هو الزاج الاحمر المحلول وسواد الليل الحديد فاعلم ذلك

them works bearing pseudo-Greek titles like *Hādīṭūs*, *Mālāṭīs*, *Isṭamākhīs* and *Isṭāmāṭis*.[17] They are part of a cycle of occult writings known as the Pseudo-Aristotelian Hermetica, which likely date to the ninth century and are often framed as conversations between Alexander the Great and Aristotle about the wisdom revealed by Hermes.[18] Ibn al-Nadīm also lists several well-known figures from the Graeco-Egyptian and Byzantine alchemical corpus, including Ostanes, Agathodaimon, Democritus, Zosimos, Maria, Pelagius, Christianos (*al-naṣrānī*) and Stephanus.[19] However, it is uncertain whether these books derived from Greek originals or were simply attributions.

Another source about the transmission of alchemy to Islamic lands comes from legendary narratives. While these offer little by way of historical data, they provide insights into how medieval Arabic authors imagined the dissemination and early development of alchemy. These legends often revolve around Christian monks who supposedly shared alchemical knowledge with worthy disciples. One example is the Byzantine monk Maryānūs, said to have instructed the Umayyad prince Khālid ibn

Maryānūs meets Khālid ibn Yazīd. From *Symbola Aureae Mensae Duodecim Nationum* by Michael Maier, 1617.

Yazīd (d. 704 or 708).[20] Their master–disciple relationship is entirely fictional, and there is no evidence of the Umayyads' interest in alchemy. Nevertheless, these legends could have arisen from a later perception among Muslim writers of Christians as important translators and conveyors of knowledge in the early Islamic period. Although there is scarce evidence for state patronage of translation before the Abbasid period, Khālid ibn Yazīdis said by Ibn al-Nadīm to have commissioned Arabic versions of several alchemical works.[21] Once again, these are all pseudo-epigraphs without Greek originals.

The topos of the monk as transmitter of alchemical knowledge is found elsewhere, such as the Jābirian Corpus, a collection of writings attributed to Jābir ibn Ḥayyān and composed between the eighth and tenth centuries. In one treatise, the author seeks the tutelage of an unnamed monk who succeeded Maryānūs.[22] The idea of monks as bearers of alchemical knowledge also occurs in Syriac literature. The *Chronicle of Zuqnīn* (*c.* 775/6) and that of Dionysius of Tel Maḥrē (d. 845) tell the story of Isaac of Ḥarrān, who was visited by a wandering ascetic at his monastery in northern Syria. Wishing to repay Isaac's hospitality, the monk produced a small amount of gold created by a mysterious elixir. However, after the monk refused to reveal his secrets, Isaac killed him, stole the elixir and was consequently appointed Patriarch of Antioch at the urging of the caliph al-Manṣūr (incidentally, the caliph credited with initiating the Translation Movement).[23] This story is likely apocryphal but may echo the idea of Christians as mediators of occult science in an era of translation.

Alchemical associations were also projected onto Byzantine emperors. A number of lost and fragmentary works in Greek are attributed to Theodosius (r. 379–95), Justinian (r. 527–65) and Heraclius (r. 610–41).[24] The reason for these attributions is unclear, though it may have served to lend alchemy some measure of respectability.[25] Similar attributions to one Heraclius (Hiraql) are found in the Arabic tradition, though their connection (if any) to Greek counterparts is as yet unclear.[26] Additionally, there are works attributed to a Stephanus (Isṭafan, possibly Stephanus of Alexandria), who is sometimes described as a monk and – reminiscent of Maryānūs's association with Khalid – a teacher of Heraclius.[27] Fainter echoes of Heraclius as occult magus occur in the tale of *Salāmān and Absāl*, a philosophical fable allegedly translated from Greek by Ḥunayn ibn Isḥāq.[28] In this story, a sorcerer-king named Hirmānūs ibn Hiraql (Heraclius) begets a son through artificial generation[29] – a concern of Jābirian alchemy, as we shall see in the following section. Later Byzantine emperors also enjoyed alchemical associations. A chronicle by Ibn al-Faqīh al-Hamadhānī (fl. 902) preserves an account attributed to ʿUmāra ibn Ḥamza, ambassador of the caliph al-Manṣūr, of his visit to the imperial court in Constantinople. There, the emperor showed him various wonders, including steam-driven automata and elixirs that could transmute copper into gold and lead into silver. At the end of his account, the narrator declares that al-Manṣūr became concerned with alchemy only after learning of these marvellous things.[30] Once again, there is no evidence to back this story's historicity, but its presence may reveal how medieval authors imagined the spread of alchemy in Muslim domains.

Alchemy among the sciences

Alchemy often intersected with other areas of scientific activity. While practitioners of this craft were undoubtedly curious about the workings of matter, their engagement with the broader sciences tended to be fluid. From its beginnings, alchemy drew its theory of matter from several adjacent intellectual traditions.[31] This eclecticism continued among Arabic authors and developed in new directions.

Medieval Arab alchemists inherited key principles from earlier Graeco-Egyptian and Byzantine principles. These included a corporealist outlook inspired in part by Stoic cosmology.[32] Whereas Aristotle believed substance to be comprised of form and matter, the Stoics held that only matter qualifies as substance, since only bodies exist. Upon creation, all prime matter is identical in substance and devoid of distinguishing qualities. These unqualified bodies are then endowed with qualities

Tables demonstrating Jābir ibn Ḥayyān's 'science of balance' from *Kitāb al-Aḥjār 'alā ra'y Balīnās* ('Book of Stones according to Apollonius').

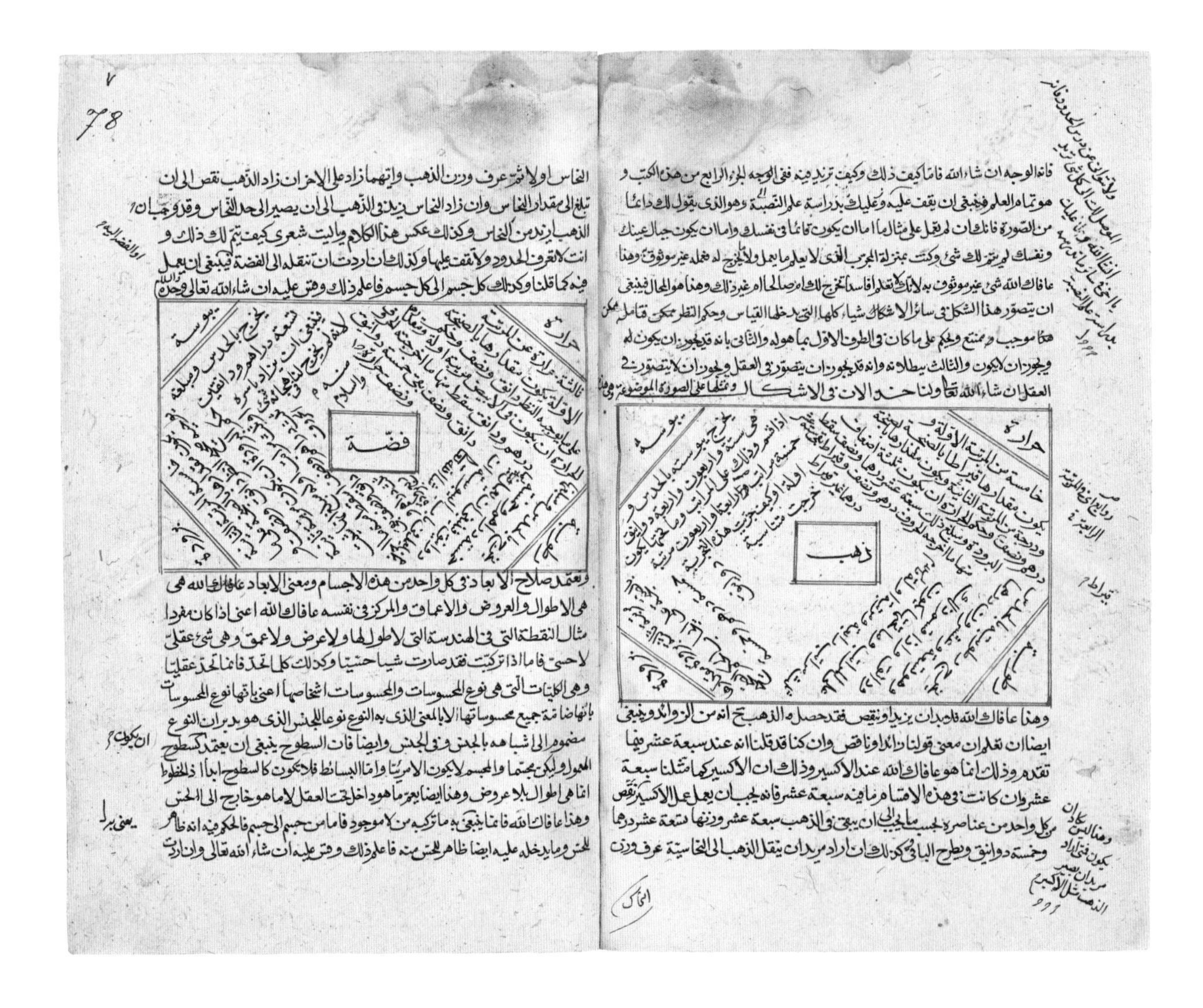

after being acted on by a spirit (*pneuma*). This meant that all matter is of one body, differing only in their spirit-imbued properties.[33] Early alchemists believed that this process could be replicated by reducing metals to prime matter and providing them with 'spirit' and 'life'.[34] In addition to possessing a spirit, metals were also said to have a soul, which Zosimos describes as having a 'sulphurous' nature.[35] By the Byzantine period, alchemists commonly understood that metals possess three 'hypostases' of body, soul and spirit.[36] These principles are taken up in the Jābirian Corpus, which maintains a rigorously physicalist conception of the universe.[37]

Connected to this outlook was an approach to metallic bodies influenced by Greek medicine. According to Galen (d. 200) and others, human bodies are made up of humours and qualities that cause illness when unbalanced and good health when balanced. The Jābirian Corpus applies this principle to metals, asserting that they require the correct balance of qualities to become gold. In order to do this, one has to determine the quantitative structure of qualities in a particular metallic body. Once again, inspiration is found in Greek medicine. In Galen's pharmacology, every medicine has four degrees of intensity that must match the severity of *dyskrasia* (imbalance) in a human body.[38] Building on this notion, the Jābirian Corpus formulates a complex theory of balance predicated on four degrees of hot, cold, wet and dry, the intensities of which are conceived of as numbers. The purpose of this system was to 'treat' metallic bodies by establishing the exact proportion of their qualities and modifying them. By determining the precise constitution of metals, one could alter them and produce new ones. Jābir premised this idea on the relationship between these intensities and the 28 letters of the Arabic alphabet. These letters are divided into four groups of seven, each corresponding to one of the four qualities. The numerical value of these degrees follows a progression of 1:3:5:8 – the second degree is three times the intensity of the first, the third five times and so on. Altogether, they add up to 17, a number held in great esteem by the Pythagoreans. In Jābir's scheme, the number 17 signifies the equilibrium that governs not only bodies but the entire cosmos. In practice, the position of a letter in a word determines the intensity of the quality of hot, cold, wet or dry that it represents.[39]

Alchemy's connection to medical theory would continue among later alchemists. In his *Rutbat al-ḥakīm* ('Rank of the Sage'), the Andalusian philosopher Maslama al-Qurṭubī (d. 964) states that a good practitioner of the craft, like a physician, must be a 'servant of nature' (*khādim al-ṭabīʿa*). Just as knowledge of nature can balance humours, so the alchemist must observe nature when balancing metals.[40] Abū Bakr al-Rāzī (d. 925 or 935), another renowned alchemist of the tenth century, was himself a practising physician.

Alchemists often sought to replicate natural phenomena through human artifice. This is evident in the theory of metallic formation, influenced notably by Aristotle's *Meteorologica*. Aristotle proposed that solar heat causes the earth to release two types of exhalations: vaporous from earth's moisture and smoky from its heat. Interactions with cold or heat generate minerals with water-like and earth-like traits, yielding fusibility and solidity respectively. Therefore, fusible or malleable metals such as copper and iron are said to have a water-like quality.[41] While Aristotle never conceived of any notion of alchemy, Greek alchemists linked vapours and smoke produced in the laboratory to such natural processes.[42] Arabic alchemists continued to advance theories of metallogenesis based on creative adaptations of Aristotle's minerology, often within the framework of Hermetic science. An early work in which this occurs is the *Sirr al-khalīqa* ('Secret of Creation'), attributed to the Greek Neo-Pythagorean philosopher Apollonius of Tyana, but actually a composite work from the ninth century. Central to its cosmology is a theory of micro- and macrocosm, which is encapsulated in the legend of the *Emerald Tablet*, a fictional narrative found at the beginning and end of the work. Here, Apollonius encounters a statue of Hermes Trismegistus, beneath which lies a tablet containing the opening lines:

> Truth! Certainty! That in which there is no doubt!
> What is above is from below, and what is below is from above.[43]

In a section on minerology, Pseudo-Apollonius states that all fusible metals have their origin in mercury. Recall Aristotle's claims that fusible metals are water-like substances caused by vapour. In the *Secret of Creation*, mercury is said to begin life as water, but when subjected to the sun's heat it turns to vapour and rises to the top of its subterranean cavern. With no place to escape, the vapour cools and returns to water, having trapped the heat and dryness within itself. These hot and dry qualities combine with the subtle nature of vapour, giving rise to a substance capable of dissolving sulphur. Through a naturally occurring process of 'roasting', sulphur's dryness is broken down and concealed within mercury. The result is a composite that has a dry, 'red' (i.e. sulphurous) soul and a moist, 'white' (i.e. mercurial) body. When metals are first created from this 'coupling' of sulphur and mercury, they are initially gold, since the two are in balance. However, this balance is disturbed when imperfections (*a'rāḍ*, literally 'accidents') appear in these metals during their formation. The resulting unbalanced proportions cause the creation of baser metals such as lead, tin, copper, iron, quicksilver and silver. All of these are of a single body that is gold in substance, differing only in accidents such as colour.[44]

The above theory is foundational to the Jābirian Corpus. In the *Kitāb al-Īḍāḥ* ('Book of Elucidation'), Jābir states that all metals are composed of various degrees of mercury and sulphur that intermingle and coagulate to produce various metals according to alterations of cold moisture and dry heat during their generation. Those metals produced from the most refined and balanced combinations of sulphur produce gold and those of mercury produce silver, while the wrong ratio of each produces baser metals. This principle applied not only to nature but also to the workings of the laboratory, where a practitioner could perfect base metals by balancing their proportions of mercury and sulphur.[45]

The idea that all metals were of a single body was developed in a short treatise by the famous Baghdad philosopher Abū Naṣr al-Fārābī (d. 950/1). Though unlikely to have been a practising alchemist, al-Fārābī asserted, on the authority of Aristotle's 'book on minerals' (*fī kitabihi fī al-ma'ādin*), that metals were of one species, albeit with interchangeable accidents:

> Aristotle first investigated [alchemy] in this book by way of dialectical reasoning. He established its validity with one syllogism and refuted it with another, as was his custom when faced with several points of contention. In the end, he established alchemy's validity with a syllogism he composed from two premises, which he explained at the outset of his book. The first was that gold, silver, and all non-combustible, fusible substances are one in species (*wāḥida bi-l-naw'*), and that the difference between them is not in their bodies but in their accidents, some of which are essential, others superficial. The second premise is that any two things classified under one species differ [only] in accidents. Transferring one to the other is therefore possible.[46]

It should be noted that Aristotle nowhere states these premises in his *Meteorologica*, nor is he known to have written a book solely devoted to metals.[47] It is unclear whether the work al-Fārābī cites is authentic or merely attributed. Yet the notion that all metals were of the same species would prove influential in later discussions about transmutation.

As to alchemy's place among other disciplines, al-Fārābī asserts that it can only be practised by philosophers, since mathematics and natural philosophy must be studied prior.[48] Al-Qurṭubī also attempts to situate alchemy within a broader body of disciplines, particularly in the context of other occult sciences. According to him, alchemy is of a pair with *sīmīyā'* – a term encompassing magic (*siḥr*) and the making of talismans. One must be trained in both alchemy and *sīmīyā'*n order to learn the 'secrets of nature' and gain mastery over the upper and lower realms. This is because alchemy attends to the extraction of earthy spirits in the sublunar world, while *sīmīyā'*concerns the bringing down of spirits from the celestial world.[49] Yet al-Qurtubī cautions that one cannot turn their hand to alchemy or *sīmīyā'* without first mastering the four propaedeutic disciplines of the quadrivium, i.e. arithmetic, geometry, music and astronomy.[50] The Ikhwān al-Ṣafā ('the Sincere Brethren'), an elusive fraternity of philosophers active in Baṣra in the tenth century, held that alchemy and magic were themselves propaedeutic sciences, to be studied before mathematics, logic, natural sciences, metaphysics and other subjects.[51]

So far, we have mentioned discussions among alchemists concerning theory and practice. However, some thinkers rejected alchemy's theoretical claims, chief among them the famous philosopher Ibn Sīnā (d. 1038), known in the West as Avicenna. While Avicenna accepted that metals comprise varying degrees of sulphur and mercury, he rejected the possibility that their species could be changed through artifice. This is because species are immutable in nature and unknowable to the senses. As such, alchemists can only change the outward appearances of metals.[52] Avicenna's refutation prompted responses from later alchemists. Among them was Mu'ayyad al-Dīn al-Ṭughrā'ī (d. 1121), who argues that the four elements possess no inherent species. Rather, they receive a species compounded into a body – an action that alchemists are capable of reproducing.[53] Fakhr al-Dīn al-Rāzī (d. 1209/10), a Sunnī Ash'arite theologian, also penned a response to Avicenna.

Citing examples from medicine, he asserts that a substance's actions and effects emerge not from its species but from its mixture. As such, it is not a metal's species that makes it gold but rather its mixture and balance of qualities.[54]

Finally, something should be said about alchemy and the artificial generation of life. Underlying this idea was the belief that natural processes could be recreated through human art. Just as metals could be generated by replicating conditions in the ground, so too could animals be created by simulating conditions in the womb. The principle also rested on the widely held notion of spontaneous (i.e. seedless) generation, as articulated by Aristotle in his *De Generatione Animalium*. Here, Aristotle writes that certain animals arose from earth and moisture when exposed to 'vital heat'.[55] According to the *Kitāb al-Tajmīʿ* ('Book of Gathering') of the Jābirian Corpus, this heat is achieved by means of putrefaction or fermentation (*taʿfīn*). This heat recreates the incubating warmth of the womb and the heat generated by the perpetual motion of the heavenly spheres, which is the cause of generation in the sublunar world.[56] In order to create this effect, one uses putrefying agents such as vinegar, soil or dung, which must be frequently nourished to maintain their incubating warmth. The creature generated by this procedure is determined by the length of 'cooking' and the nutrients used to nourish the decomposing matter, in accordance with the elements and qualities associated with that creature. The capable alchemist can generate any number of beings from this process, including hornets, snakes, scorpions and even rational human beings.[57] The quest for artificial life places the Jābirian Corpus in the company of works on natural and astral magic. For example, the *Kitāb al-Nawāmīs* ('Book of Secrets', known as the *Liber Vaccae* in the Latin West) and *al-Filāḥa al-nabaṭiyya* ('The Nabatean Agriculture') of Ibn Waḥshiyya (fl. 930) discuss the creation of living beings – including humans and, in some cases, human–animal hybrids – in the context of both talisman-making and spontaneous generation.[58]

The elixir and transmutation

As noted, alchemists conceived of metals as possessing a single body, each differing only in accidents such as colour. As such, every metallic body contained an inner essence – or substance – of gold that could be brought to its surface through transmutation. In Greek, this process was called *ekstrophē* ('inversion').[59] Similarly, alchemists in the Islamic world tended to refer to transmutation as *taqallub* ('inversion').[60] One achieved this through an agent known as an elixir, *al-iksīr* in Arabic. The term derives from the Greek *xērion* (possibly via the Syriac *ksīrīn*), originally a medicinal powder used in the treatment of eyes. As discussed above, alchemists often applied medical concepts to their craft, and so additional terms for the transmuting agent included 'medicine' (*dawāʾ*, pl. *adwiya*), while its ingredients were often referred to as 'drugs' (*ʿaqqār*, pl. *aqāqīr*).[61]

Various explanations and descriptions were given for the elixir's characteristics and function, often with recourse to wordplay, false etymology and encoded language (on which more below). In Graeco-Egyptian and Byzantine alchemy, the transmuting agent was often referred to as 'divine water', the 'philosophers' egg', 'a stone that is not a stone' and, later, the 'philosophers' stone'.[62] Sometimes Greek writers would elaborate on these names, such as when Zosimos writes that the transmuting agent is a 'yeast', because a small quantity yields a large amount.[63] Arabic writers took up many of these descriptions: in the Jābirian Corpus we encounter the terms 'yeast of the philosophers' (*khamīr al-ḥukamā'*), the 'philosophers' stone' (*ḥajar al-falāsifa*), and the 'philosophers' egg' (*bayḍ al-falāsifa*).[64] Jābir explains the etymology of the word elixir (*al-iksīr*) as deriving from the Arabic root *k-s-r*, because it is broken down and crumbles (*yankasiru wa-yanfatitu*), possibly a reference to its powdered state.[65] In a similar vein, al-Qurṭubī asserts that the elixir's name derives from its 'breaking power' (*al-quwwa al-kāsira*) over metals into their constituent parts, just as digestion breaks down food into nourishment.[66]

A common notion about the elixir is that it is triple and quadruple in nature. Jābir states that the elixir must possess (i) a 'body' that acts as a vessel; (ii) a 'soul' that protects the metal from heat; and (iii) a 'spirit' that 'resurrects' bodies and endows them with their distinctive qualities.[67] Al-Qurṭūbī elaborates on this scheme, explaining that the elixir is three (or a triangle) in nature and four (or a square) in qualities. The four natures signify the elixir as the sum of four elements. The three, meanwhile, signifies (i) the spirit's imparting of 'whiteness' to the metal through moistening and cooling; (ii) the soul's 'redness' on account of its hot and dry qualities; and (iii) the body's 'blackness', which hardens malleable substances through coldness and dryness. Accordingly, the elixir turns metals to gold by 'whitening' copper and 'reddening' silver.[68] A later author, this time of a text attributed to Avicenna, reports that the elixir possesses a further four actions: (i) withstanding fire; (ii) tincturing substances; (iii) mixing with and immersing in substances; (iv) providing stability and fixation to substances.[69]

While most alchemists could broadly agree on the function of the elixir, many were divided about its composition. The earliest Greek alchemists tended to stipulate that the starting point of the tincture was a mineral substance. However, the Jābirian Corpus insists that the most essential and noble of substances in the manufacture of the elixir are those from animals. This is because they break down more readily through distillation, making it easier to extract their qualities for use in metals.[70] The identity of these 'animal stones' would later be specified by al-Rāzī, who lists hair, crania, brains, gall, blood, milk, urine, eggs, mother-of-pearl and horn.[71] The alchemist Muḥammad ibn Umayl (fl. tenth century) rejected the use of animal products in the creation of the elixir. His main contention was that alchemists misapprehend the codes and symbols of the ancients, interpreting references to things like eggs and gall in a literal manner.[72] He further asserts that the elixir must be composed of that which it seeks to transform. While he does not specify its exact composition (other than that it is made from a mysterious substance called 'magnesia'), Ibn Umayl tells us that the elixir functions in the way that a sperm or plant seed acts upon material of the same species. With its 'nutritive spirit' (*al-rūḥ al-nāmiya*), the elixir 'nourishes' the incipient body.[73] The implication here is that for the elixir to work on metallic bodies, it must itself be composed of something metallic. Al-Qurṭubī, on the other hand, adopts something of a *via media*, insisting that the elixir is a 'world' (*ʿālam*) – that is, a microcosm composed of animals, plants and minerals.[74]

In addition to ingredients, the preparation of the elixir required various operations involving technical equipment. It is here that we observe alchemy as both a technical and theoretical science. The working of the 'stone' was conceived of in stages. In his *Kitāb al-Asrār* ('Book of Secrets'), al-Rāzī stipulates seven operations (possibly on the model of the seven metals and planets). These are: (i) calcination (*taklīs*); (ii) ceration (*tashmīʿ*); (iii) dissolution (*taḥlīl*); (iv) mixing (*tamzīj*); (v) coagulation (*ʿaqd*); (vi) sublimation (*taṣʿīd*); and, finally, (vii) extracting the 'reddening waters' (*al-miyāh al-muḥammira*), that is, the transmuting agent.[75] There were further procedures such as assation (*tashwiya*), oxidation (*taṣdiʾa*) and the making and application of 'philosophers' clay' (*ṭīn al-ḥukamāʾ*), used for luting pots, vials and other vessels. These included the famous alembic (*al-anbīq*, from the Greek *ambix*), the cucurbit (*qarʿ*) and the aludel (*al-uthal*, from the Greek *aithalīōn*).[76] Their manufacture is set

A group of men discover a statue of a figure holding a tablet containing alchemical symbols inside the temple of Abū Sīr, Egypt. From *The Silvery Water* by Ibn Umayl, 1339.

out by al-Rāzī and described by later encyclopaedists. However, individual sources are rarely sufficient in building a complete picture, since alchemists were guided not only by texts but also practice and observation, the results of which could often differ from one case to another.

Alchemy and secrecy

A hallmark of alchemical writing is its symbolic language. One reason for this was alchemy's artisanal nature, passed down from one craftsman to another. As previously mentioned, alchemy in the medieval Islamic world lacked a centralised curriculum, resulting in significantly varied interpretations and practices. We have already observed debates about the use of animal products in the elixir-making process. One driver of these debates was the prevalence of symbols (*rumūz*) and enigmas (*alghāz*) in descriptions of procedures, ingredients and recipes, the interpretation of which often led to widely differing approaches.[77] Moreover, the use of encoded language introduced an intriguing paradox. While alchemical authors frequently underscored the hidden nature of their knowledge, they would simultaneously proclaim its revelation to those they deemed deserving. Consequently, many texts adopt an initiatory tone, portraying the author as a mentor guiding a pupil.[78] This literary approach takes diverse forms, most significantly fictional dialogues between a legendary sage and their student. In addition to Maryānūs and Khalid ibn-Yazid, a classic example of this literary framework comes from the *Sirr al-asrār* ('Secret of Secrets') attributed to Aristotle and addressed to Alexander the Great.[79]

Historians have long debated the origins of alchemy's esotericism. Some have posited a link between the images and writing inside ancient Egyptian temples and the early use of alchemical ciphers.[80] This is suggested by such texts as *On the Working of Copper*, a work lost in Greek but preserved in Syriac. Here, Zosimos states that the method of modifying silver was a mystery that had been interpreted by the ancient teachers and prophets from various 'priestly idols' (*ṣalmē d-kumrē*).[81] Similar practices are found elsewhere in the late antique Mediterranean world; the Neoplatonist philosopher Porphyry of Tyre (d. 301), for example, is known to have written a treatise on the hidden meaning of statues.[82]

Medieval writers were also aware of alchemy's associations with Egypt. Ruins of Egyptian temples were visible and accessible throughout the Middle Ages, remaining open to inquisitive visitors who speculated about the meaning of their images.[83] Additionally, Arab authors were informed about early alchemy's background through translated Greek texts, many of which retained references to Graeco-Egyptian themes. Chief among these themes was the temple priest as guardian of arcane mysteries. For example, Zosimos complains of practitioners who put their names on works by Hermes but are censured by priests who keep the originals in their temple sanctuaries.[84] By the tenth century, authors invoked the image of Egyptian temples in their alchemical writing. Ibn Umayl begins one work with an account of his visit to an ancient temple (referred to in Arabic as *birbā*, pl. *barābī*, from the Coptic *p'erpe*) in Giza.[85] Inside he encounters statues, images and inscriptions in *birbāwī* ('temple-writing', possibly hieroglyphs).[86] In a passage reminiscent of the *Emerald Tablet*, Ibn Umayl recounts his discovery of a statue bearing a tablet on its lap. The tablet's images and symbols, Ibn Umayl asserts, epitomise what the philosophers have hidden away.[87] He later states that these symbols constitute the 'hieratic speech' (*al-kalam al-kāhinī*) of the Egyptian temple priests, which contemporary alchemists have misconstrued, leading them to erroneously prescribe impure animal substances for the preparation of the elixir.[88] These ancient Egyptian associations also emerge in Ibn al-Nadīm, who mentions chambers within a temple dedicated to various alchemical operations like mixture, grinding, dissolution and distillation.[89] Al-Qurṭubī emphasises the superiority of alchemical images over written text, asserting that these symbols were initially discovered in the temple (*birbā*) of Ikhmīm (Panopolis in Egypt, Zosimos's birthplace).[90]

In practice, alchemical symbols functioned as *Decknamen* ('cover names') for the elixir and its ingredients. We have already noted the varied names ascribed to the elixir. In addition to these, one also encounters poetic language that conveys both the elixir's nobility and modesty, referring to it as 'expensive-cheap', 'lowly-precious', 'known-unknown' and 'present-absent'.[91] A connected strategy in the Jābirian Corpus involves *tabdīd al-ʿilm* ('scattering of knowledge'). This is characterised by a haphazard treatment of topics, which contrasts with the systematic approach one ordinarily finds in philosophical texts. This method was likely employed to safeguard secrecy and render alchemical doctrines less accessible to a wider readership.[92] Elsewhere, one encounters a diverse array of symbols and pseudonyms for alchemical ingredients, or 'drugs'. These designations often hint at a substance's characteristics while masking its true nature. Examples include the 'escaped slave' (*al-ʿabd al-ābiq*) for mercury, denoting its tendency to 'flee' when subjected to heat, and 'scorpion' (*ʿaqrab*) for sulphur, an allusion to its penetrating qualities.[93]

In addition to recognising the importance of alchemical symbols, medieval intellectuals also theorised their use. In defence of encoded language, al-Fārābī states that obscure expressions shield knowledge from those unversed in philosophy. He then asserts that unless the alchemical art is hidden from all but a few, gold and silver would cease to have any value, leading to social collapse.[94] Other writers sought to defend the use of symbols even as they attempted to decode them. According to Ibn Umayl, while symbols may sometimes appear contradictory, their principles remain consistent.[95] He later bemoans practitioners who refuse to invest the necessary effort into deciphering the mysteries of the ancients, which leads them to a superficial understanding of their meanings. Among them were those who believed they possessed direct access to divine secrets.[96] Al-Qurṭubī, meanwhile, complains of critics who claim that alchemical codes are devoid of meaning.[97] To establish the credibility and utility of ciphers, al-Qurṭubī sets out to explain them. In doing so, he enumerates three species of alchemical symbol: (i) adornment (*tamlīḥ*), which is employed by way of literary artifice; (ii) obfuscation (*ikhfāʾ*), which is intended to throw the reader off the trail; (iii) and disclosure (*iẓhār*), which nudges the reader towards the symbol's meaning.[98]

Illustration of the parts of the distillation apparatus in *Kitāb al-aqālīm al-sabʿa* ('Book of the Seven Climes') by Abū al-Qāsim al-ʿIrāqī *c.*1300 CE.

ثم اغوزه في الوسط وركب الغطا عليها وطينها واجعلها في بنيه مثل عمل
النشادر واوقد عليه بنار لينه نصف يوم حتى تذهب الرطوبه ثم قوي
عليها النار تمام ثلاثه ايام بلياليها ثم اضعها يبرد يوم اخر وافتحها تجد
قد صعد على الوجه جوهرا كانه الخفية البيضا فخذها واعلم انك قد حزت
ملك الدنيا فاخزنها في اناء زجاج واحكم الوصل بكلما تقدر عليه فان
الحكمه بالشد الجيد لئلا يروحن ويهربن منك فاعلم ذلك ثم خذ
من الحجر الاول طرى فاغسله واجعله في قرعه وانبيق الى ثلثها او نصفها بلا زياده
وركب عليها الانبيق الواسع المزراب واحكم وصلها واوقد عليها بنار لينه
مثل حرارة الشمس يطلع الما صافيا

انبيق
الوصل
قرعه
قابلة التقطير

فاعلم يا ولدي ان كانت نارك
شديده طلع الما اصفر مطرب الى الحمره فيكون مفسد
فيكون نارك برشد تنال ما تريد بسرعه بمشية الله وعونه
حتى اعزل الثفل حتى تحتاج اليه ثم خذ من ذلك الما
الابيض عشر دراهم التي منها ثلاثة دراهم ونصف من ذلك النشادر فانه يتحلل فيه
ويصير في اشد بياض من اللبن الحليب وهو الذي يقال له لبن العذرى فاجعله في
قدح العقد واحكم وصلها باللطف ثلاثة ايام بلياليها باين ما تقدر عليه وعلامة
انعقاده ليس يطلع في القدح الفوقاني عرق البته فاعلم انه انعقد ثم ضعه

Alchemy and Islam

We have already noted alchemy's intersection with adjacent disciplines such as Hermeticism, Aristotelian philosophy and Galenic medicine. It should also be noted that alchemy came into contact with various modes of religious discourse. This encounter occurs as early as the Jābirian Corpus. While the core of these texts may have been written by Jābir ibn Ḥayyān, they contain later accretions by writers with markedly Shī'ite commitments. Jābir himself is said to have been a companion of Ja'far al-Ṣādiq (d. 765), the sixth Shī'ite imam.[99] One Jābirian work recounts a story about 'Alī ibn Abī Ṭālib (d. 661), the first imam and fourth of the 'rightly guided' caliphs. In it, a crowd of people demand to know from 'Alī whether alchemy exists as a legitimate science. 'Alī responds in the affirmative but does so in a highly gnomic manner. When pressed for clarification, he sternly reminds the crowd that such secrets cannot be disclosed to the masses.[100] Moreover, the Jābirian Corpus contains elements of Shī'ite esotericism inspired by the doctrine of the imam's occultation and apparition. This is notable in discussions about metals possessing a hidden (*bāṭin*) and manifest (*ẓāhir*) quality.[101]

Religious discourse also featured as a rhetorical device among alchemists, particularly those who wished to lend legitimacy to their theories and methods. This approach is notable in Ibn Umayl, who views alchemy as a primordial science revealed by God to Adam. While the idea of alchemy as a *donum dei* was already present in Byzantine alchemy,[102] Ibn Umayl marshals this strategy against those alchemists who sully their elixirs with animal products. How, he asks, can something so holy emerge from impure substances like rotten eggs, blood and urine? Moreover, how can a discipline bearing the imprimatur of the imam 'Alī deal in such filth?[103] Al-Qurṭubī also incorporates religious ideas into his work. As has been mentioned, he situates alchemy within a broader system of occult science that encompassed magic and the talismanic arts. On this scheme, alchemy, like magic, brings an effect into existence *ex nihilo*, just as God creates with the word 'be' (cf Quran 3:59 and 19:35). Echoing Plato's *Theaetetus* (176 B 1), al-Qurṭubī explains this principle as the philosopher imitating the divine.[104] To this effect, he employs the term *tarjīḥ* ('preponderation') from Islamic theology (or *kalām*), which denotes God's causation of temporal change.[105]

It is important to bear in mind that there existed no centrally defined notion of orthodoxy that addressed alchemy or the occult sciences in general. Rather, Islamic theologians and jurists tended to hold a variety of views on the subject. Famous among those who engaged constructively, if not positively, with the occult sciences was the Sunnī Ash'arite theologian Abū Ḥāmid al-Ghazālī (d. 1111). In his *Tahāfut al-falāsifa* ('Incoherence of the Philosophers'), he counts astral determinations, physiognomy, dream interpretation, talismans and *nīranjāt* ('spells') among the natural sciences.[106] Although he elsewhere condemns some of these practices as being disapproved by law,[107] al-Ghazālī seizes on others for polemical ends. Against the Avicennans, who believe that matter is only receptive to its specific forms, al-Ghazālī points to the wondrous workings of talismans, which combine terrestrial and astral properties to produce effects contrary to their material dispositions.[108]

It is perhaps for this reason that he is portrayed as being favourably disposed to alchemy, an adjacent practice, in a pseudo-epigraphic work entitled *Sirr al-ʿālamayn* ('The Secret of the Two Worlds'), which discusses transmutation as an allegory for spiritual renewal.[109] The occult sciences are also present in the thought of Fakhr al-Dīn al-Rāzī, a later Ashʿarite of similar renown. In addition to a treatise on astral magic, he also wrote in defence of alchemy, as mentioned above.[110] Moreover, al-Rāzī was known to have practised this craft, at least according to his later biographers.[111]

Alchemical themes also featured prominently in Islamic mysticism. When Jungian psychoanalysis held sway over the historical study of alchemy, it was widely believed that descriptions of chemical processes were merely accounts of mystical transformation. The field has long since moved on from such reductive and ahistorical interpretations.[112] However, this does not mean that medieval Sufi writers did not draw inspiration from alchemical themes. Mention has already been made of the *Sirr al-ʿālamayn*, a work attributed to al-Ghazālī that discusses spiritual development through the language of transmutation. Al-Ghazālī adopts a similar approach in an authentic Persian work entitled *Kīmīyā'yi saʿādat* ('The Alchemy of Happiness'). A later author, Najm al-Dīn Kubrā (d. 1221), the founder of the Kubrawiyya Sufi order, developed a system of spiritual progression on the model of colours associated with the stages of the elixir's preparation.[113] Similarly, the philosopher-mystic Ibn ʿArabī (d. 1222) referred to the goal of spiritual purification as 'red sulphur', a cover name for the elixir.[114]

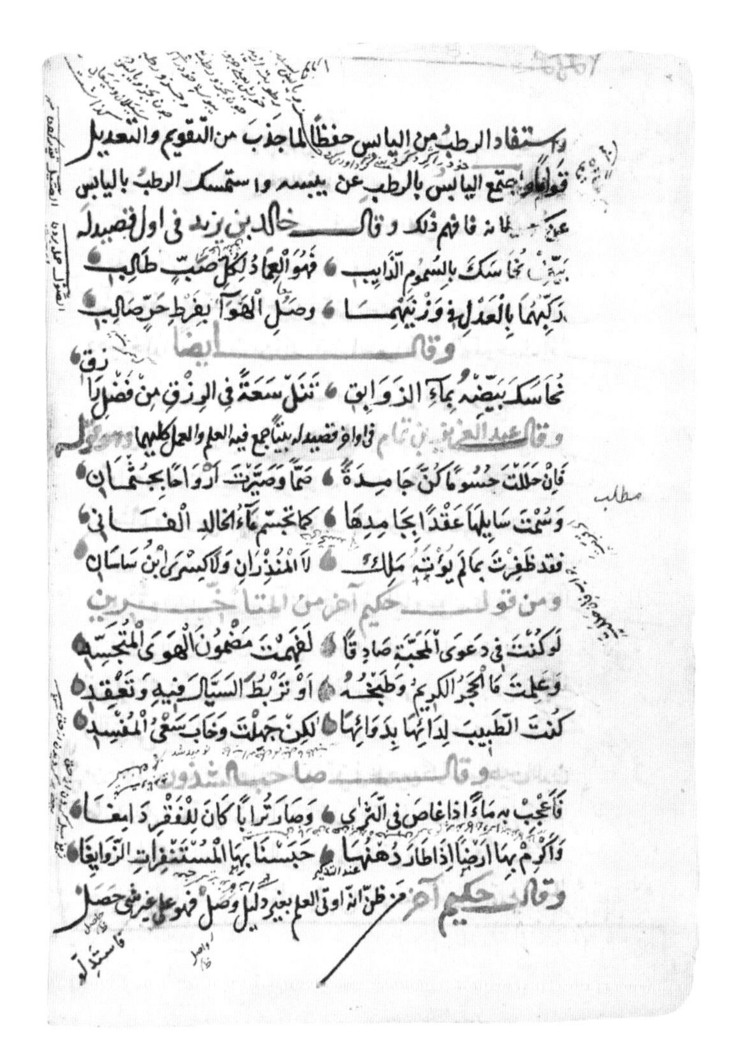

Alchemical compendium in the British Library, London, copied in 1764 in Algiers. Pictured here is a poem attributed to the Umayyad prince Khalid ibn-Yazid.

Other topics and avenues of research

In the foregoing I have attempted to contour some of the most salient topics in the study of medieval Arabic alchemy. However, there remain many others that I have been unable to cover in this modest survey. Before ending this chapter, I will give a brief account of some further aspects of medieval Islamic alchemy that would repay further study.

I have surveyed the history of Arabic alchemy from about the eighth century until the twelfth. Yet it is important to note that alchemy was a *continuous* tradition, developing through the Middle Ages and into modern times. Important authors of a later pedigree include the Malmluk-era alchemist ʿIzz al-Dīn Aydimir al-Jildakī (d. 1342). Al-Jildakī produced several systematic works covering all aspects of alchemy, in addition to commentaries on earlier writers.[115] He would later influence writers such as the Ottoman alchemist ʿAlī Çelebī.[116] Other centres of alchemical activity included seventeenth-century Morocco, where many intellectuals practised the craft alongside astronomy, logic and medicine.[117] A further glimpse of Arabic alchemy's modern afterlife comes from the historian of science and chemistry teacher Eric Holmyard. Writing in the first half of the twentieth century, he relates a personal anecdote about one Al-Haj Abdul-Muhyi Arab, the mufti of Shah Jahan Mosque in Woking, England. Abdul-Muhyi wrote him an introduction to a well-known alchemist in Fez, who would later show him 'a subterranean alchemical laboratory in the old part of that city'.[118]

Another emerging area of enquiry is alchemy's literary tradition. In this chapter I have briefly discussed various discovery legends that provided the literary frame for much alchemical writing, particularly those in the Hermetic mould. In addition to these, one encounters Arabic poetry as a site of alchemical knowledge. In many ways, poetry served as a fertile medium for alchemical writing, favouring as it did wordplay, allegory and riddle. Such literary devices were central to alchemy and could also be found in Arabic poetry on medicine and other learned subjects.[119] A collection of poems is attributed to the Umayyad prince and alchemist of legend Khālid ibn Yazīd, as well as later figures such as al-Ṭughrāʾī and Ibn Arfaʿ Raʾs (fl. twelfth century).[120] The poetic writings of these authors would occasion several commentaries by later writers such as al-Jildakī and others.[121] An examination of these poems and their associated commentary traditions reveals the innovative methods through which alchemy continued to develop.

Page from *al-Miṣbāḥ fī asrār ʿilm al-miftāḥ* ('Elucidation on the Secrets of the Science of the Key') by ʿIzz al-Dīn Aydimir al-Jildakī. This book is a compendium on alchemy that occasionally employs a secret alphabet (in red).

وتعلم ان الاشيا تقوى باشكالها وتضعف باضدادها فلا تضاف المدبرات المنسوبة الى زحل
الا الى المدبرات المنسوبة الى زحل وكذلك لا تضاف المدبرات المنسوبة للمشتري الا الى المدبرات المنسوبة
الى المشتري وكذلك القول في المدبرات المنسوبة للمريخ فيمكن الجمع بينها فتقوى باشكالها وكذلك المدبرات
المنسوبة للشمس ايضا وكذلك المنسوبة للزهرة وكذلك المنسوبة لعطارد وكذلك المنسوبة للقمر
فمقتضى ما ذكرناه يصير لكل كوكب مفتاح اعظم كبير معلق في سلسلة وله عدة اصابع طوال
وفي كل اصبع منه عدة مفاتيح صغار وكل مفتاح صغير فيه عدة اصابع قصار وفي كل اصبع من
الاصابع الطوال والقصار عدة اسنان فافهم ما نقول وبالله المستعان واعلم ان المفاتيح المنسوبة
لزحل تؤثر التاثير البالغ فيما ينا سب زحل من الاشياء المعدنية التي هي موضوع الصناعة الالهية وفي
جسده المنسوب اليه فتقومه وتصالحه وتعدله الى ما يقارب مزاج الشمس او القمر وكذلك القول في
المفاتيح وكذلك القول في المدبرات المنسوبة للمشتري فانها تفعل افعال الكوكب السعد الذي هو المشتري
في الاجسام المعدنية المنسوبة للمشتري لا سيما في [illegible] وكذلك القول في
المدبرات المنسوبة للمريخ فانها تفعل افعال المريخ القوية اذا كان سعيدا مسعودا في الاجسام المعدنية
بالسرعة والقوة لا سيما في [illegible] فتكون مدبراته كالصواعق والبرق المتلاحق من القوة والنفوذ
باذن الله عز وجل الفعال لما يريد وكذلك القول في المدبرات المنسوبة للشمس المنير فان فيها افعال انقلاب
صورة الاكسير فيجعل الذهب المنير كثير الضيا قوي النورانية والشعاع و [illegible]
[illegible] الذي يذوب بايسر الحمي من غير امتناع وينتقل بحمد الله الى طبيعة [illegible]
في المنظر والمخبر الله اكبر الله اكبر وكذلك القول في المدبرات المنسوبة للزهرة فانها تحيل الاجسام المنسوبة
اليها من المعادن وتظهر من اسرارها ما هو مخفي وكامن وتوصلها الى طبيعتي النيرين من غير شك
ولا ريب ولا مين وكذلك القول في المدبرات المنسوبة لعطارد فانها تفعل في الاجسام المنسوبة اليه
فعل الصلاح وكذلك تفعل في [illegible] وفي المغنيطس وفي [illegible]
وفي [illegible] وفي سائر الارواح من حل وعقد وتمكين واصلاح وكذلك ايضا في تعديل المزاج

ويجعل الذهب

مگر قوت را چاره پیدا کنیم
ز خلق جهان بی نیازی کنیم
زن کارپرداز روشن ضمیر
بدان خوی اپتکه گشت خواهش پذیر
یکی منظری بود با آب و رنگ
مقرنس بر آورده از خاره سنگ
عروسانه بر شد بران جلوه گاه
بر ان پری سینه بسته کرد ماه
براموده چون کپس و مشک سپید
بموی سیه مهرهای سفید
صلیبی دو گیسوی مشکین کمند
دران مهره آورده پاپیچ و بند
بنظارگان گفت گیسوی من
ببینید در طاق ابروی من
نمودار اکسیر پنهانیم
بجویید در صحن پیشانیم
۳۹

ALCHEMY IN INDIA

DAGMAR WUJASTYK

A rare depiction of women in alchemy. Miryam, a Christian maiden versed in the art, is consulted by other alchemists. From *Khamsa* ('Five Poems'), by Nizami. Early 1590s CE.

Several historical disciplines that developed on the Indian subcontinent could be understood as alchemical traditions in the sense that they deal with the transmutation of matter. The first is called the 'doctrine of elements' (Sanskrit: *dhātuvāda*) and is associated with the making of gold. The second is the doctrine of mercury (Sanskrit: *rasavāda*, also *rasavidyā* and *rasaśāstra*), which is focused on the making and application of mercurial elixirs for the transmutation of metals and the human body. The third is the South Indian Tamil Siddha tradition, which combines yogic with alchemical and medical thought, using both mercury and other substances as agents of transmutation.

References to these disciplines are preceded by stories of spiritual masters turning various substances into gold. The earliest of these are found in Buddhist and Jain literature. The Buddhist sources describe miraculous events in which a Buddhist master turns stones, or even mountains, into gold.[1] In some sources, this change is effected through the application of another substance, which is usually undefined or generically referred to as 'drugs', 'juice', 'medicine' or 'decoction'.[2] None of these sources mention mercury as an agent of transmutation. In the seventh century CE, the Chinese pilgrim Xuanzang told stories of a Nagarjuna in his *Record of the Western World*, in which Nagarjuna is described as being able to make gold but also 'well practised in the art of compounding medicines; by taking a preparation (pill or cake) he nourished the years of his life for many hundreds of years so that neither the mind nor appearance decayed'.[3]

There are also similar stories in Jain sources. For example, the *Vasudevahiṇḍī* ('Vasudeva's Wanderings', fifth century CE), a narrative text written in Prakrit by the Jain monk Saṅghadāsa, describes an ascetic making gold from iron pyrites and *rasa*, which may or may not be mercury.[4] The Shvetambara monk Haribhadra's Prakrit version of the *Dhūrtakhyāna* ('Stories of Rogues', eighth century CE) offers a new topic: it describes a well full of mercury that has the power to transform 1,000 times its weight of base metals into gold.[5]

These stories introduce some of the themes that later become characteristic of the alchemical disciplines, namely (1) the transmutation of materials, and (2) the prolongation of life and rejuvenation through potions. The sources do not, however, make reference to an organised, authoritative body of knowledge: there is no mention of a doctrine (*vāda*), science (*vidyā*) or discipline (*śāstra*) with generally applicable techniques in any of these sources. Nor is there any reference to groups of practitioners dedicated to or versed in the craft. Rather, the stories are bound to the advanced spiritual achievements of the individual effecting transmutation. Here, spiritual advancement precedes the ability to effect transmutation, rather than spiritual experience being effected through alchemical practice. The *Stories of Rogues* is alone in ascribing special transmutational properties to mercury: a theme that becomes the central focus of *rasavāda*, the mercurial alchemical discipline.

A young woman entices mercury from a well, as men, waiting to harvest it, hide nearby. Detail of a miniature painting illustrating a popular fable, from an album painted in India, Provincial Mughal style, 1740–60.

The doctrine of elements (dhātuvāda)

Early references to the doctrine of elements as a discipline concerned with the transmutation of minerals and metals into gold are first found in Jain literature and in Sanskrit plays dating to the second half of the first millennium CE. The Sanskrit word *dhātu*, rendered as 'elements' here, may refer to any metal or mineral. The perhaps earliest reference to a metallurgical science occurs in Kauṭilya's *Arthaśāstra* ('Treatise on Success'), where the term *dhātuśāstra* – 'the science of metals/elements' – is used to denote knowledge about mining, ores, smelting metals, etc. This science was not concerned with transmutation or the making of gold.[6] What may possibly be the first mention of the term *dhātuvāda* is found in a somewhat surprising source, namely Vātsyāyana's *Kāmasūtra* ('Treatise on Pleasure', *c.* third century CE). In the third chapter of Book One, a list of 64 arts that should be studied by women along with the *Kāmasūtra* includes *dhātuvāda*, which Doniger and Kakar simply translate as 'metallurgy'.[7] However, the text offers no further explanation, so we do not know whether this denotes the alchemical or the more basic metallurgical art. However, the seventh-century Sanskrit play *Kādaṃbarī* by Bāṇabhaṭṭa mentions a *dhātuvāda* that probably refers to making gold. The play offers a first, satirical take on the figure of the alchemical practitioner:

> He had a tumor growing on his forehead that was blackened by constantly falling at the feet of the mother Goddess...and was blind in one eye from a batch of invisibility salve [*siddhāñjana*] given him by a quack...He had brought a premature fever on himself with an improperly prepared mercurial elixir [*rasāyana*] used as a vermifuge...He had a collection of palm-leaf manuscripts containing material on conjuring, tantra, and mantra, which were written in letters of smoky-red lacquer. He had written the doctrine of Mahākāla as such had been taught to him by an old Mahā-pāśupata. He was afflicted with the condition of babbling about buried treasure and had become very windy on the subject of transmutational alchemy [*dhātuvāda*]...He had increased his grasp on the *mantra-sādhana* for becoming invisible, and knew thousands of wonderful stories about Śrīparvata.[8]

While *dhātuvāda* is named as a discipline in this story, the person undertaking it is not (just) an alchemist, but a South Indian Shaiva ascetic dabbling in a variety of alchemical and other esoteric activities.[9] The author's assessment is clear: the practitioner is foolish, greedy and gullible, and the practices are ridiculous. Nonetheless, this passage is the first instance in which we see alchemy linked with Shaiva tantric thought, the excavation of buried treasure and rejuvenation and longevity therapies (*rasāyana*). All these elements play a role in mercurial alchemy.

In the *Kuvalayamālā* ('Garland of Blue Lotuses'), a Prakrit-language novel written by the Jain monk Uddyotana-sūri in the late eighth century, alchemists receive a mixed evaluation, but the craft itself (Prakrit: *dhāuvāo*) is presented as one of the arts an educated person should have mastered. The protagonist of the novel, the prince Kuvalayacandra, encounters a group of alchemists unsuccessfully trying to make gold in the forests of the Vindhya mountains. The prince steps in and helps them. His success in making gold is explained as being due to knowing the technical aspects of the operations, as well as the spiritual ones, such as performing the

Dhātuvāda, or metallurgy, the 39th of the 64 traditional arts (*Chathusashti Kalas*) for women listed in the *Kāmasūtra*. Painted scroll from Odisha, 20th century.

correct prayers to Jinas and Siddhas.[10] Later in the story, Kuvalayacandra explains that there are different kinds of alchemists:

> The man who is familiar with the procedure of mixing (the substances) together is (the man known as) one 'conversant with alchemical operations'; he who has skill in fixing (all substances right up to) mercury is (the man who is called) a 'master alchemist'.
>
> The man who knows how to take the metal out of a soil (propitious to this end) and blows it with the aid of alkalis is called, obviously, by all an 'alchemist', so it is said.[11]

This story depicts successful alchemists as persons with both theoretical and technical expertise, but also portrays spiritual and ethical virtue as a crucial part of the craft. When asked by his wife why 'alchemists are constantly seen to err and to fail to achieve their purpose', the prince explains that:

> There are indeed some master alchemists who are doomed always to fail. These are men devoid of virtue, men who are completely lacking in purity and who do not respect chastity; they are under the yoke of greed and are greedy to possess everything; they are prepared to deceive their own friends, prove themselves devoid of gratitude, and take refuge in no god; they have succeeded in assimilating no magic spell; nor do they have any helper; they are ignorant and devoid of the quality of perseverance; they criticize their master, lack trust in themselves, and are lazy.[12]

Conversely, those who are virtuous and spiritually advanced will be able to achieve the goals of the practice. This passage thus describes an established metallurgical discipline with skilled practitioners who have complementary areas of expertise. The aim is once again the making of gold, with no mention of longevity, special powers or spiritual ends.

The doctrine of mercury (rasavāda)

Around the tenth century CE, a new literature emerged that described the doctrine of mercury (*rasavāda*), i.e. alchemical practices that centre on the uses of mercury. This literature represents an insider perspective with alchemists describing their own craft. The two earliest texts of the genre, the tenth-century *Rasahṛdayatantra* ('Book of the Heart of Mercury') and the *c.* eleventh-century *Rasārṇava* ('Ocean of Mercury'), lay out an alchemical programme in which liquid quicksilver is cleansed, distilled, mixed and amalgamated with other organic and inorganic substances, calcined and compounded into an elixir. The descriptions of these procedures, which make up the bulk of the alchemical works, are technical and practical, though not always detailed enough to be used as prescriptions. Consider, for example, these verses from the Book of the Heart of Mercury (Chapter 2, verses 3–13) describing the first nine procedures for making a mercurial elixir:

1. Steaming (*sveda*)
   The steaming of mercury is carried out with a sixteenth each of mustard, salt, the three pungent substances, leadwort, fresh ginger and radish together with sour gruel for three days over a mild fire.
2. Trituration (*mardana*)
   Trituration with sour gruel is carried out for three days with molasses, burnt wool and salt, together with soot, brick dust and mustard, each in the amount of a sixteenth part to the mercury.
3. Thickening (*mūrchā*)
   The three inherent faults of mercury are called 'dirt', 'fire' and 'poison'. Through dirt, it produces fainting; through fire, a burning sensation; through poison, death.
   Aloe removes dirt, the three myrobalans remove fire and leadwort removes poison. Therefore, one should thicken it seven times with a mixture of these.
4. Condensation (*ūrdhvapātana*)
   Having prepared a copper paste, a condensation procedure is conducted because of suspicion of remaining lead and tin. And, thus freed of its faults, the cleansed mercury condenses.
5. Evaporation and condensation (*utthāpana*)
   Through the trituration, it becomes completely cleansed and freed from lead and tin. The mercury is brought to a rise from a sour gruel decoction in a condensation apparatus.
6. Revivification (*nirodhā*)
   It is rendered useless through trituration, thickening and condensation because of its weak potency. Once it has become replenished though revivification with salt water, it will no longer be impotent.
7. Fixation (*niyamana*)
   When it has thus properly regained its potency, the unsteady [mercury] will subsequently be fixed through a steaming with betel, garlic, salt, false daisy, spiny gourd and tamarind.
8. Stimulation (*dīpana*)
   Mercury that has been stimulated with a steaming for three days with alum, ferrous sulphate, borax, black pepper, salt, mustard, moringa and sour gruel becomes 'one who desires a morsel'.
9. Feeding (*cārana*)
   Thus, when it has been stimulated and cleansed, the king of essences becomes radiant as a thousand strings of flashes of lightning and then it should be fed a seed of essences, etc.

   First, during the gold procedure, gold should be placed in its mouth and silver in the silver procedure; ground in a mortar together with divine herbs, it swallows as 'one without a mouth'.[13]

The further steps are described in the following chapters, with much attention given to the preparation of mica before it is added to the elixir. Not all alchemical works follow the exact sequence or number of procedures laid out in the Book of the Heart of Mercury. There is also some variation in the ingredients used, though mercury features in all of the works. The final product is a mercurial elixir, which is first applied to metals to transmute them into silver or gold. The Ocean of Mercury (Chapter 17, verses 165–6) established a key paradigm in this context:

A 15th-century copy of the Jain text *Kuvalayamālā* ('Garland of Blue Water Lilies') by Jain monk Uddyotana-sūri, 779 CE.

> Mercury must always be applied to the body in the same way it is applied to metals.
> It equally enters into the body and metals, O Goddess.
> First, observe it in metal, then apply it to the body.[14]

Accordingly, in a final lengthy procedure, the elixir is applied to the human body through ingestion or by placing a mercurial pill under the tongue.[15] The expected outcomes of consuming the elixir range from a prolonged lifespan, halted or reversed ageing and improved general and reproductive health to the attainment of special powers (*siddhi*), immortality, the experience of godhood and liberation while living (*jīvanmukti*):

> One should eat mercury that has the transformative power of one to a hundred, a thousand, one hundred thousand, ten million, or one hundred million or more, ground with sulphur: it will give special powers.
>
> Then the mercury penetrates and one will produce children radiant as divine children, and one becomes invariably loved by women, and free from wrinkles and grey hair.
>
> The intelligence, strength and power of the one undergoing the elixir regimen grows, together with his lifespan. The divine intelligence and divine qualities of one who has attained these grow further.

One who has become fully perfected through mercury, who has left behind troubles, ageing and death, and is possessed of good qualities, continually roams all the worlds through moving in the air.

He will also become a giver and a creator here in the triad of worlds, like the Lotus-born; one who maintains, like Vishnu, and a destroyer, like Rudra.[16]

Religious and spiritual elements

While the technical aspects of alchemical practice take up the largest part of the works, some of them dedicate substantial attention to religious and spiritual activities. There is a divide in terms of narrative frameworks, which is already found in the two earliest alchemical works. The Book of the Heart of Mercury's author Govinda dedicates his summary of the craft in 506 verses to the king of the Kirātas, whom he refers to as a master on the subject of mercury (*rasācārya*). Here, one expert is speaking to another, soberly delivering a learned digest on a practical discipline with stated soteriological aims. These include liberation in a living body (*jīvanmukti*) and the attainment of the Absolute or Supreme Spirit, here named *brahman*. Notably, the Book of the Heart of Mercury does not display strong sectarian affiliation: it is the only early work within this body of literature that is not emphatically Shaiva or Shakta.[17] The Ocean of Mercury, on the other hand, is firmly placed in a Shaiva, or more specifically Kaula,[18] setting: it is presented as a dialogue between the god Shiva and his consort, the goddess Parvati, in which Shiva instructs Parvati on alchemy as esoteric knowledge. Here, mercury is a manifestation of Shiva as the Absolute in the form of his semen and must be worshipped.[19]

The Ocean of Mercury includes many elements entirely missing from the Book of the Heart of Mercury, such as the auspicious positioning and set-up of the laboratory; lists of instruments used in alchemical operations; the initiation of the alchemist; the desired characteristics of the persons involved in the alchemical endeavour; and the use of mandalas, mantras and propitiations of deities to ensure the success of all alchemical operations. Many of these elements reflect a Shaiva tantric worldview.[20] Several works carry forward the narrative device of a dialogue between various forms of the god and goddess, portraying alchemy as an esoteric science only accessible to initiated adepts. These typically feature tantric elements as well. Later texts emphasised the medicinal qualities of the elixirs rather than their contribution to spiritual experience, gradually omitting reference to tantric or other religious elements. Works such as the sixteenth-century *Rasendrasārasaṃgraha* ('Collection of the Essence of Mercury') focused almost entirely on iatrochemistry, replacing traditional herbal ayurvedic formulations with mercurials.[21] Conversely, from approximately the eleventh century, ayurvedic works started to integrate recipes for mercurials and other mineral formulations into their therapeutic frameworks.[22] To this day, the medical elements of the doctrine of mercury survive in India under the name of *rasashastra* (*rasaśāstra*), as part of government-regulated ayurvedic education and practice.

The alchemical practitioner

While the alchemical works offer ample information on the techniques of the craft, they give rather less information on the practitioner. There is even some uncertainty as to whether the person producing the elixir is the same as the one consuming it.[23] However, several works describe the desired characteristics of the persons involved in alchemy. This passage in the Ocean of Mercury (Chapter 2, verses 2–11) provides a glimpse of the alchemical team, consisting of the teacher, apprentice and assistants:

The first procedure: steaming mercury, a reconstruction by the author of the first of 18 procedures for making a mercurial elixir, as described by Govinda in *Rasahṛayatantra* ('Book of the Heart of Mercury').

> The venerable Goddess spoke:
>
> Please tell me [this], O Lord! What sort [of person] should the alchemical preceptor be? And, O God, what so with the apprentice who makes the practice of alchemy his supreme goal?
>
> The venerable Bhairava spoke:
>
> [Such a person] is considered [to be a] guru: [he is] free from desire, egoless, devoid of avarice or duplicity, devoted to the path of the Clan (*kula-marga*), ever devoted to the veneration of his [own] guru, self-restrained, pedagogically savvy, energetic, unselfish, upright, truthful, dexterous, good-natured, possessed of every good quality, pure, knowledgeable in multiple alchemical teachings, possessed of laboratory experience, [and] well-versed in the precepts of alchemical initiation. He is [also] someone who has full knowledge of apparatus, botanicals, the primary reagents, the nuances of color, the components of the mineral seeds (*bījas*), the conjoining of opposites, calcinating salt, coloring, flowing, oil, metallic 'leaf' preparations,

progression, the extraction of color and pliability, [and] the 'youthful' and 'aged' as well as the 'sky-going' and 'earth-going' [types of] calcination.

The [prospective] apprentice is knowledgeable about what is to be done according to place and time. [He is] compassionate and dexterous, entirely free of avarice and duplicity, having mantra practice as his highest goal, endowed with [auspicious] bodily markings and attributes [and] depth [of character], beloved by his teacher, ever devoted to the veneration of the gods, to fire, to the circle of Yoginīs, and to the Clan, an educated connoisseur of Tantra, a speaker of truth, and firm in his vows.

Men who are equipped with hands that bear the marks of water-pots, spades, banners, conches, and so forth, are, O Goddess, to be employed in 'treasure' (*nidhi-*) practice.

[Men who are] ever ferocious, powerful [and] massive brutes with dark bloodshot eyes and bent noses are recommended for 'pit' (*vila-*) practice.

In smelting (*dhātuvāda*), one should employ vigorous red-headed [men] with emaciated upper bodies, and hot and hardened feet.[24]

In this description, and in particular in the formulation of the alchemist's ethical characteristics, we can see some continuity with the alchemist portrayed in the Garland of Blue Lotuses several centuries earlier, albeit brought into a Hindu Shaiva tantric context rather than a Jain one.

Women in alchemy

Women's roles in Indian alchemy are not clearly defined. Generally speaking, all the central actors seem to be male. However, there is mention of women in one specific context, namely in the initiation of the alchemical disciple. Her role is never fully spelled out; instead, she is described in general terms, with a focus on her physical appearance and, notably, her menstrual cycle. The relevant passages, such as the one below, suggest that the woman was a kind of vessel for processing mercury, with her menstrual fluid providing one of the essential ingredients. However, the latter concept is only featured in those alchemical works with tantric elements.

Now, a woman who is in the full bloom of youth, fair of form, of pleasant laughter, [and] lovely hair, a perennial lover of dairy products, agreeable in her conversation, [and] ever fond of the tales and teachings of Shiva; whose mouth is lotus-shaped [and each of] whose eye[s] has the appearance of a blue lotus flower, whose teeth are like diamonds, whose lower lip resembles a tender shoot, whose breasts are even, prominent and full; and whose vulva – the two parts of which are fleshy, smooth, [and] rounded, with inward curling black pubic hair – is symmetrical and resembling a fig leaf; and who is slim at the waist and supreme in [her] devotion to Shiva: [such a] slender, gracious and doe-eyed woman who menstruates on the full moon or the new moon, from fortnight to fortnight, is to be known, assuredly, as a 'lotus woman'.

> The aforementioned [woman] called 'cowrie shell' is of six types. She is a provider of the essential fluid. O Mistress of the God of Gods, [it is] with her, precisely, that one should perform the alchemical procedures.[25]

The early alchemical works are unequivocal in envisioning the recipients of alchemical products as male. However, as the later works redefined their aims from elixir alchemy to iatrochemistry, women were included as recipients of medicinal formulations, especially those for sexual health, to increase fertility and for rejuvenation. Alchemical works from the *Rasaratnākara* ('Jewel Mine of Mercury', thirteenth to fifteenth century CE) onward included sections or even whole chapters on virility, fertility and sexual-stamina treatments (*vājīkaraṇa*). This is one of the eight traditional subject areas of the much older ayurvedic tradition, and there is much overlap in how medical and alchemical texts treated the subject. However, alchemical works also included methods of treatment drawn from tantric or magical rather than ayurvedic milieus in their *vājīkaraṇa* sections. They also added contraceptive and abortive measures, which are only featured relatively late in ayurvedic literature.[26]

Tamil Siddha alchemy

Tamil Siddha alchemy was formulated as part of wider Tamil Siddha thought. The Sanskrit term *siddha* ('perfection') has been applied to several groups or unorthodox sects practising tantra, yoga, medicine and alchemy with the objective of attaining perfection and immortality. The practitioners of the Tamil Siddha sects oriented themselves on practices and writings attributed to the authoritative figures of Siddhars (a Tamilised version of the Sanskrit Siddha, 'perfected being', also Tamil *cittar*, pl. *cittarkaḷ*). Tamil Siddhars' writings do not sharply differentiate between

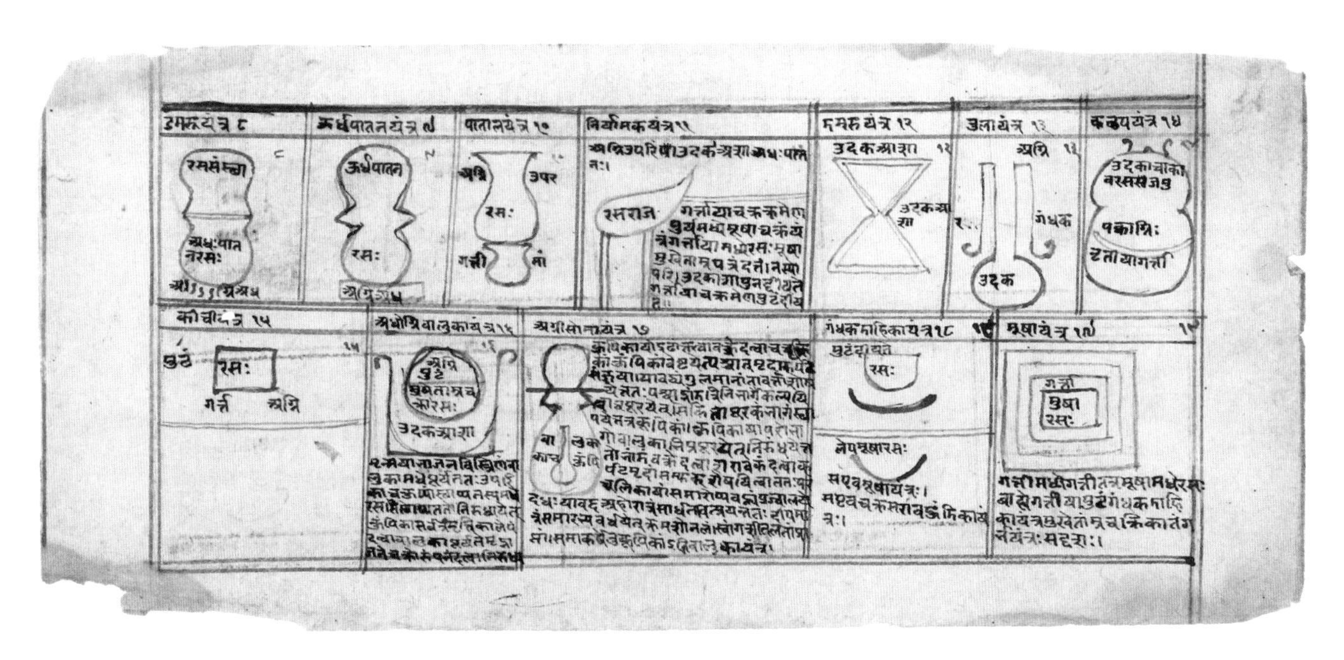

A rare, illustrated folio of the Sanskrit text *Rasendramaṅgala* ('The Benediction of Mercury') by the alchemist philosopher Nagarjuna.

yoga, medicine and alchemy: they are presented as interrelated parts of a single body of knowledge and practice.[27] The alchemical elements of the Tamil Siddha tradition clearly draw on the Sanskritic tradition.[28] However, there are also significant differences, beginning with the language of Tamil Siddha literature. Unlike the Sanskrit sources, Tamil alchemical works often employ coded language that obscures the meaning of the text for the uninitiated. At the same time, colloquial expressions may give access to persons outside the literary elite.[29] The texts promote the use of mercury and describe similar procedures for processing it as found in the Sanskrit sources. However, other substances receive as much or even more attention. Among these are special salts and soils, such as the 'triple salt' (*muppu*), the 'tied salt' (*kaṭṭuppu*) and fuller's earth (*pūnīr*, 'a water of earth').[30] *Muppu* in particular takes a prominent place in Tamil Siddha recipes, as in the following:

> The one who knows the methods of calcining will become a doctor. Regard the one who calcines as an alchemist who immobilizes [substances]. [In order to calcine metals] it is necessary to know the *muppū* which has been spoken about. Listen about the killing [i.e. calcination] of metals. The powder of *muppū* which is called 'the power' is necessary [for the process].[31]

Both the use of language and the importance of local substances bind the tradition to its geographical location. However, some of the later sources tell stories of Siddhars travelling and both importing and exporting alchemical knowledge.[32] Today, Siddha alchemy survives as a subdiscipline of Siddha medicine, just as the alchemy of the Sanskrit works survives as ayurvedic *rasashastra*. However, there are also still those, including women, who operate outside the more official forms of Siddha medicine. These practitioners are still deeply involved in esotericism and alchemy and are considered proficient in the transformation of mercury.[33]

An 18th- or 19th-century copy of the 13th-century *Rasaratnākara* ('Jewel Mine of Mercury') by Nityanātha.

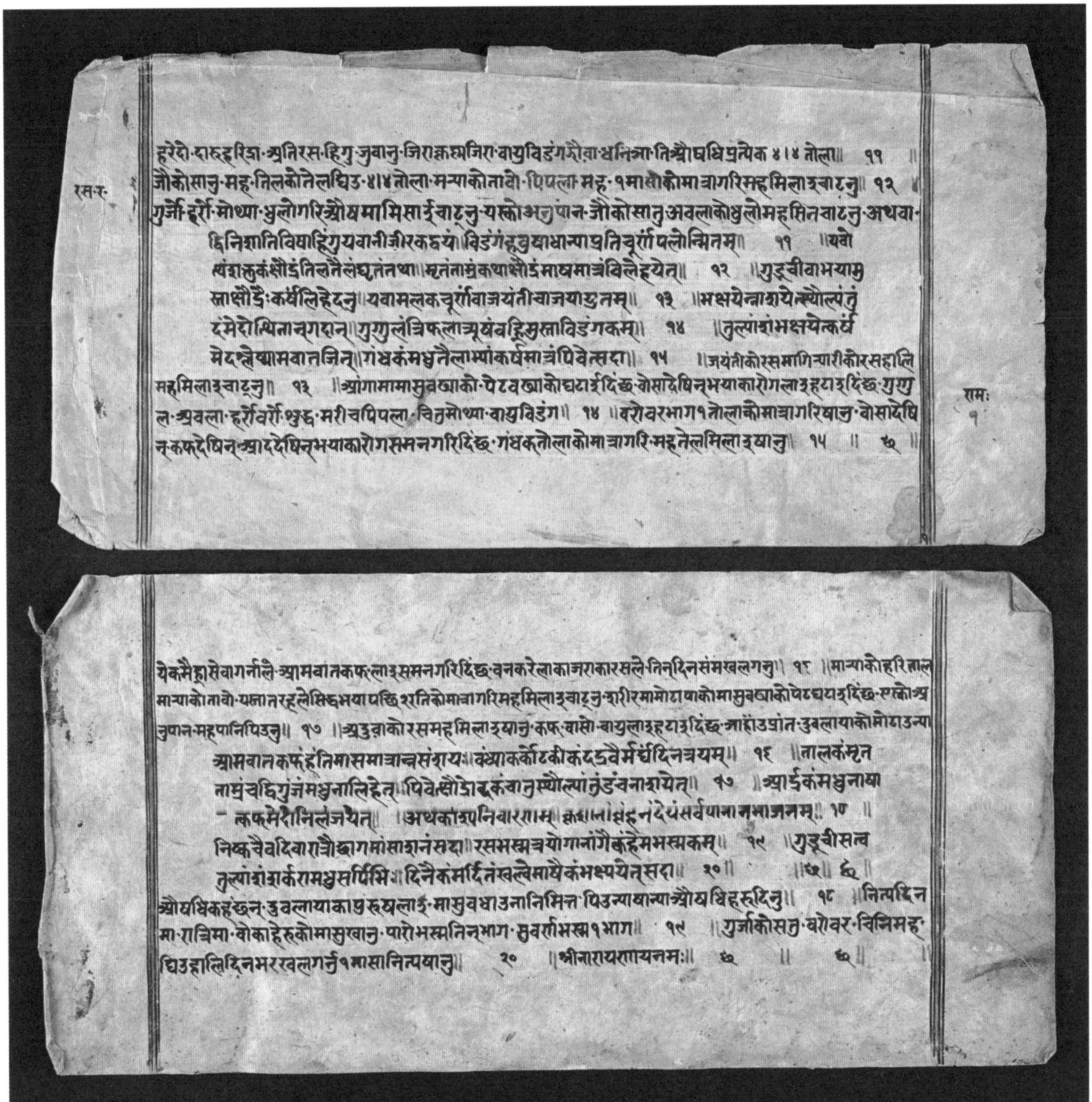

रस·र·

हरेदो·दारुहरिद्रा·अतिरस·हिंगु·जुवानु·जिराकृष्णजिरा·वायुविडंगसैंधा·धनिआ·तिन्औषधिप्रत्येक ४।४ तोला॥ ११ ॥
जौकोसातु·मह·तिलकोतेलघिउ·४।४तोला·मन्याकोतावो·पिपला·मह·१मासाकोमात्रागरिमहमिलाइखानु॥ १२ ॥
गुर्जो·हर्रो·मोथ्या·धुलोगरिऔषमामिसाई खानु·यस्कोअनुपान·जौकोसातुअवलाकोधुलोमहसित खानु·अथवा·

द्विनिशातिविषाहिंगुयवानीजीरकद्वयं॥ विडंगंह्रुषाधान्याप्रतिचूर्णंपलोन्मितम्॥ ११ ॥यवो
त्थंशक्तुकंक्षौद्रंतिलतैलंघृतंतथा॥ मृतंतानुंकथाक्षौद्रंमाषमात्रंविलेहयेत्॥ १२ ॥गुडूचीवाभयामु
स्ताक्षौद्रैःकर्षंलिहेदनु॥ यवामलकचूर्णंवाजयंतीवाजयायुतम्॥ १३ ॥भक्षयेन्नाशयेत्स्थौल्यंतु
दंमेदोस्थितानगदान्॥ गुग्गुलंत्रिफलात्र्यूषंवह्निमुस्ताविडंगकम्॥ १४ ॥तुल्यांशंभक्षयेत्कर्षं
मेदश्लेष्मामवातजित्॥ गंधकंमधुतैलाभ्यांकर्षमात्रंपिबेत्सदा॥ १५ ॥जयंतीकोरसमाणिन्यारीकोरसहालि
महमिलाइखानु॥ १३ ॥आंगामामासुवख्याको·पेटवख्याकोघटाईदिंछ·वोसादेषिनभयाकारोगलाइहटाइदिंछ·गुग्गु
ल·अवला·हर्रोवर्रो·शुद्ध·मरीचपिपला·चितुमोथ्या·वायुविडंग॥ १४ ॥वरोवरभाग१तोलाकोमात्रागरिखानु·वोसादेषि
न·कफदेषिन·आदर्देषिनभयाकारोगसमनगरिदिंछ·गंधकतोलाकोमात्रागरि·महतेलमिलाइखानु॥ १५ ॥ ८७ ॥

रामः
१

येकमैहासेवागर्नाले·आमवातकफलाइसमनगरिदिंछ·वनकरेलाकाजराकारसले·तिनदिनसंमखलगनु॥ १६ ॥मान्याकोहरिताल
मान्याकोतावो·यस्ताटरहलेसिद्धभयापछि शतिकोमात्रागरिमहमिलाइखानु·शरीरमामोटापाकोमासुवख्याकोपेटघटाइदिंछ·यस्कोअ
नुपान·महपानिपिउनु॥ १७ ॥अदुवाकोरसमहमिलाइखानु·कफ·श्वासो·वायुलाइहटाइदिंछ·आहांउप्रांत·दुवलायाकोमोटाउन्या

आमवातकफंहंतिमासमात्रान्नसंशयः॥ वंध्याकर्कोटकीकंदढ्वैर्मर्द्यंदिनत्रयम्॥ १६ ॥तालकंमृत
ताम्रंचवह्निगुंजंमधुनालिहेत्॥ पिबेत्क्षौद्रोदकंवानुस्थौल्यांतुंडंवनाशयेत्॥ १७ ॥आर्द्रकंमधुनाश्वा
- सकफमेदोनिलंजयेत्॥ ॥अथकार्श्यनिवारणम्॥ कृशानांबृंहणंदेयंसर्वपानान्नभाजनम्॥ १८ ॥
निष्कंचैवदिवारात्रौछागमांसाशनंसदा॥ रसभस्मत्रयोगानांगैकंहेमभस्मकम्॥ १९ ॥गुडूचीसत्व
तुल्यांशंशर्करामधुसर्पिभिः॥ दिनैकंमर्दितंखल्वेमाषैकंभक्षयेत्सदा॥ २० ॥ ॥८२॥ ८०॥
औषधिकहंछन्·दुवलायाकाप्रुरुषलाई·मासुवधाउनानिमित्त·पिउन्याखान्याऔषधिहरुदिनु॥ १८ ॥नित्यदिन
मा·रात्रिमा·वोकाहेरुकोमासुखानु·पारोभस्मतिनभाग·सुवर्णभस्म१भाग॥ १९ ॥गुर्जाकोसतु·वरोवर·चिनिमह·
घिउहालिदिनभरखलगर्नु१मासानित्यखानु॥ २० ॥श्रीनारायणायनमः॥ ८२ ॥ ८० ॥

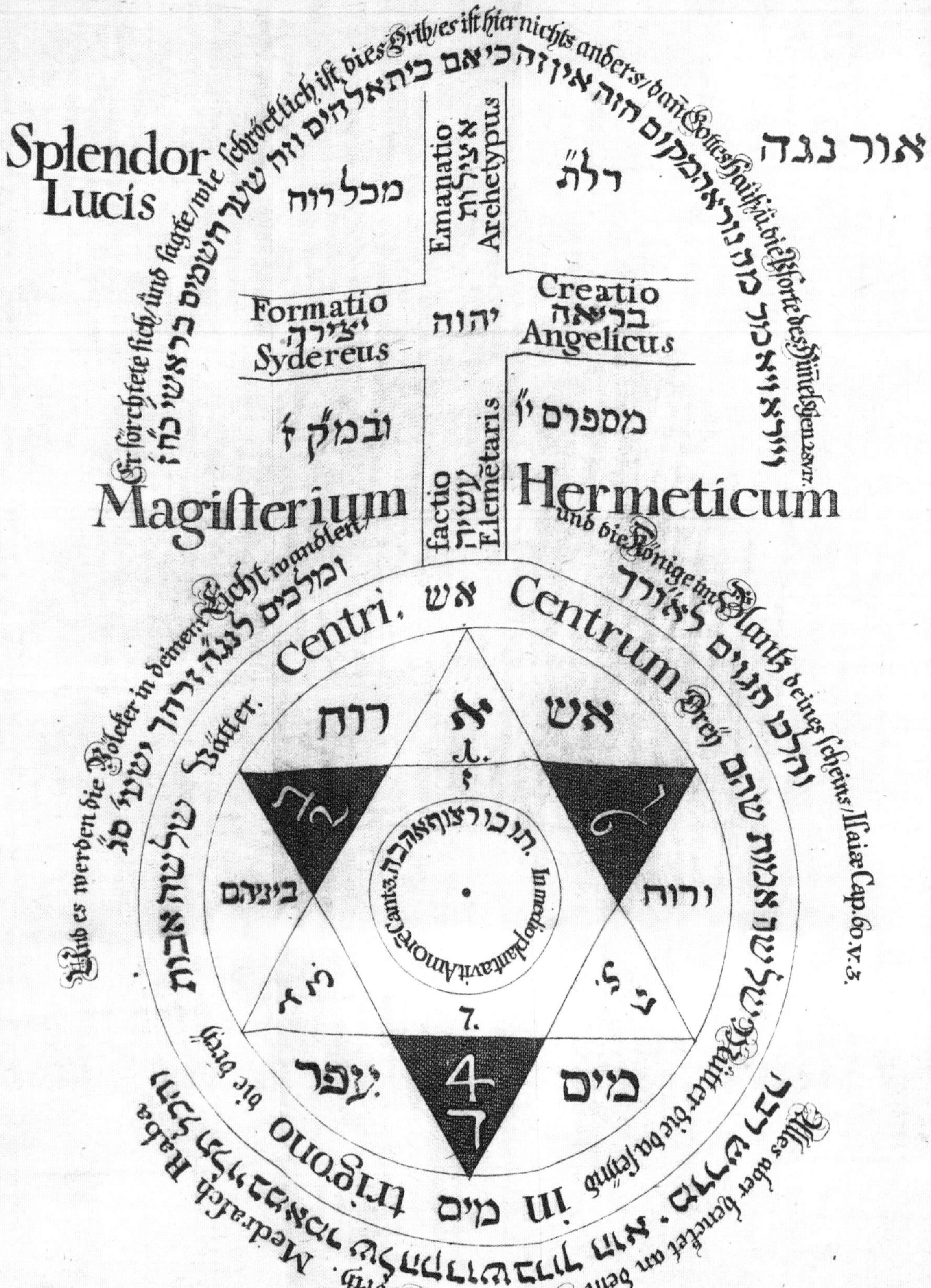

Splendor Lucis

אור נגה

Er förchtete sich/und sagte/wie schröcklich ist dies Orth/es ist hier nichts anders/dan Gottes Hauß/und die Pforte des Himels Gen. 28. v. 17.

ויירא ויאמר מה נורא המקום הזה אין זה כי אם בית אלהים וזה שער השמים בראשי׳ כח׳

Emanatio
אצילות
Archetypus

מכל רוח

דלת

Creatio
בריאה
Angelicus

Formatio
יצירה
Sydereus

יהוה

ובמק׳ ז׳

מספרם י׳

factio
עשיה
Elemētaris

Magisterium Hermeticum

Und es werden die Völcker in deinem Licht wandlen/ und die Könige im Glantz deines scheins/ Isaiae Cap. 60. v. 3.

והלכו גוים לאורך ומלכים לנגה זרחך ישעי׳ ס׳

Centri. אש Centrum

Vätter. Drey

שלשה אבות שלשה אמות

רוח א אש

בינהם ורוח

עפר מים

7.

תוכו רצוף אהבה

In medio plantavit Amorē Cant. 3.

Mütter die da seynd in מים trigono die drey

Alles aber hencket an dem Göttlichen Worth. Medrasch Raba

והכל תלוי במאמר של הקדוש ברוך הוא · מדרש רבה

JEWISH ALCHEMY AND KABBALAH

JOHN M. MACMURPHY

Diagram from *Or Nogah*, better known by its official Latin title, *Splendor Lucis*, by Alois Wiener von Sonnenfels, 1745.

The notion of 'Jewish Alchemy' has baffled scholarship for decades.[1] Some claim that the phenomenon was widespread. Others asserted it was minimal, incidental or even non-existent.[2] In the following analysis, we shall explore not only the world of alchemy within Jewish circles but also expose for the first time its hidden relevance within kabbalistic and other related European esoteric milieus, including the unique metaphysical alchemical practices of the highly influential, yet mysterious, brand of Kabbalah known as the Lurianic school.

An important issue that must be addressed relates to the term 'Jewish Alchemy' itself. After all, the concept seems to convey that alchemical operations have a certain faith-based quality, or in this case, pertain somehow to Judaism. Naturally, such formulae, as we shall soon see, lend themselves nicely to religious metaphors. However, the methods themselves, which generally involve the manipulation of substances, appear to be free of any theological dogma. Scholars in the field of Jewish Magic, such as Gideon Bohak and Yuval Harari, have faced a similar conundrum concerning magical techniques and recipes. The offered solution, which can be applied also to alchemical praxis, suggests that it is the religious identity of the adepts themselves that can lend itself to the classification of the area of study.[3] As such, for our purposes, 'Jewish Alchemy' is the engagement with alchemical thought and its practice by Jews.

In this respect, one pioneering approach was established by the eminent Religious Studies researcher Raphael Patai (1910–96) in his monograph, cleverly titled *The Jewish Alchemists* (1994).[4] This work, while containing some misleading analysis and assumptions, demonstrates an impressive survey of Judaic sources, personalities, practices and influences surrounding the topic of alchemy. Despite laying the foundation for future research, *The Jewish Alchemists* was highly criticised when it came out.[5] Apart from perhaps over-reaching with its historical accuracy, the main issue appeared to have been that its vision went against the perspective of Gershom Scholem (1897–1982), the father of the modern study of Jewish

Mysticism, who maintained negative views on the subject of 'occult' praxis[6] within a context that relates to Judaism. This explains why such areas of study, until recently, were largely ignored in academia.

Evidence of the association of alchemy with the Jews can be gleaned throughout history, starting from antiquity. In fact, in the writings of Zosimos of Panopolis (fl. *c.* 300), one of the earliest records about alchemical theory and praxis, it is revealed that his mentor was a certain Maria the Jewess, an alchemist who lived in Hellenistic Egypt in the early third century. Among her various teachings, she is credited with inventing a double boiler configuration known as *a bain-marie*, from the Latin *Balneum Mariae*, or 'The Bath of Mary'. This method, which allows for one container to be heated evenly through a temperature transfer from boiled liquid in a secondary vessel, has persevered until today and can be seen in factories, laboratories and even modern-day kitchens.[7]

However, this early association did not end in antiquity. During the Middle Ages, while Arabic alchemy made a name for itself through prominent figures such as Jābir ibn Ḥayyān (*c.* 721–815), it may be surprising to note that certain related professions such as goldsmiths, metalworkers and especially jewellery-makers were associated with and occupied by Jews.[8] In addition, disdain for such crafts was not only exhibited by pre-Islamic Arabs but also after the rise of Islam. This is supported by archaeological evidence that demonstrates scant findings of Islamic jewellery.[9] It has been debated whether Muslim decrees in religious texts against both creating and wearing such items correlate to the small number of artefacts found.[10]

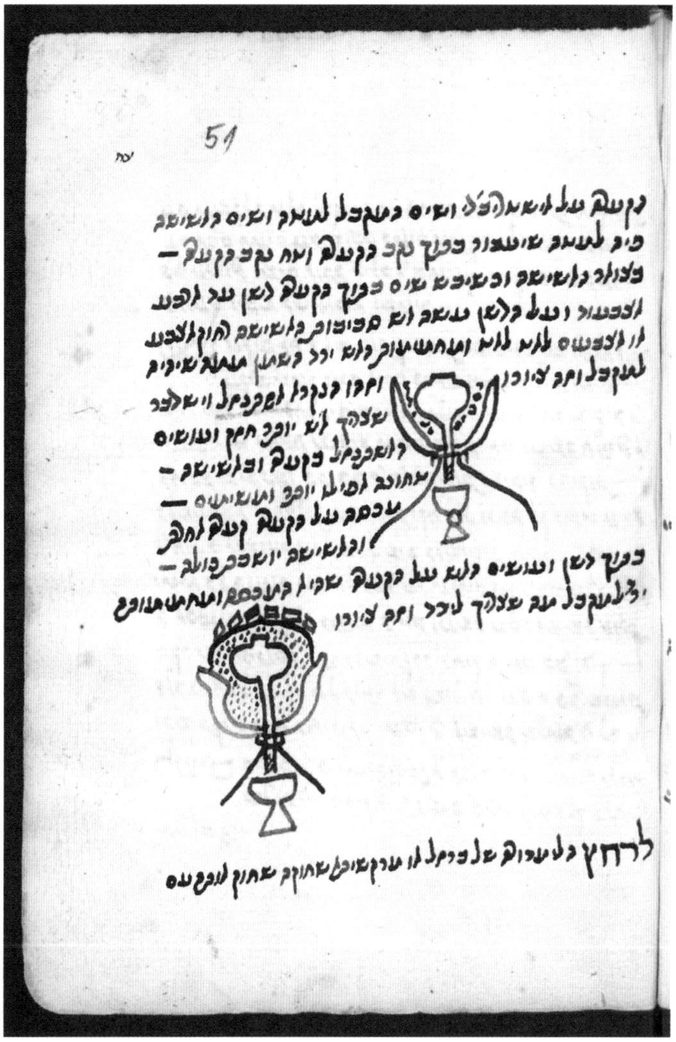

Moreover, we also have documentation that supports Judaic involvement with alchemy during the Middle Ages. In 1896–97, the Cambridge scholar Dr Solomon Schechter (1847–1915), aided by twin sisters Agnes S. Lewis (1843–1926) and Margaret D. Gibson (1843–1920), uncovered what would turn out to be the largest depository of medieval Jewish manuscripts and documents from a long-standing synagogue in Cairo. This collection is usually denoted as the 'Cairo Genizah', named after the 'storeroom', or *genizah*, of the Ben Ezra synagogue in Fustat (old Cairo). Jewish religion and customs assert that holy writings, especially ones that contain divine names, should never be destroyed. As such, this repository consists of witnesses that stretch back as far as the early eleventh century. Within this archive, we find further confirmation of jewellery-making and metal-smithing being labelled as a 'Jewish craft'.[11] In addition, certain discovered material is

Alchemical operations and drawings in a Hebrew manuscript, *c.* 16th–18th centuries.

Opposite:
'Maria the Jewess'. In *Symbola Aureae Mensae Duodecim Nationum* by Michael Maier, 1617.

specifically associated with occult practices. Most recently, these sources, written predominantly in Arabic with Hebrew letters (i.e., Judaeo-Arabic), were surveyed by Gideon Bohak and his team.[12] Their findings demonstrate the presence of content consisting of magic (1,026 fragments), divination (247 fragments) and astrology (349 fragments). Relevant to our present discussion, 68 fragments were identified as dealing with alchemy.[13] In a previous study by Yosef Yinon (Paul B. Fenton), some of the alchemical compositions from the Cairo Genizah were pseudoepigraphically attributed to Hermes Trismegistus[14] – the father of alchemical thought and the purported author of the *Emerald Tablet*, one of the most important and enigmatic texts in Hermetic philosophy.

This being said, the positive attitudes of medieval Jews towards Hermetic thought in an alchemical context may not have been unequivocal, as evidenced in *Sefer ha-Tamar* ('The Book of the Palm Tree' – ספר התמר).[15] The work was most likely composed around the eleventh century by a Muslim Arab based in Sicily. The authorship has been attributed to enigmatic and celebrated figures such as Abufalah of Syracuse (Saragossa), an elusive yet influential alchemist, and even the famous mathematician Jabir ibn Aflah (d. *c.* middle of twelfth century),[16] known in the Latin world as Geber – not to be confused with the aforementioned Jābir ibn Ḥayyān (identified by the same latinised nomenclature). According to Patai, the fact that the text was translated into Hebrew and was even incorporated into Jewish writings demonstrates its influence on the Jews.[17] The content explicitly rejects the teachings of Hermes and his followers. However, this dismissal appears suspect. As Scholem points out, much of the theoretical notions including alchemical terminologies are 'very very close to this [Hermetic] literature'.[18]

The relationship between 'alchemical Hermeticism' and the Jews was not just one-sided. In the early modern period, we see special reverence towards Jewish lore and personalities from esoteric practitioners who aligned themselves with Hermetic thought. For example, on the cover plate of the work *Symbola Aureae Mensae Duodecim Nationum* ('Symbols of the Golden Table of the Twelve Nations', 1617),[19] the famous German physician and alchemist Michael Maier (1568–1622) displays portraits of all influential alchemists throughout time. At the very top, right next to the image of Hermes himself, is Maria the Jewess. As such, if Hermes was the 'father' of alchemical thought, Maria can certainly be credited as its 'mother'. In addition, we also see her associated with a character from the Hebrew bible. In the body of Maier's book, she is depicted performing an alchemical operation, as the text identifies her with none other than Miriam, the sister of Moses.[20] Though ahistorical, such correlation is understandable considering the murky details about Maria the Jewess and the overwhelming familiarity of early modern thinkers with biblical narratives and personalities.

The fascination with Judaic lore was not incidental. Rather, it was consistent with Renaissance thought that championed the notion of *prisca theologia*, or ancient theology – the idea that many traditions are simply various permutations of an age-old truth. This approach did not escape Jewish circles, as we can see in the case of Yohanan Alemanno (*c.* 1435–*c.* 1504), an Italian humanist, Kabbalist and alchemist.

Title page. *Symbola Aureae Mensae Duodecim Nationum*, by Michael Maier, 1617.

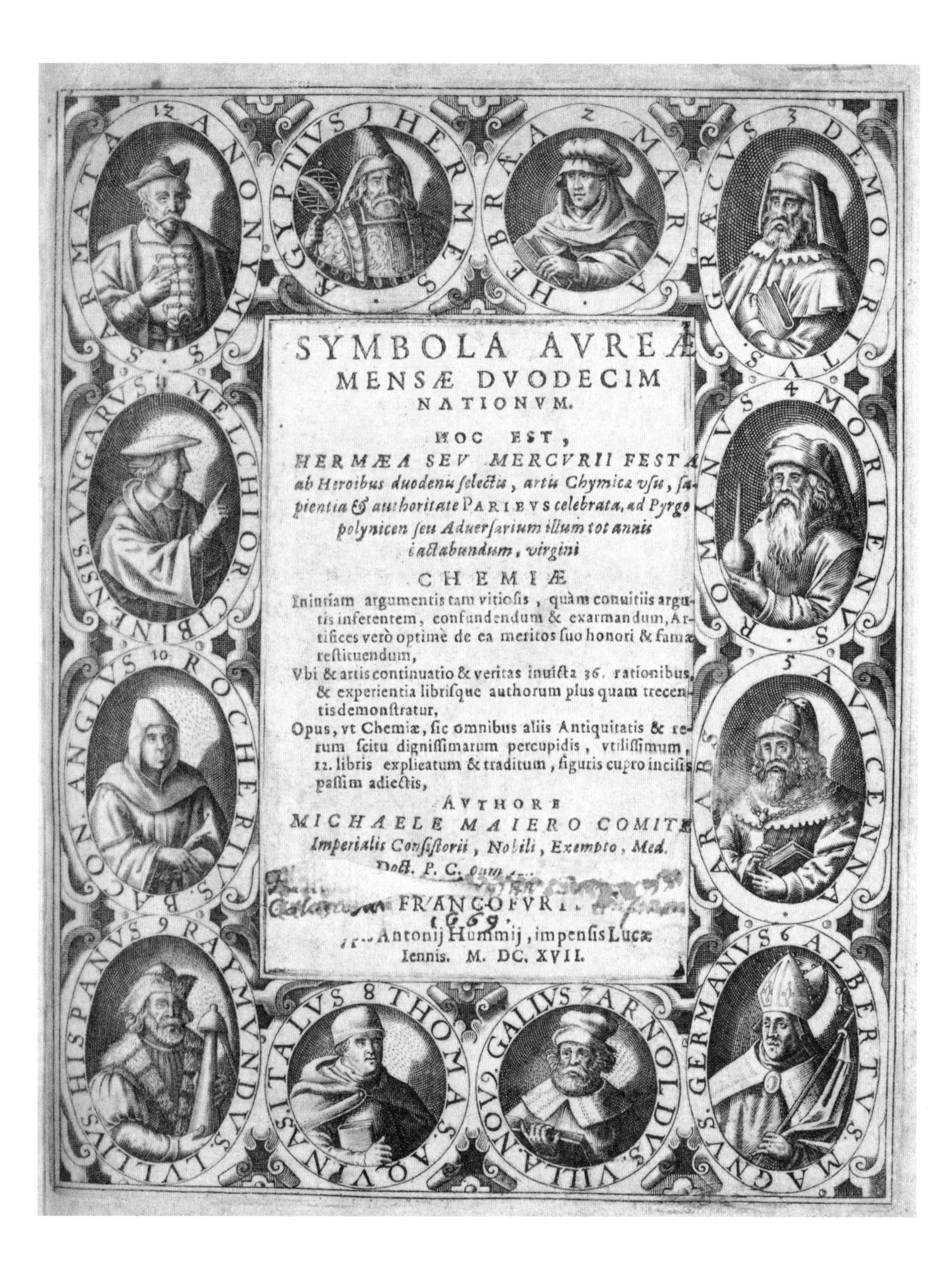

SYMBOLA AVREÆ MENSÆ DVODECIM NATIONVM.

HOC EST,

HERMÆA SEV MERCVRII FESTA ab Heroibus duodenis selectis, artis Chymicæ usu, sapientia & authoritate PARIBVS *celebrata, ad Pyrgopolynicen seu Aduersarium illum tot annis iactabundum, virgini*

CHEMIÆ

Iniuriam argumentis tam vitiosis, quàm conuitiis argutis inferentem, confundendum & exarmandum, Artifices verò optimè de ea meritos suo honori & famæ restituendum,

Vbi & artis continuatio & veritas inuicta 36. rationibus, & experientia librisque authorum plus quam trecentis demonstratur,

Opus, vt Chemiæ, sic omnibus aliis Antiquitatis & rerum scitu dignissimarum percupidis, vtilissimum, 12. libris explicatum & traditum, figuris cupro incisis passim adiectis,

AVTHORE

MICHAELE MAIERO COMITE Imperialis Consistorii, Nobili, Exempto, Med. Doct. P. C. [illegible]

FRANCOFVRTI

1669.

[illegible] Antonij Hummij, impensis Lucæ Iennis. M. DC. XVII.

1 HERMES ÆGYPTIUS · 2 MARIA HEBRÆA · 3 DEMOCRITUS GRÆCUS · 4 MORIENUS ROMANUS · 5 AVICENNA ARABS · 6 ALBERTUS MAGNUS GERMANUS · 7 ARNOLDUS VILLANOVUS GALLUS · 8 THOMAS AQUINAS ITALUS · 9 RAYMUNDUS LULLIUS HISPANUS · 10 ROCHERIUS BACON ANGLUS · 11 MELCHIOR CIBINENSIS UNGARUS · 12 ANONYMUS SARMATA

Alemanno's involvement with alchemy is well-attested. In one of his many obscure works, he expounds on his version of the ideal educational curriculum which he recommends for posterity.[21] The programme contains both exoteric fields such as mathematics, philosophy and the sciences as well as esoteric knowledge such as magic, kabbalah and alchemy – including studies in using an alembic, a common still[22] and the study of one of the most recognised European alchemical texts, *Turba Philosophorum* ('The Assembly of the Philosophers', 1572),[23] a translation of an Arabic work claiming to transmit alchemical dictums of ancient Greek philosophers. In his unpublished book *Chay ha-Olamim* ('The Eternal Life' – *חי העולמים*), a composition describing human development from birth to the point of recognising the divine, Alemanno mentions more alchemical influences such as the aforementioned Abufalah of Syracuse, as well as Ibn Rushd (1126–98), also known as Averroes – the medieval doctor, alchemist and philosopher best known for criticising Aristotelian philosophy. According to Moshe Idel, one of the foremost contemporary authorities in the academic study of Jewish Mysticism, while Alemanno criticised Aristotle, he was deeply influenced by Abufalah and his attributed *Sefer ha-Tamar*.[24] In fact, in *Sefer Sha'ar ha-Cheshek* ('The Book of the Gate of Desire' – *ספר שער החשק*), Alemanno lists Abufalah as a transmitter of knowledge that can be traced back to King Solomon, one of his greatest inspirations.[25]

ספר
שער החשק
יוחנן אלימן
יעקב
משה חיים ברוך
פה ליוורנו יע״א
שנת החכמ״ה
IN LIVORNO)(Per Gio. Vinc. Falorni)(con Ap.

Title page. *Sha'ar ha-Cheshek* ('The Gate of Desire'), by Yochanan Alemanno, 1790.

Abraham Eleazar the Jew. Hand-coloured frontispiece from *Uraltes chymisches Werk*, pasted into *Theatrum chymicum*, c. 1780.

Overleaf left:
Rosicrucian diagram from *Geheime Figuren der Rosenkreuzer aus dem 16ten und 17ten Jahrhundert* ('Secret Symbols of the Rosicrucians from the 16th and the 17th centuries'). Anonymous compiler.

Overleaf right:
Detail from *Il Metamorfosi Metallico et Humano*. By Giovanni Battista Nazari, 1564.

Alemanno's uniqueness is not only attributed to his interest in kabbalah and alchemy, but he is also best known for being, alongside the Sicilian Jewish convert and humanist Flavius Mithridates (fl. late fifteenth century), among the primary Kabbalistic mentors of Giovanni Pico della Mirandola (1463–94), one of the early founders of Christian Kabbalah. While this approach mainly focuses on Christological hermeneutics of kabbalistic and Judaic literature, its Jewish counterpart, in contrast, generally centres around the mystical exegesis of the Torah based on the configuration of hypostatic emanations known as *sefirot* (ספירות), or a corresponding system of divine names and their applications.

Alemanno's syncretic occult interests, including magic and kabbalah, were undoubtedly passed on to Pico and, in turn, his students – most notably, his adherent Johannes Reuchlin (1455–1522). Quite ironically, Pico himself is not known for his interest in alchemy, unlike his mentor and especially the followers of his legacy within the European esoteric traditions, as attested by enigmatic works such as *The Book of Abraham the Jew*;[26] personalities like Paracelsus (1493–1541), Giovanni Agostino Pantheo, John Dee (1527–1608/9) and Heinrich Khunrath (1560–1605); and initiatic groups, including the seventeenth-century German-based Rosicrucians.[27]

The popular interest among the early modern thinkers in this syncretic approach to the various systems has led to the proliferation of correspondences tables in the occult literature that allow adepts to 'translate' knowledge from one system to

R. ABRAHAMI ELEAZARIS

Uraltes Chymisches

Werk,

Welches ehedessen von dem Autore
Theils in Lateinischer und Arabischer, theils auch in Chaldäischer und Syrischer Sprache geschrieben,
Nachmals von einem Anonymo
in unsere deutsche Muttersprache übersetzet,

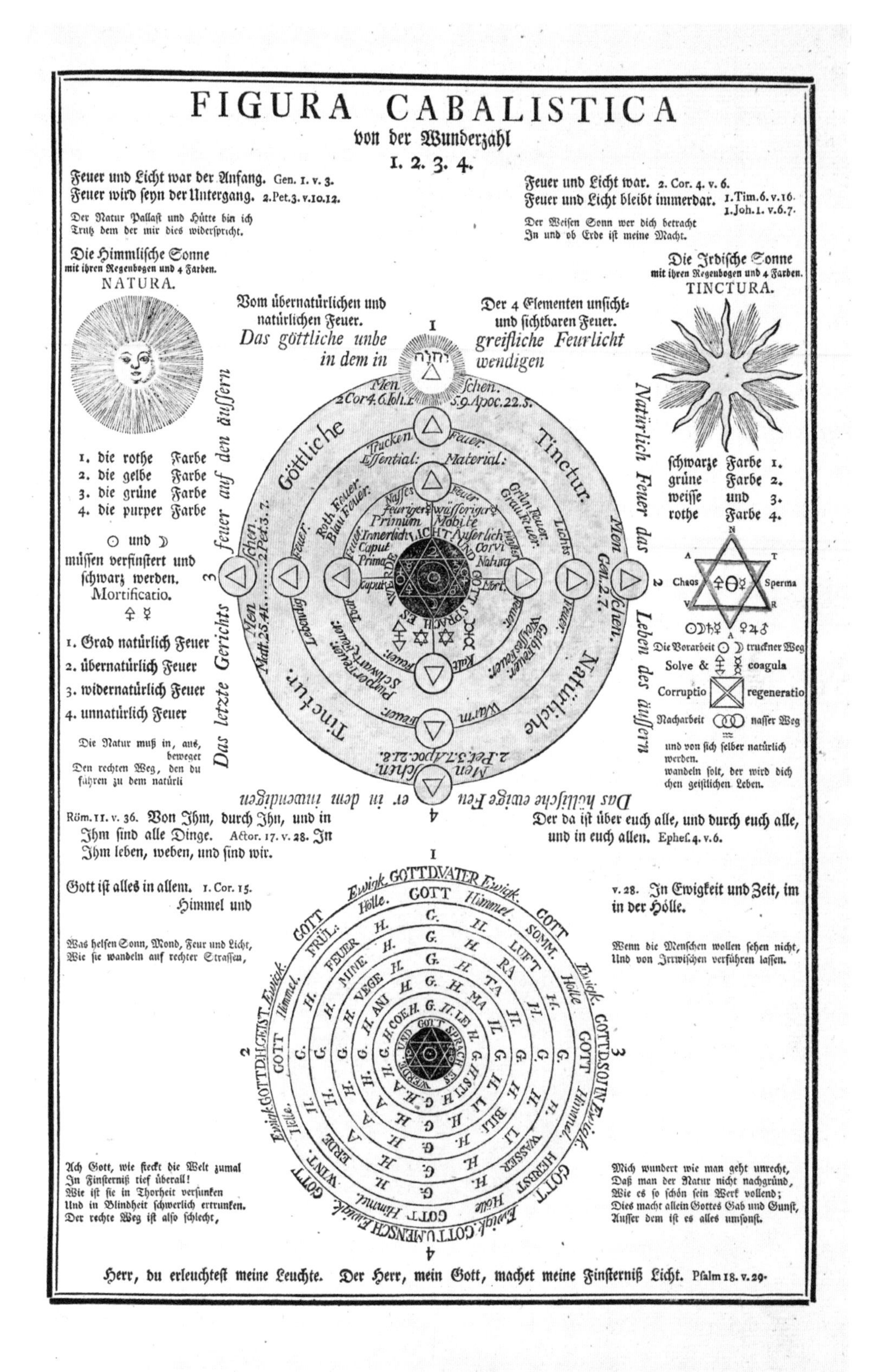

FIGURA CABALISTICA
von der Wunderzahl
1. 2. 3. 4.
Feuer und Licht war der Anfang. Gen. 1. v. 3.
Feuer wird seyn der Untergang. 2.Pet.3. v.10.12.
Der Natur Pallast und Hütte bin ich
Trutz dem der mir dies widerspricht.
Feuer und Licht war. 2. Cor. 4. v. 6.
Feuer und Licht bleibt immerdar. 1.Tim.6. v.16.
1.Joh.1. v.6.7.
Der Weisen Sonn wer dich betracht
In und ob Erde ist meine Macht.
Die Himmlische Sonne
mit ihren Regenbogen und 4 Farben.
NATURA.
Die Irdische Sonne
mit ihren Regenbogen und 4 Farben.
TINCTURA.
Vom übernatürlichen und natürlichen Feuer.
Der 4 Elementen unsicht- und sichtbaren Feuer.
Das göttliche unbe greifliche Feurlicht
in dem in wendigen
Men schen.
2 Cor.4.6.Joh.1. 5.9.Apoc.22.5.
Göttliche
Tinctur.
Natürliche
Tinctur.
Natürlich Feuer das Leben des äussern
Das letzte Gerichts feuer auf den äussern
Men schen. Matt.25.41. 2 Pet.3.7.
Men schen. Gen.2.7.
Men schen. 2.Pet.3.7.Apoc.21.8.
Das höllische ewige Feu er in dem inwendigen
1. die rothe Farbe
2. die gelbe Farbe
3. die grüne Farbe
4. die purper Farbe
schwarze Farbe 1.
grüne Farbe 2.
weisse und 3.
rothe Farbe 4.
und
müssen verfinstert und schwarz werden.
Mortificatio.
Chaos
Sperma
Die Vorarbeit truckner Weg
Solve & coagula
Corruptio regeneratio
Nacharbeit nasser Weg
1. Grad natürlich Feuer
2. übernatürlich Feuer
3. widernatürlich Feuer
4. unnatürlich Feuer
Die Natur muß in, aus, beweget
Den rechten Weg, den du fahren zu dem natürli
und von sich selber natürlich werden.
wandeln solt, der wird dich chen geistlichen Leben.
Röm. 11. v. 36. Von Ihm, durch Ihn, und in Ihm sind alle Dinge. Actor. 17. v. 28. In Ihm leben, weben, und sind wir.
Der da ist über euch alle, und durch euch alle, und in euch allen. Ephes. 4. v. 6.
Gott ist alles in allem. 1. Cor. 15. v. 28.
Himmel und Hölle.
In Ewigkeit und Zeit, im in der Hölle.
Was helfen Sonn, Mond, Feur und Licht,
Wie sie wandeln auf rechter Strassen,
Wenn die Menschen wollen sehen nicht,
Und von Irrwischen verführen lassen.
Ewigk. GOTT D. VATER Ewigk.
Ewigk. GOTT D. H. GEIST Ewigk.
GOTT D. SOHN Ewigk.
Ewigk. GOTT U. MENSCH Ewigk.
GOTT SPRACH ES WERDE
Ach Gott, wie steckt die Welt zumal
In Finsterniß tief überall!
Wie ist sie in Thorheit versunken
Und in Blindheit schwerlich ertrunken.
Der rechte Weg ist also schlecht,
Mich wundert wie man geht unrecht,
Daß man der Natur nicht nachgründ,
Wie es so schön sein Werk vollend;
Dies macht allein Gottes Gab und Gunst,
Ausser dem ist es alles umsonst.
Herr, du erleuchtest meine Leuchte. Der Herr, mein Gott, machet meine Finsterniß Licht. Psalm 18. v. 29.

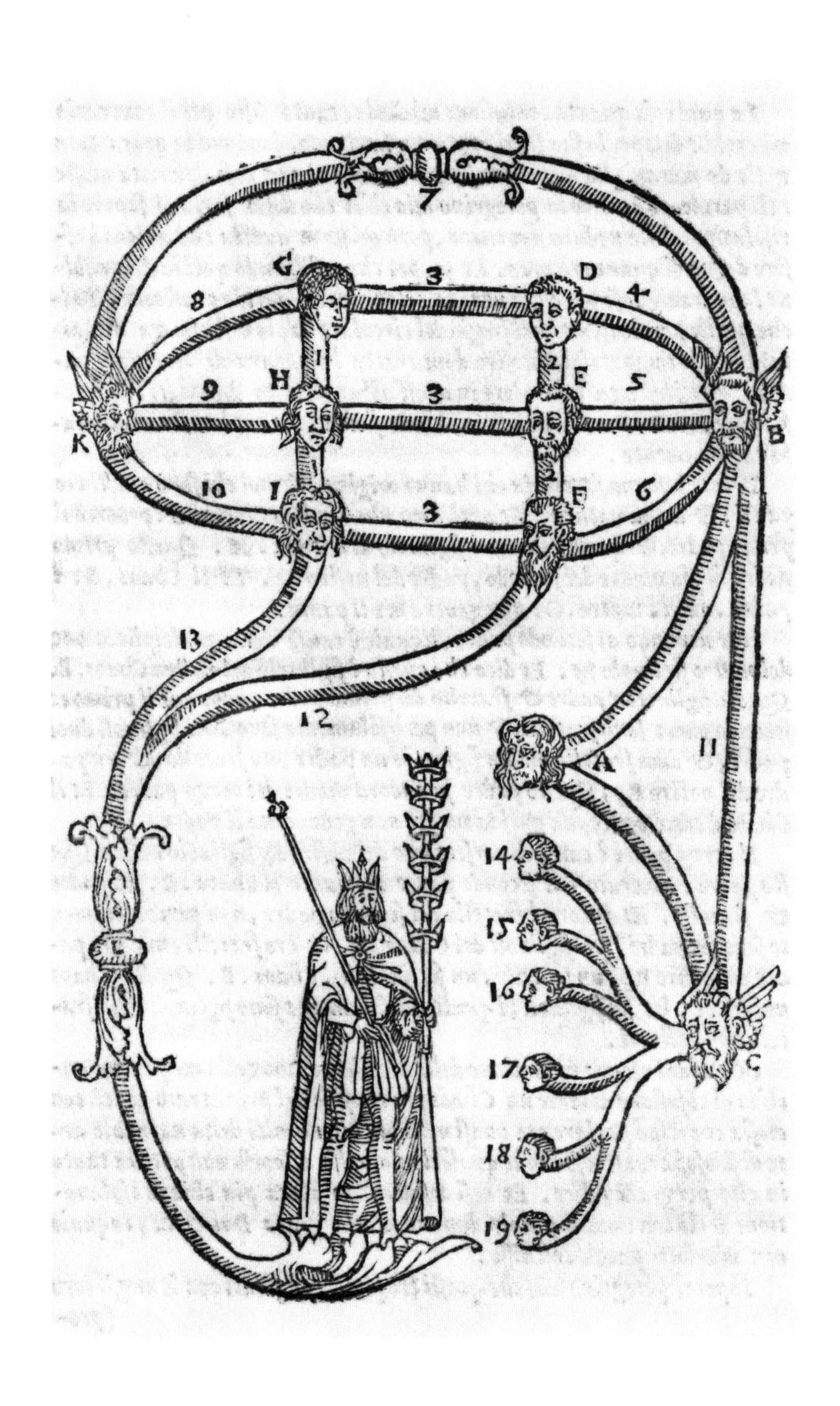
7
G
3
D
4
8
1
2
H
9
3
E
5
K
B
1
2
10
I
3
F
6
13
12
A
11
14
15
16
17
C
18
19

Kabbalistic diagram from *Kabbala Denudata*, by Christian Knorr von Rosenroth, 1677–84.

another. Thus, for example, the planets used in astrology could be associated with alchemical metals and even with the kabbalistic *sefirot* system of divine emanations. In fact, according to Pico such usage was not only useful but actually necessary. This is exemplified in his *Conclusions* (1486),[28] where it is stated that one must use kabbalah in order to effectuate magical operations. As it is explained: 'No magical operation can be of any efficacy unless it has annexed to it a work of Cabala, explicit or implicit.'[29] Considering the close link between, for instance, 'natural magic' and alchemy, we can see how such theses served as the inspiration for the emerging esoteric literature that soon followed.

In addition to the tables, we see treatises that allow for a theoretical synthesis between alchemy and kabbalah. One such case is *Or Nogah* ('The Brilliant Light' – *אור נגה*),[30] better known by its official Latin title, *Splendor Lucis* ('The Splendour of Light', 1745).[31] This book was written by Alois Wiener von Sonnenfels (1705–68), formerly Chayim Lipmann Perlin Sonnenfels, the son of Wurzbach Lipmann, chief rabbi of Brandenburg between 1713 and 1725 under King Frederick William I of Prussia. In 1735, Sonnenfels converted to Catholicism and, shortly after, he composed *Or Nogah* – an alchemical kabbalistic work written in Hebrew and German (with additional Latin text). Much like other Christian kabbalistic writings, this composition contains a vast amount of Christological content. In addition, the text includes ample discussions concerning the Philosophers' Stone – a key component in the process of the transmutations of metals.[32]

However, one of the best known examples of this genre is *Esh Metzaref* – אש מצרף (sometimes written as *Aesch Mezareph*), which can be translated as 'The Purifying', or 'Refiner's Fire'.[33] This composition survives, albeit in fragmented form, in Christian Knorr von Rosenroth's (1636–89) *Kabbala Denudata* ('Kabbalah Unveiled', 1677/8, 1684).[34] At the time, this arrangement represented one of the largest compendiums of kabbalistic dictums translated into Latin. Unlike other writings in this compilation, *Esh Metzaref* has been divided up and peppered across the first part of the collection, which consists of a kabbalistic lexicon. The syncretic approach is very clear in this work, as not only do we see the association of alchemical metals with the *sefirot*, but the text also presents evidence of astral magic and astrology. This is exemplified by the presence of magical squares, numerical configurations that are embedded in equal-sided grids, that can be related to specific planets. This enigmatic composition and its unique approach to alchemy was not entertained exclusively by occult adepts. In fact, it also piqued the interest of Sir Isaac Newton (1643–1727), the father of modern science. Newton, who also conflated Maria the Jewess with the biblical sister of Moses,[35] actually owned a copy of the *Kabbala Denudata*. During my research, I discovered that his particular method of folding in pages of interest reveals that he was specifically attracted to content containing excerpts from *Esh Metzaref*.[36]

As for authorship, the original composition is said to have been written by a certain 'Rabbi Mordechai', who is mentioned briefly in the text.[37] However, the identity of this individual is not clear. Assuming he was a real person, one possibility is that the name is a reference to Mardochaeus (or Mordechai) de Nelle. This

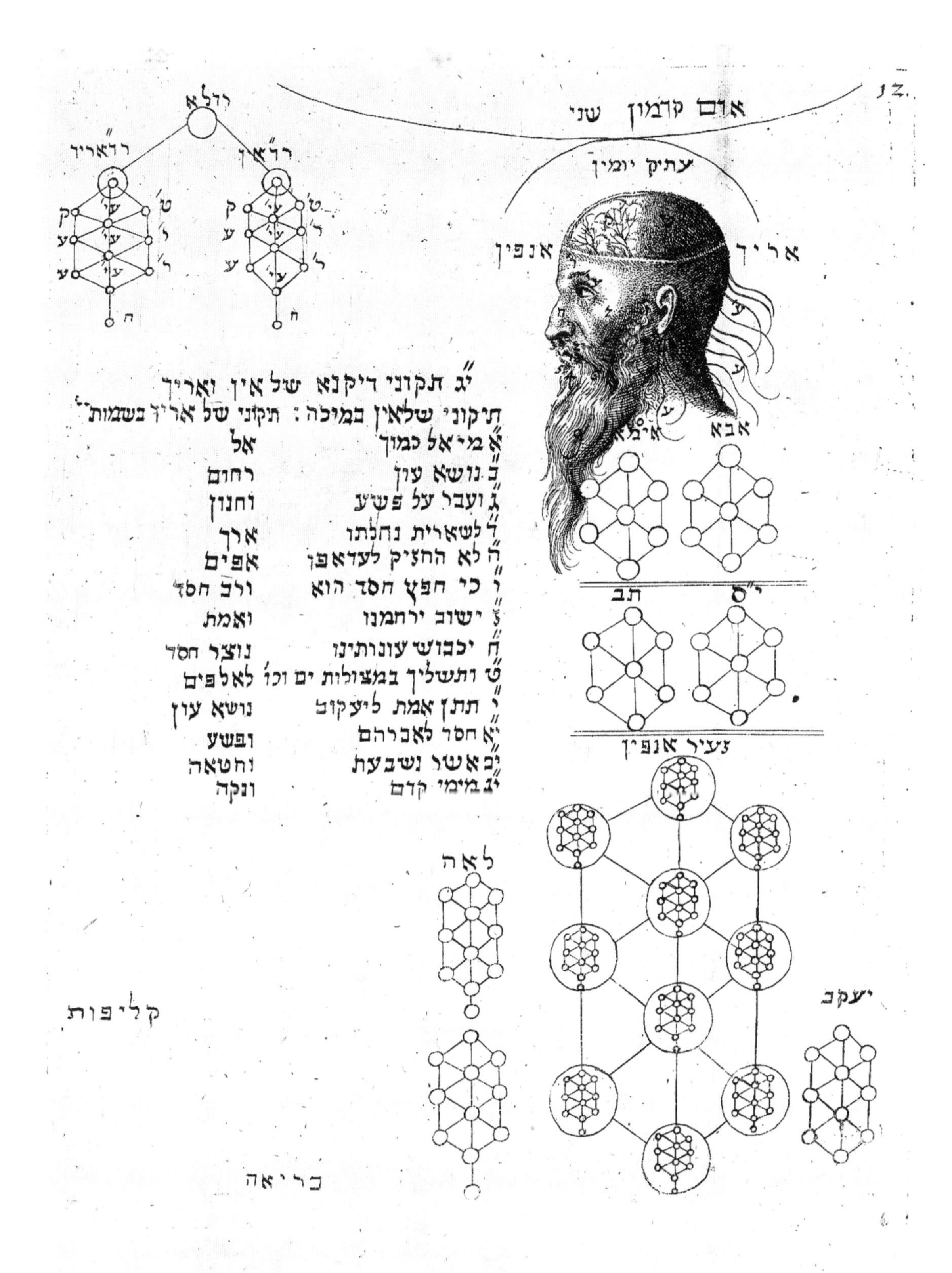

אדם קדמון שני
עתיק יומין
אריך אנפין
ידלא
רד"אין
רדאריך
י"ג תקוני דיקנא של אין ואריך
תיקוני שלאין במיכה : תיקוני של אר"ך בשמות
א" מי אל כמוך — אל
ב" נושא עון — רחום
ג" ועבר על פשע — וחנון
ד" לשארית נחלתו — ארך
ה" לא החזיק לעד אפו — אפים
ו" כי חפץ חסד הוא — ורב חסד
ז" ישוב ירחמנו — ואמת
ח" יכבוש עונותינו — נוצר חסד
ט" ותשליך במצולות ים וכו' — לאלפים
י" תתן אמת ליעקב — נושא עון
י"א חסד לאברהם — ופשע
י"ב אשר נשבעת — וחטאה
י"ג מימי קדם — ונקה
אבא
אימא
יש"ס
תב
זעיר אנפין
לאה
יעקב
קליפות
בריאה

mysterious personality is known from several sources that attest to his activities and travels in Eastern Europe, including a visit to Emperor Rudolph II (1576–1612) in Prague, one of the foremost patrons of occult disciplines, earning him the nickname the 'German Hermes Trismagistos'.[38] However, most recently, Esotericism scholars Mike Zuber and Raphael Prinke have cast doubt over the existence of de Nelle as a historical figure.[39]

Scholem and Patai proposed another candidate for the elusive author of *Esh Metzaref*, namely Mordechai de Leon of Modena (d. 1615), the son of (Yehudah) Aryeh of Modena (1571–1648), known for his disdain and harsh polemics against the kabbalists.[40] While Aryeh dabbled in alchemy, Mordechai was devoted to this endeavour full-time. Moreover, based on his father's testimony, we see that he practised it alongside a Catholic cleric named Giuseppe (Joseph) Grillo in Venice.[41] Having both a rabbi and a priest practising alchemy together is direct evidence of how such praxis transcended religious differences – a trend that continued well into modernity.[42] However, Mordechai's enthusiasm may also have led to his demise as it is reported that he died from a mishap with, or perhaps as a result of, his alchemical activities. The operation in question involved specifically the transmutation of lead into silver by the use of arsenic and certain salts. His father speculated that the inhalation of such vapours was most likely the culprit, as Mordechai exhibited heavy nosebleeds before his passing.[43] Scholem finds parallels between this account and certain methods involving using arsenic to produce silver described in *Esh Metzaref* – thus strengthening the association and attribution of authorship to Mordechai de Leon of Modena.

Portrait of Rabbi Yehudah Aryeh Leon of Modena. On the title page of his *Historia de Riti Hebraici* ('History of Jewish Rites'), 1638.

The term *Esh Metzaref*, or more specifically, the verb root *tzaraf* (צרף) and its various meanings, was of unique interest to the kabbalists. Commentaries on the proto-kabbalistic text *Sefer Yetzirah* ('Book of Formation'), one of the most influential works in Jewish Mysticism, posits that magical manifestation or 'creation' is possible by the combination (*tzeruf*) of the Hebrew characters.[44] Another interpretation can be seen in the genre of literature that includes ecstatic techniques championed by Abraham Abulafia (1239–d. after 1291), one of the early proponents of Prophetic Kabbalah. In his writings, we see meditations designed to produce altered states of consciousness and even divine union by utilising the permutation (*tzeruf*) of letters.[45] The understanding of *tzeruf* as 'refinement' took an interesting turn around the sixteenth century, as another alchemical nomenclature with the same meaning became dominant, namely the concept of *tikkun* (תיקון), or rectification.

Up until now, we have explored the physical aspect of alchemy. However, we are now well-equipped to delve into its more cryptic transcendent counterpart. Alchemical thought posits that the non-volatile essence of all metals, or its 'body' (*sôma* – σῶμα), is the same. However, properties such as colour, known as its volatile 'spirit' (*pneûma* – πνεῦμα), can be altered and even refined. Such processes help facilitate the transmutation of lesser, more common metallic substances like lead, into ones that are rarer and purer such as gold. As can be imagined, the concept of the 'purification of spirits' can lend itself quite nicely to symbolism involving rectifying one's soul.[46] In fact, examples of such metaphors exist in the Jewish literature as far back as the Bible. Though, around the period of the golden age of alchemy in early modern Europe, we begin to see more literal representations of metaphysical alchemical operations in the kabbalistic writings. The most representative example of this is in the Lurianic school. Though some treatises were written by Isaac Luria (1534–72), the founder of this system, much of his oral teachings were captured by his students – especially his trusted note-takers Moshe Yonah (fl. sixteenth century) and then later Chayim Vital (1543–1620). The latter is of particular interest to our present enquiry as, apart from being a kabbalist, he was known for his engagement with occult arts, especially alchemy.[47]

Current academic scholarship asserts that Vital and especially Luria opposed the practical applications of kabbalah, sometimes labelled *Kabbalah Ma'asit* (קבלה מעשית), including alchemy. As such, any alchemical content in Vital's writings is seen as simply incidental – knowledge that was filtered through the lenses of the alchemical education he acquired during, his (much regretted) past.[48] However, this attitude was championed by later editors, and both Luria and Vital held respect for such wisdom, which may have even been the basis for their special relationship.[49]

Chayim Vital's diary, which was recently published under the title *Sefer ha-Pe'ulot* ('The Book of Operations' – ספר הפעולות),[50] demonstrates a myriad of formulae across various subjects ranging from medical recipes to metallurgic manipulation, including a successful gold-making transmutational procedure. However, we also find in the teachings of his master metaphysical alchemical operations that involve rectifying divine influx both in the macrocosmic universe and its parallel microcosmic human domain.

Similar (allegorical) processes can be gleaned from the treatises of other European esoteric thinkers. For example, in the teachings of Michael Maier we see allusions to the cosmos as an alchemical vessel, particularly as it pertains to the biblical narrative of the first six days of creation, sometimes referenced as *septimana philosophica*, or the 'philosophical week'.[51] In addition, we also have the work of Giovanni Battista Nazari titled *Il Metamorfosi Metallico et Humano* ('The Metallic and Human Metamorphosis', 1564),[52] where man's sinful nature represented in his physical form can, much like metallic transmutations, transform into the perfect being, symbolised as Christ's body.[53] As mentioned, such religious symbolisms were very common in alchemical literature.[54]

In the Lurianic myth, impurities originate in a primordial catastrophe known as the 'breaking of the vessels' (*Shvirat ha-Kelim* – *שבירת הכלים*), where during the creation of the universe, emanations deriving from the ineffable source called ***ain sof*** *אין סוף* (literally 'without an end') could not contain the infinite light and thus simply broke, leaving shards, identified as the infamous ***klipot*** (husks – *קליפות*), with residual sacred luminance trapped inside them. The biblical Adam was created in order to rescue these holy sparks through the process of alchemical rectification. However, given the mishap in the Garden of Eden, Adam's soul, paralleling the cataclysm of the macrocosmic emanations, also shattered, leaving the task of ***tikkun*** to humanity.

Rectification in the Lurianic system involves spiritual praxis that sublimes the divine sparks and purifies them from the impure realm of the *klipot* – the lowest part of the metaphysical cosmos which also includes our earthly domain. While this can be achieved through general religious practices such as the study of the Torah or liturgical prayers and meditations, we also find that the human body can function as an 'alchemical apparatus' in certain cases such as fasting, eating and even sexual intercourse. For example, by abstaining from nourishment the body burns the bodily fat and blood. Much like burnt animal offerings, an adept may sublime the resultant influx back to the transcendent. Eating also allows for souls, believed to be trapped in food, to liberate, rise up within the individual, purify and return to the source. Lastly, similar processes occur during lovemaking. In Luria's own writings, we find that sexual desire originates in the hypostatic domain, condenses down the spine and then becomes sexual fluids. These liquids, in turn, are rectified back up during the orgasm – a process that resembles the alchemical double pelican vessel that allows for cyclical sublimation and is sometimes pictured in the literature as two people embracing.[55]

As we can see, the above illustrations demonstrate and serve as prime examples that the interaction between the Jews, alchemy and kabbalah is both very much present and also quite complex. In fact, given this perspective, it would appear that alchemical notions were not only implemented by the kabbalists but may also have served as the inspiration for their precepts. While some preliminary research has already been established, the intricate layers of these relationships and connections are just now being uncovered and unravelled. Should this trajectory continue, we may find further evidence of praxis and additional affirmation of the integration of alchemical ideas in both exoteric and esoteric Jewish thought, as well as the metaphysical systems of other related European occult milieus.

Double pelican vessel. Image from *De Distillation* by Giambattista della Porta, 1608.

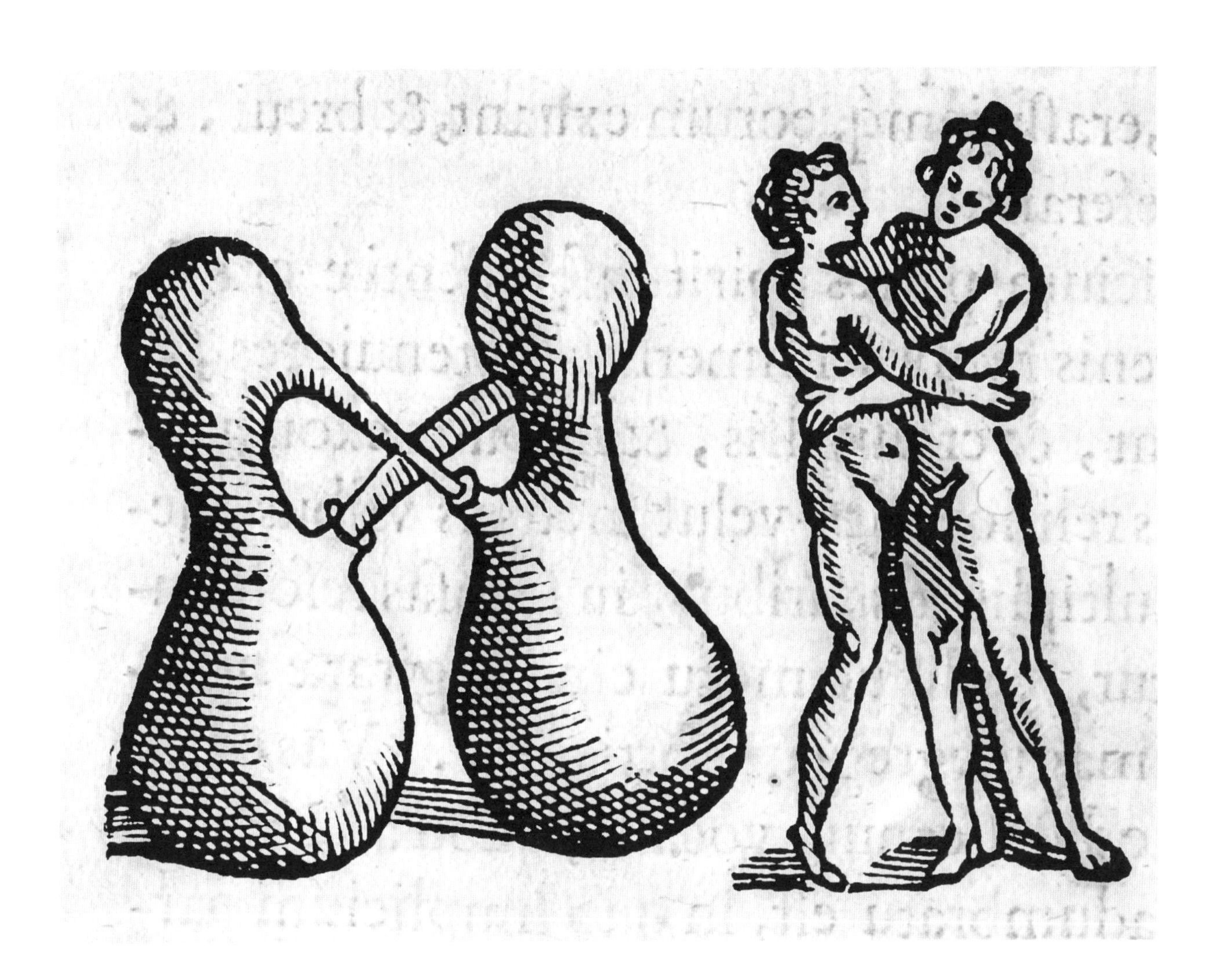

IOÁNES
STRATENSÍS
FLANDRVS
1570

MEDIEVAL AND EARLY MODERN ALCHEMY IN EUROPE

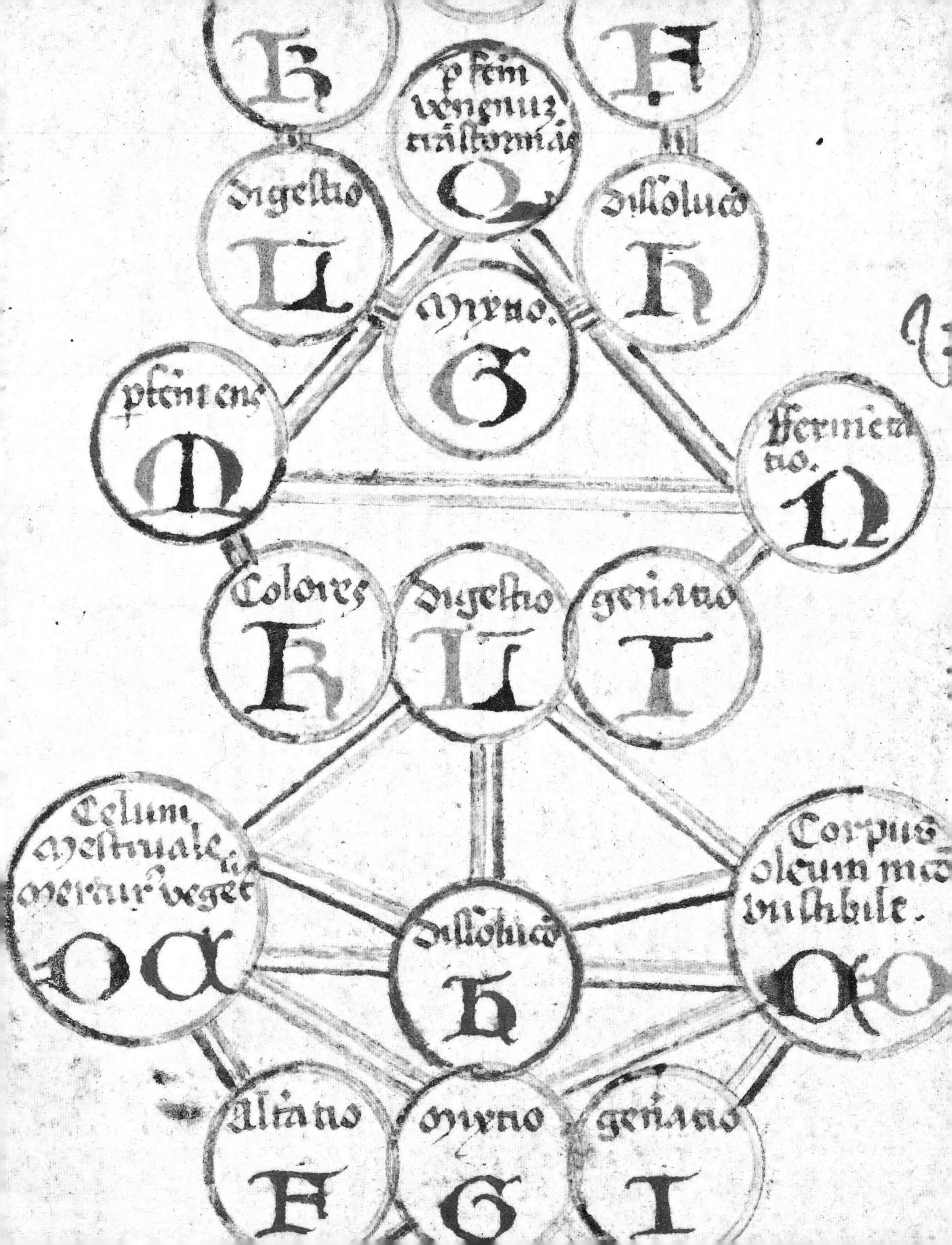

R
H
digestio
L
H
mixtio
G
M
fermentatio
N
Colores
R
digestio
L
T
Celum
Corpus
h
F
G
I

MEDIEVAL LATIN ALCHEMY

PETER J. FORSHAW

Detail of *Tertia distinctio* ('Third Distinction'), 1351–52.

Previous vignette: The Alchemist's Workshop. Oil on canvas by Jan Van Der Straet, 1570.

'For this book is divine and most full of divinity, the true and perfect proof of the Old and New Testaments.'[1] So writes the English monk Robert of Chester, who had been involved with the first translation of the Quran into Latin the previous year.[2] He is not talking here about a religious work, however, but about a new form of knowledge that he is ushering into Western consciousness: the art of alchemy, which first enters the Latin West, as he records, on Friday 11 February, 1144. The book in question has been given various titles over the years, *De compositione alchemiae* ('On the Composition of Alchemy'), *De transfiguratione metallorum, & occulta, summaque antiquorum Philosophorum medicina* ('On the Transfiguration of Metals, and the Hidden and Greatest Medicine of the Ancient Philosophers'), or simply the *Liber* or *Testamentum Morieni* ('The Book or Testament of Morienus').[3] It is an account of how the Christian Morienus first travels from Rome to Alexandria to study alchemy with the venerable scholar Adfar (Stephanos of Alexandria)[4] and then, after his master's death, moves to Jerusalem to live as a hermit, until he is persuaded to go to Egypt to instruct the Umayyad prince Khalid ibn-Yazid (*c.* 665–704) in the secrets of the art.[5]

Although the standard etymology of 'alchemy' is usually said to be from the Greek term *chēmeia*, supposedly based on the verb *χέω* (*cheō*), 'to melt' or 'make fluid', with the addition of the Arabic definite article *al-* as a prefix, that is, 'the art of melting', in this first encounter we receive a rather unexpected explanation of *alchymia*, in language surely alluding to the *Emerald Tablet*, as 'a corporeal substance composed from the One and through the One, joining the most precious things together...and naturally converting the same things'.[6] *Alchymia* is introduced not as the knowledge necessary for the work, but as its sought-for product, the philosophers' stone or elixir, which will transmute silver into purest gold.[7] Morienus speaks of primal matter and the four elements, of the Latin and Arabic names of minerals, the properties of the different metals, of the colour changes seen at various stages in the work, and compares the generation of the stone to that of a human being, through the states of coitus, conception, pregnancy, birth and nourishment.[8]

He seldom uses allegorical language, although he does mention the Green Lion,[9] but some statements are nevertheless ambiguous and open to various levels of interpretation, especially the declaration that the matter necessary for the philosophers' stone 'comes from you, who are yourself its source' (literally 'its mine').[10] Ultimately, Morienus informs Khalid, 'God confers this divine and pure knowledge to his faithful', for it is nothing less than a 'donum Dei' (a gift of God).[11]

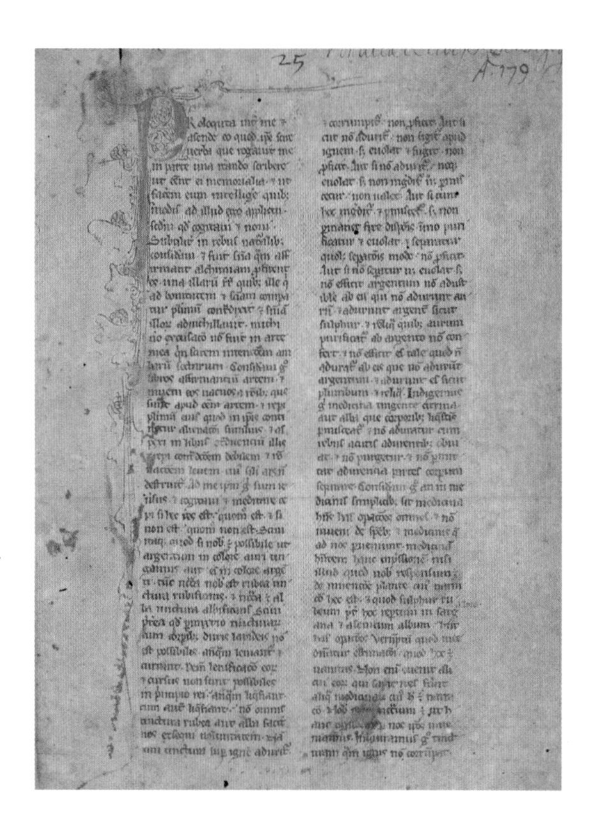

Robert of Chester's translation of *Liber de compositione alchemiae*, by Morienus. 13th century.

Evidently a degree of uncertainty existed about how to classify this new knowledge. Two early references to the 'science of alchemy' can be found in the works of the philosophers Dominicus Gundissalinus (fl. 1150) and Daniel of Morley (*c.* 1140–1210). In *De divisione Philosophie* ('On the Division of Philosophy'), the former introduces the 'sciencia de alquimia' as one of the eight parts of natural science, while in the latter's *Liber de naturis inferiorum et superiorum* ('Book on the Natures of Lower and Upper Things') the 'scientia de alckimia' is listed as one of the eight parts of astrology.[12] This new science, nevertheless, had an enthusiastic reception, with the famous German Dominican theologian and natural philosopher, the *Doctor universalis* Albertus Magnus (*c.* 1200–80) writing in *De mineralibus* ('On Minerals') of testing alchemical silver and gold and defending the theory of transmutation.[13] His contemporary the English *Doctor mirabilis* and Franciscan friar Roger Bacon (*c.* 1219–*c.* 1292) takes a different tack in his *Opus tertium* ('Third Work'), displaying an interest in potential medical applications of alchemy, in purifying existing apothecary medicines and in the prolongation of human life.[14] Albertus's student the Italian *Doctor angelicus* Thomas Aquinas (1225–74), like him later canonised as a saint, is said to have declared that 'This art either finds a man Holy, or its discovery makes him Holy!'[15]

By far the best-known exponent of laboratory alchemy in the Middle Ages is the author known as Pseudo-Geber, identified as the Italian Franciscan friar Paul of Taranto,[16] who drew heavily on Arabic sources, including works attributed to the famous Persian alchemist Abū Mūsā Jābir ibn Ḥayyān (*c.* 721–*c.* 815).[17] Geber's *Summa perfectionis magisterii* ('The Summation of the Perfection of the Magistery', *c.* 1300) is arguably the most influential medieval work of transmutational alchemy. It starts with a discussion of the physical and mental prerequisites for anyone intending to be an 'Artificer', bluntly stating that 'whoever does not know the natural principles in himself, is already far removed from our art, since he does not have the true root upon which he should found his goal'.[18] The *Summa* discusses the feasibility of gold-making and the possible transmutation of one species of metal to another (with the analogy, for example, of how a worm mutates into a fly)[19] and enters into a detailed summary of current knowledge about the minerals and the six metals (lead, tin, iron, copper, silver and gold) used in the work, particularly the relative proportions and grades of sulphur and mercury in their composition, the former being the cause of their corruption and the latter, the 'true medicine of alterable things',[20] the cause of their perfection.[21] Gold, the most perfect of metals, is described as

> a metallic, yellow, heavy, silent, brilliant body, temperately digested in the womb of the earth, and washed for a very long time by a mineral water, extensible under the hammer, fusible, and able to withstand the tests of cupellation and cementation.[22]

It is 'made from the most subtle, fixed and brightest substance of quicksilver, and from a little of the substance of clean, fixed sulfur of little redness', its heavy weight being due to the fineness of its particles, which are greatly compressed.[23] Geber works systematically through the other metals, stating, for instance, that copper is 'midway between *Sol* [gold] and *Luna* [silver]' and is easily converted to either;[24] tin and lead, however, have too much liquid mercury, hence their low melting points; iron and copper, on the other hand, possess too much inflammable dry sulphur, which is why their filings burn when dropped into a fire.[25]

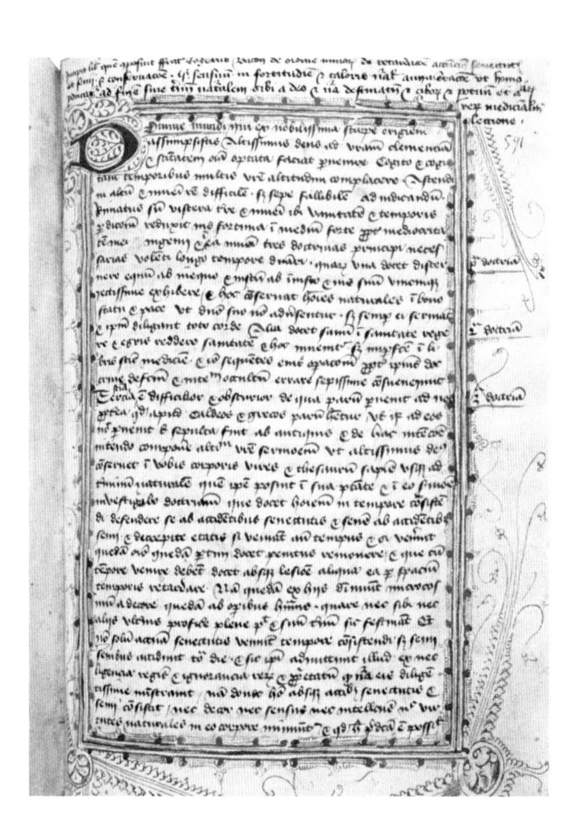

Decorated page from *Opus tertium* ('Third Work'), by Roger Bacon. Beginning of the 15th century.

Geber also introduces his reader to eight different alchemical processes (sublimation, descension, distillation, calcination, solution, coagulation, fixation and ceration),[26] together with the colour changes one should expect during these activities, and goes into some detail in his descriptions of laboratory apparatus: a thick-walled furnace, we learn, gives a strong, concentrated fire, compared to the weak fire provided by a thin-walled furnace.[27] In a similar manner he discusses different kinds of baths (ash, water), various glass vessels (aludels, alembics), 'subtle waters' for dissolving substances, and different kinds of fuel (wood and horse dung) for heating.[28] He reassures his reader that he speaks from personal experience: 'We have seen with our eye and touched with our hand the sought-for goal of this, with our magistery',[29] and that he has not written 'anything except that discovered by us ourselves'.[30] The voice of common sense sounds in his advice to check the accuracy of all results with metallurgical assaying techniques. All this, the reader is informed, is only accomplished with 'the greatest application of labor and with a lengthy spell of intense meditation. For with this you will find it, and without it not.'[31] One final element of Geber's writing that is particularly noteworthy concerns the transmission of this knowledge. Early in the book he states that 'this magistery does not need a hidden discourse, nor a wholly open one',[32] but it is at the very end of the work that he lays his cards on the table, in the concluding section: 'the Author relates how he has hidden the Science'. There he writes of the practice of 'dispersa intentio' (dispersion of knowledge), in order to protect alchemy's secrets from the profane:

> And lest we be attacked by the jealous, let us relate that we have not passed on our science in a continuity of discourse, but that we have strewn it about in diverse chapters...And we have also hidden it where we have spoken more openly.[33]

Ultimately – and encouragingly, if somewhat ironically – having dutifully pored over this technical book on alchemical theory and practice, the 'son of the doctrine' learns that he will make discoveries 'not by inquiring of doctrine', nor (at least not quickly) by 'pursuance of books', but 'by inquiring of the motion of his own nature. For he who seeks knowledge through the goodness of his own diligence will find it.'[34]

Although the Catalan philosopher Ramon Llull (*c.* 1232–1315/16) is renowned today for his Art of combinatorial logic, and even considered a precursor of modern computing, he was not known for any interest in alchemy. Yet this did not

Handwritten index in an early edition of *Summa perfectionis magisterii* ('The Summation of the Perfection of the Magistery') by Pseudo-Geber (attributed to Paul of Taranto), 1486–90.

discourage alchemical writers from issuing works under his name. One particularly influential Pseudo-Lullian work is the *Testamentum* (*c.* 1332), which spoke, like Geber, of metallic transmutation, as well as the production of artificial and precious stones.[35] Although Geber had discussed 'solar' and 'lunar' medicines, these were in the context of 'healing' impure metals, with little sense of treating human beings. Pseudo-Lull, however, like Roger Bacon, recommends alchemical medicine, specifically the *elixir vitae* (elixir of life) for the comfort and treatment of human infirmities, in addition to discussing alchemy's ability to purify all precious stones and transmutate all metallic bodies into true silver and gold.[36] He also engages in a theo-alchemical discussion of God's creation of the macrocosm or Greater World with 'a certain pure substance called the quintessence', divided into three parts: the purest, for the highest angels; then the heavens, planets and stars; and, finally, the least pure part for the terrestrial realm, including the four elements.[37] He describes the alchemical magistery as the creation of a microcosm, 'by way of miracle'.[38] He also introduces a reference to Christ's shedding of his blood on the cross 'in order to exalt us',[39] as well as to the creation of Adam and the subsequent 'transcreation' of Eve, with a consideration of how 'our magistery is assimilated to the work of the creation of man'.[40] Presumably to avoid accusations of transgressing the boundaries of religion, Pseudo-Lull takes care to state that he wishes his reader to understand this 'with a scientific and not a hagiographic spirit'.[41]

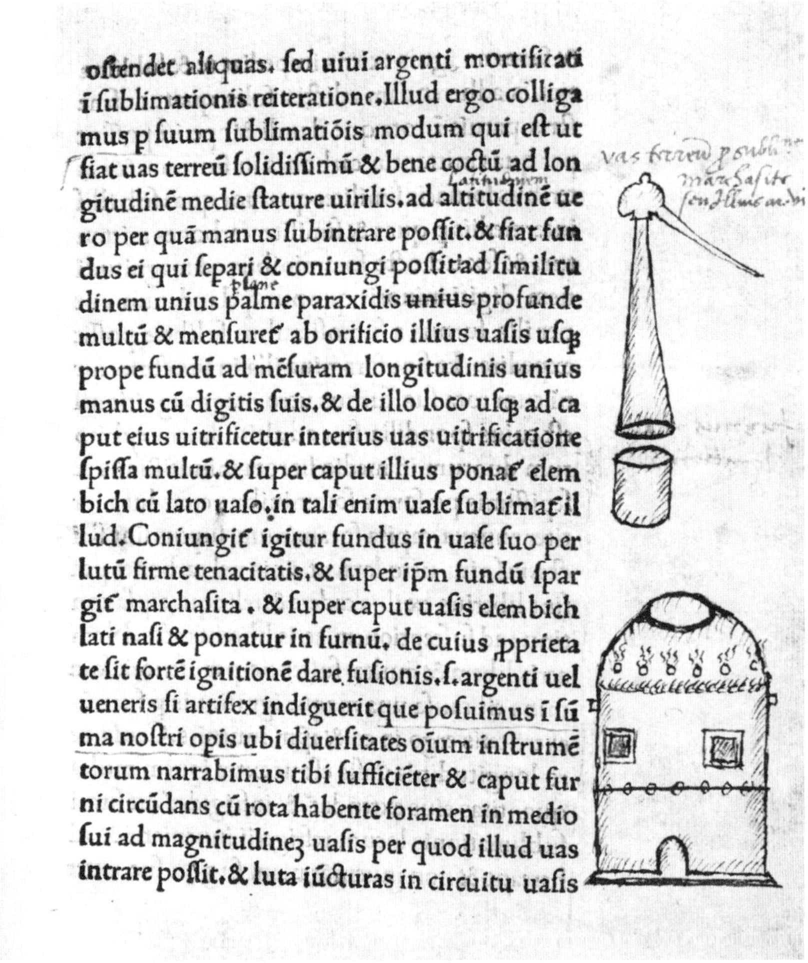
ostendet aliquas. sed uiui argenti mortificati
i sublimationis reiteratione. Illud ergo colliga
mus p suum sublimatiōis modum qui est ut
fiat uas terreū solidissimū & bene coctū ad lon
gitudinē medie stature uirilis. ad altitudinē ue
ro per quā manus subintrare possit. & fiat fun
dus ei qui separi & coniungi possit ad similitu
dinem unius palme paraxidis unius profunde
multū & mensuret ab orificio illius uasis usq
prope fundū ad mēsuram longitudinis unius
manus cū digitis suis. & de illo loco usq ad ca
put eius uitrificetur interius uas uitrificatiōe
spissa multū. & super caput illius ponat elem
bich cū lato uaso. in tali enim uase sublimat il
lud. Coniungit igitur fundus in uase suo per
lutū firme tenacitatis. & super ipm fundū spar
git marchasita. & super caput uasis elembich
lati nasi & ponatur in furnū. de cuius pprieta
te sit fortē ignitionē dare fusionis. s. argenti uel
ueneris si artifex indiguerit que posuimus i sū
ma nostri opis ubi diuersitates oīum instrumē
torum narrabimus tibi sufficiēter & caput fur
ni circūdans cū rota habente foramen in medio
sui ad magnitudinē uasis per quod illud uas
intrare possit. & luta iūcturas in circuitu uasis

Marginal drawings in Pseudo-Geber's *Summa perfectionis magisterii* 1486–90.

Inspired by the genuine Lullian Art, as found in the *Ars Brevis* ('The Short Art', *c.* 1307/8), with its 'use of letters of the alphabet, combined on geometrical figures, for the working out of problems',[42] the *Testamentum* communicates by various triangular, square and circular diagrams, which assist in reflecting on the operative relationships between substances, as well as serving as aids to 'memorative, recordative or recollective' practices.[43] For example, he includes a wheel of substances with the categories (A) Hyle or Primal Matter, (B) the Elements, (C) the Vapours of the Elements, (D) Clear Water, (E) Vitreous Azoc (described elsewhere as the 'earth and

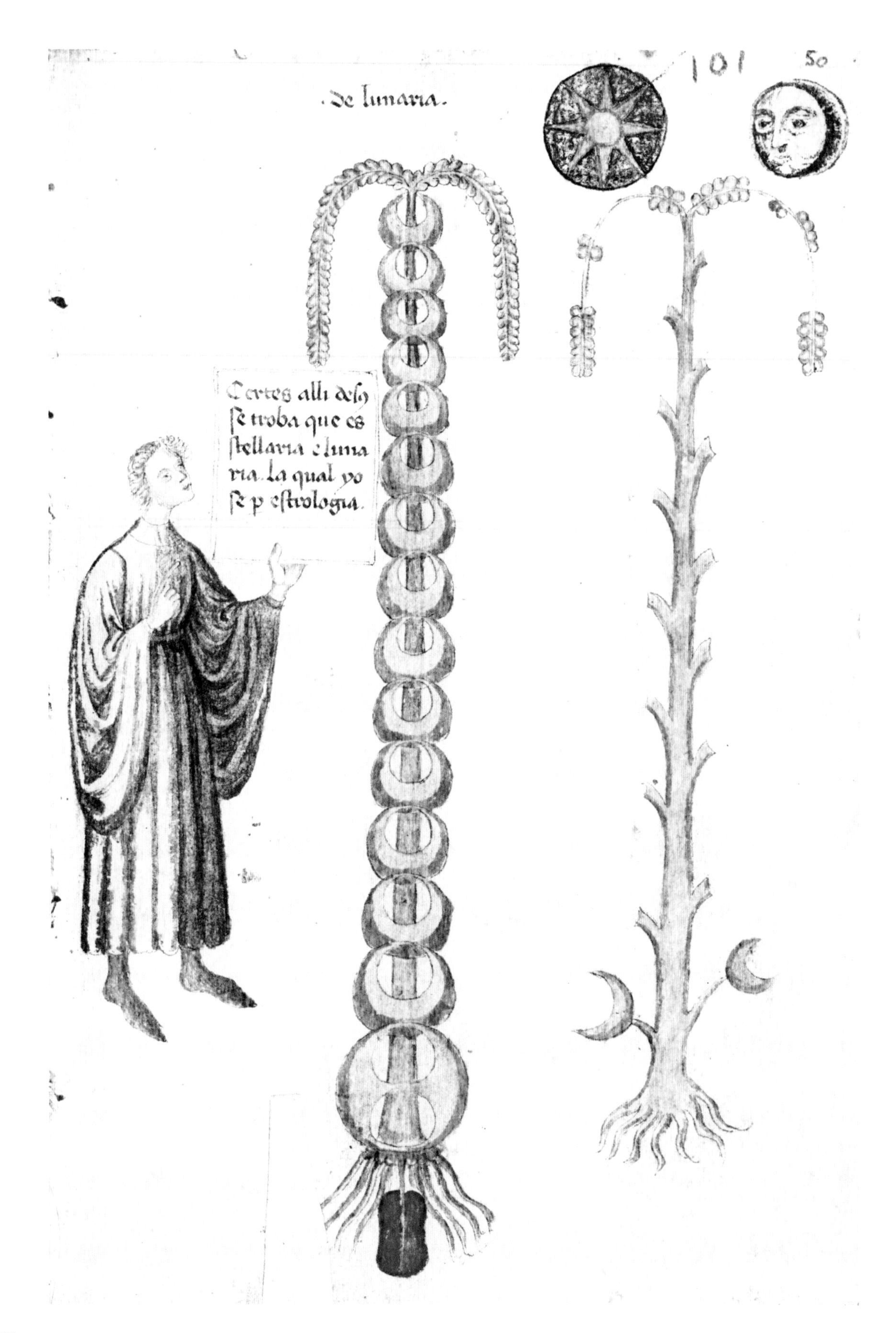

101
50
.De lunaria.
Certes alli desy
se troba que es
stellaria e luna
ria. La qual yo
se p estrologia.

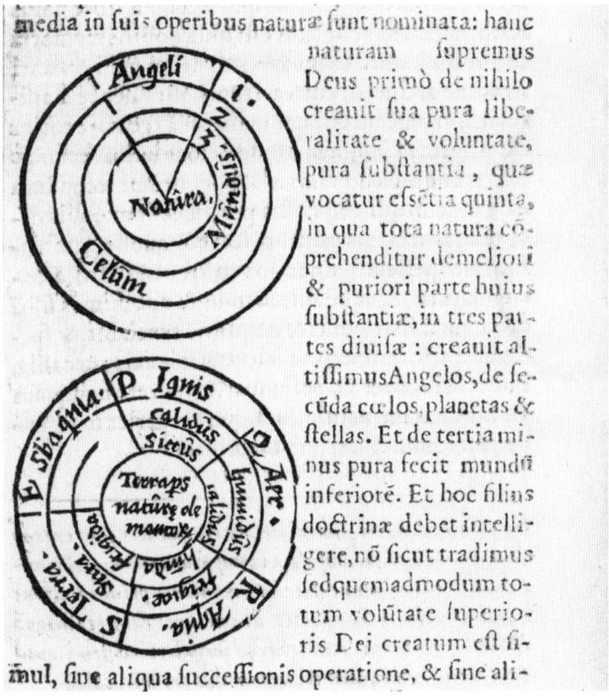
media in ſuis operibus naturæ ſunt nominata: hanc naturam ſupremus Deus primò de nihilo creauit ſua pura liberalitate & voluntate, pura ſubſtantia, quæ vocatur eſsẽtia quinta, in qua tota natura cõprehenditur demeliori & puriori parte huius ſubſtantiæ, in tres partes diuiſæ: creauit altiſſimusAngelos, de ſecũda cœlos, planetas & ſtellas. Et de tertia minus pura fecit mundũ inferiorẽ. Et hoc filius doctrinæ debet intelligere, nõ ſicut tradimus ſedquemadmodum totum volũtate ſuperioris Dei creatum eſt ſimul, ſine aliqua ſucceſſionis operatione, & ſine ali-

Creation of the Macrocosm, from Pseudo-Lull, *Theorica testamentum*, Cologne: David Berthelin, 1573, p.15.

mother of the metals', i.e. mercury),[44] (F) Vapours of Vitriol, (G) Sulphur and (H) the Metals to assist in contemplating the connections between primal matter, the elements, minerals and metals.[45] He also provides tables explaining the significance of the letters used in the *Testamentum*'s diagrams.[46]

While Geber wrote of concealing knowledge by dispersing information, but without the use of any enigmatic language, Lull opts for allegory and we find a far more pronounced presence of the alchemical bestiary, the concealment of substances and encoding of processes by means of dragons, lions, basilisks and birds. Thus we read that 'the start of our operation is in the form of a raven's head',[47] as a mercurial vapour which 'descends from high in the air in the form of a black raven, whose head is red, its feet are white, and its eyes are black';[48] of a great dragon (the one nature which contains in itself the natures and properties of the four elements),[49] 'cast out of the great desert of Arabia, because he would immediately be suffocated by thirst, and perish in the dead sea'.[50] He also introduces new analogies for the compounding of sulphur and mercury: on a religious level as a form of matrimony (between body and spirit);[51] astrologically, as a conjunction of the sun or moon, that is, an eclipse; and mythologically (and biologically), as a hermaphrodite, all images that were to gain traction in later works.[52] It is notable that Pseudo-Lull places far more emphasis on the necessity of secrecy than Geber, warning that

> If you were to reveal in brief words that which he [God] formed over a long period of time, you would be condemned on the day of the great judgment, as a perpetrator of injury against God's majesty.[53]

Therefore, Lull paradoxically explains, 'the books of this science speak through silent voices strongly marked'.[54]

If Pseudo-Geber's focus was on the transmutation of metals, to which Pseudo-Lull briefly added human medicine, the French Franciscan friar John of Rupescissa (*c.* 1310–*c.* 1362) takes the medicine much further with the distillation of quintessences for medicinal purposes. In one of the most widespread works of the Middle Ages, the *Liber de consideratione quintae essentiae omnium rerum* ('Book on the Consideration of the Quintessence of All Things', 1351–52),[55] Rupescissa informs his reader that the first secret of his art is that one can take care of the 'inconveniences of old age', especially for religious men.[56] It is possible to save and preserve a body from decay; in fact, this is a secret that God gave to Adam when he was expelled from paradise, so that he might live for ever.[57] Rupescissa tells us that this wondrous

Oppostie:
The author with Lunaria grass ('Honesty'), the sun and the moon, from John of Rupescissa (Jean de Roquetaillade), *Liber de consideratione quintae essentiae omnium rerum* ('Book on the Consideration of the Quintessence of All Things'), Latin and Catelan, 1351–52.

remedy is given three names by the philosophers: *aqua ardens* (fiery water), soul or spirit of wine, and *aqua vitae* (water of life), and promises that any flesh, fish or fowl placed in it will be free from decay.[58]

Rupescissa's intention is to 'console poor men of the Gospel, so that they do not lose their prayer in this work',[59] and that 'they may be rendered capable, courageous, and strong, for all the labours of perfection'.[60] He gives practical advice on how to cut costs in the laboratory, such as by placing distillation vessels in horse dung, which ferments at a low heat, rather than expending precious funds on wood or coal; likewise, the distillation of *aqua ardens* does not require high-quality wine but can be done cheaply with wine that has turned.[61] This is in fact one of the secrets of the magistery, that the incorruptible quintessence can be extracted, by sublimation, from corrupt wine.[62] Furthermore, this quintessence is in all things, in animal flesh, even in human blood. Rupescissa is particularly knowledgeable about extracting the quintessence as an essential oil from fruits, roots and leaves[63] and discusses at great length which plants (and animals) are suitable for this in relation to the four Aristotelian qualities of heat, cold, dryness and moisture, each of which have their different curative properties. He extols the powers of the quintessence, which can, for example, 'extract wood or iron from wounds',[64] purge noxious humours from the body, rescue those close to death who have been abandoned by physicians,[65] or treat physical ailments like leprosy, scabies and paralysis. It also works on a mental level, curing those afflicted by 'fantastic passions, imaginations, and foolish demonic infestations and temptations',[66] or those suffering from fear or loss of courage, by restoring audacity, fortitude and courage.[67] The quintessence wards off witchcraft and demons, especally those ills associated with the malefic planets of Saturn and Mars.[68] With respect to this, Rupescissa displays far more interest in and knowledge of astrology than previous authors, writing of the influence of the stars on the magnet and iron, of the moon's influence on the sea, the sun on gold, the moon on silver, as well as the medical notion of the *melothesia* or astral man, in which each zodiac sign influences a different part of the body.[69]

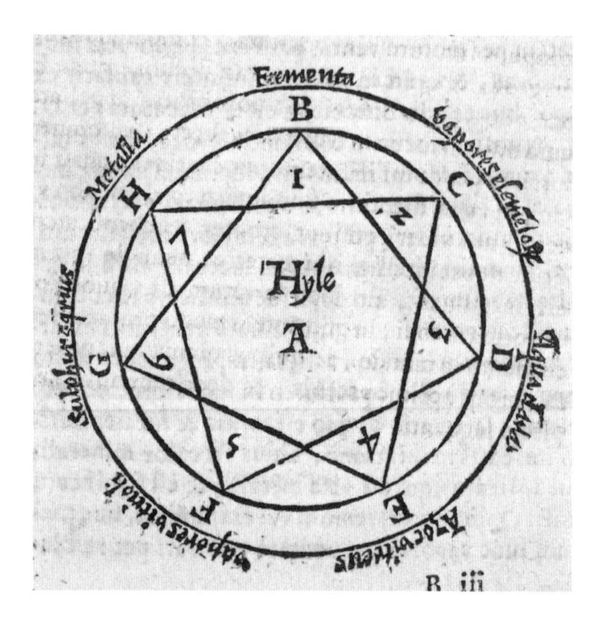

Wheel of Hyle (Primal Matter). In *Theorica testamentum* by Pseudo-Lull, 1573.

Rupescissa works through long lists of botanical materia medica, including narcotics (opium, mandrake, henbane, duckweed, psilocybe mushrooms, purslane),[70] but also writes of extracting the quintessence from gold, silver and the other metals. Compared to Pseudo-Lull's more theoretical approach, here we find echoes of Geber, with instructions for building the 'most secret furnace of the Philosophers', the athanor.[71] He admits that the art can transmute base metals in the blink of an eye but swears not to divulge such secrets as it would go against the practice of religion.[72] In truth, none of this knowledge is of any worth to one whose mind has not first been deified by highest contemplation and holiest life.[73] He makes a noteworthy and provocative statement that 'gold can be made in three ways: first, natural or mineral gold; secondly, alchemical gold; thirdly, the gold of the philosophers' stone'.

Alchemical gold, given in a drink, does not work as a cordial to make a man's heart happy; but because it is composed from corrosives, it causes wounds to swell. Natural gold, meanwhile, added to a drink achieves nothing. Gold produced by the stone is the only kind that can be digested as nourishment, heals lepers and every infirmity, and is thus rightly called 'God's gold'.[74] This is the gold that Rupescissa claims can safely be 'spread throughout the whole world, without fear of lords and tyrants, and to preserve it in all times of wars and tribulations, especially at the time of the Antichrist'.[75]

So far, although alchemy has been described as a 'gift of God' and analogies have been drawn between God's creation in Genesis or Christ's crucifixion, its practice has mostly been presented as something natural, albeit with apotropaic effects such as averting evil and exorcising demons. This was to change, however, in the influential work of the Italian Pietro Antonio Boni of Ferrara (fl. *c.* 1323–30), better known as Petrus Bonus, the *Pretiosa margarita novella* ('New Pearl of Great Price', 1330).[76] The title promotes the high value of alchemy, indeed fosters a close connection between alchemy and Christianity by associating it with Christ's parable of the merchant in Matthew 13:45–6: 'Again the kingdom of heaven is like to a merchant seeking good pearls. Who when he had found one pearl of great price, went his way, and sold all that he had, and bought it.' Bonus's focus is on supporting belief in alchemy by responding to Scholastic criticisms of its theory. It is an erudite work in which he displays not only an extensive knowledge of Aristotle, but also shows himself to be a most informed reader of alchemical literature, whose authors he cites frequently, including Morienus, Albertus and Geber. It is interesting to note whom he doesn't mention by name, such as via a damning remark that must surely be referring to Rupescissa: 'some idiots say that alchemical gold does not comfort the heart'.[77]

Alchemy is initially introduced as one of the *artes naturales* (natural arts), alongside medicine and horticulture, one that truly teaches how to know matter.[78] A few pages later we learn that it is 'the science by which the principles, causes, properties and passions of all metals are completely known, and which are...transmuted into true gold'.[79] Then, however, Bonus starts to sound like Rupescissa, stating that 'the art itself and its investigation and regimen of operations...exists above nature (*supra naturam*) by way of a miracle'.[80] He supports this with multiple citations from alchemical authors speaking of divine inspiration,[81] before making the novel and extremely bold assertion that 'this art is natural and divine, and that by means of it the ancient philosophers were seers of future divine miracles'.[82] We learn that the projection of the stone onto imperfect metals, by which they are transmuted into gold, is considered natural.[83] On the other hand, the fixation of the soul and spirit, achieved by the addition of a 'hidden stone' (*lapidis occulti*), not grasped by the senses but purely through inspiration or divine revelation, is of another order, divine and supernatural.[84]

From their observations of matter in the laboratory, Bonus suggests that the ancient pagan philosophers came to know of the Day of Judgement and of the resurrection of the dead into glorified, incorruptible bodies of almost unbelievable lucidity and subtlety.[85] Likewise, because the stone conceives, impregnates and gives birth to itself, they foresaw the coming of the miraculous virgin birth and the appearance of God in human flesh.[86] This knowledge, for Bonus, is supernatural and divine, and he believes that the prophets of old did not practise alchemy for the sake of acquiring silver or gold, nor for any other benefit save for love of the art for its own sake.[87] Indeed the transmutation or conversion he seems to have foremost in mind is less of matter and more of faith: 'I firmly believe that should any unbeliever truly know this divine art, he would necessarily become a believer affirming the trinity of God and would believe in Christ Jesus our Lord, the Son of God.'[88]

Nonetheless, this knowledge, which raises the understanding to the supernatural and divine,[89] remains restricted. Like Geber, Bonus admits that he does not transmit knowledge in continuous discourse but sows it in different chapters.[90] What is interesting is his frankness in stating that it 'can be taught and learned in one day, nay, in one hour, by someone who knows'.[91] In fact, 'if one were willing to give the whole art in all its practical necessities, but leaving out all the figures, they could write it in eight or twelve lines.'[92] Admittedly, that would not qualify the student to

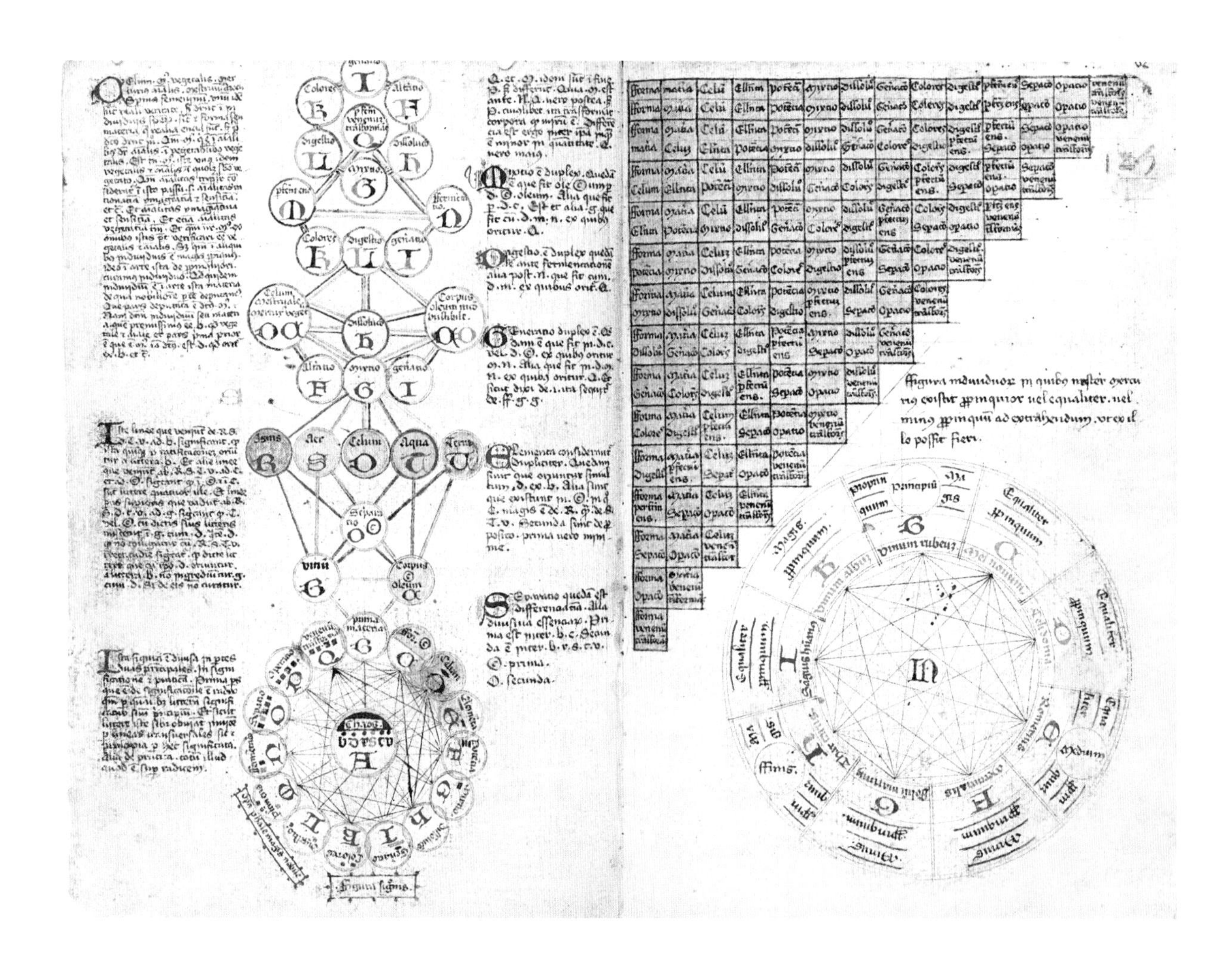

Alchemical diagrams, from Ramón Llull, *Tertia distinctio* (*Third Distinction*), Latin and Catalan, 1351–52. The Parker Library, Cambridge, Corpus Christi College, MS 395, fos 61v–62r.

Opposite:
Incipit from *Pretiosa margarita novella* ('New Pearl of Great Price') by Pietro Antonio Boni of Ferrara, *c.* 1450–80.

Despagne

Incipit margarita nouella com
posita per magistrum bonum
lombardum de ferraria philo
sophicum excellentissimum.

In quid bonus
bonus lombardus de fer
raria phie entia realia
sunt in tripliti gradu sm phm
metha. Quia aut sunt con
iuncta motui et materie et de
ipsis est sciencia naturalis. Aut
sunt coniuncta materie et separa
ta a motu et de istis est mathe
Aut separata a motu et materia
et de istis est metha i. diuina
Sed de separatis a mā et coniunc
tis motui non est sciencia cū tale
ens ee oīo sit impossibile Cum
igitur sciā alkimie sit de ente rea
li coniuncto motui et matiē neq
sub metha i. diuina continetur
quia non ē de ente reali separato
a motu neq de ente reali separato
a materia sicut sunt intelligencie
Oz ig nrio q sub naturali phi
losophia ponatur cū ipā sit de ente
reali coniuncto motui et matiē.
Similr entia nālia cernunt
sibi matiam determinatā ut sim
tas nasum et visus oculum et
ideo sine sua matiā et sine sub
iecto non possunt intelligi. En
tia aut mathe nō cernūt sibi
matiam determinatam ut linea
triangulus numerus quia talia
entia possunt ee in qualibz mā
Entia autē me i. diuina
cum careant matia nulla sibi

cernunt matiam q sit spiritualiter
intelliguntur. Cum ig ars al
kimie sit sub parte nālis phie, cū
libro de mineralib subalternet oport
nrio q cernat sibi matiam determina
tam. Et quia intentio artis al
kimie est sequi naturā oīa et per
oīa ut faciat idē aurū quod ipā
que et qualis erit matia apud na
turā talis necessario erit apud ar
tem, cū de potentia matiē determinate
extrahat forma sua et nō de matia
aliena. Si autē fuerit matia
nō eadem sz quodam modo similis
sm genus erunt generata nō eadē
sz similia sm genus et nō sm spēm
et formā oīo. Cum ig matia sit
vna et determinata oīo apud nām
et nō multiplex sq nrio illa vna
eadē matia determinata erit similr
apud artē alkimie et nō multiplex
Cū ig matia sit. hec aut mā
est argentū viuū sm oēs phos tā
nāles q alkimistas. Ig et apud
artem alkimie matia erit similr
ar. vi. Et quia matia nō pducit
se ipām in ee ad formā et genera
tionē sz ab agente proprio diri
gitur et informatur tūc nrio oport
q istud argentū vi. q ē matia diri
gatur et informetur a suo agente
proprio ad finē intentū natē. qui est
generacio metallorum et auri et aliorum,
quorumdā que nūc relinquū. Hoc
aut agens est sulphur sibi in pro
priis mineriis coniunctum hñs illud
ar. vi. a propria vtute coagulare et
digerē p calorem naturalē mineralē
sm phos nāles. ita ut in fine digonis

in actu et rubeu in potencia et albu
et imperfectu et pficit^r Rubedine
et no alio ut dr in turba phorum
Et hoc est quod dicit Rosinus
sol albus est apparicione rubeus
vero experimento ¶ Et anaxago
ras dicit solem ee lapide rubeum
ardente anima cui iungit^r sol spu
mediante e alba et est et est de
natura lune et dicitur argentu
vivu phor ¶ Et io nuc patet
veritas eius quod dicit hmes s. q
sine rubeo lapide nulla fit verax
tinctura ¶ Et pz veritas ei9 quod
dicit morien9 s. q ad effectum no
pueniunt donec sol et luna i vnu
corp9 redigant^r quod aute dei
preceptum dei evenire no pot ¶ Et
sic pz ecia quomodo ars alqmie
sine sole no pficit^r nec complet^r qm
ipm est verum fermentu ta solis
qz lune ¶ Et pz veritas ei9 quod
dicit ras in lumine luminum.
¶ Rubicund9 fui candida duxit
vxore ¶ Et simile ei9 quod dicit
haly nisi quis rubore cu candore
adiciat introducat ecia q cu labo
re et expectacione pauit ad rubo
ris fulgore accede no pot ¶ Et
pz simile q duplex e auru vnu
albu et aliud rubeu ¶ De quib9
dicit Rosinus tu aut nisi auru po
nas michi habes ¶ Et de hoc
auro albo dicit ras libro pfti magis
terij q ipm e corp9 neutru qz nec
egm nec sanum ¶ Et hoc aurum

pro.

albu est argentu vivu de quo dicit
geber ca^o de ar. vi. q nullu metal
loru submergit^r in ipo nisi sol et
q e mediu coniungendi tincturas
¶ Et istum de hoc sole et argento vi^o
dicit ca^o de na solis cum ipo sile
quiescent^r spus et fugiunt^r q ipm
maximo ingenio quod no puenit
ad artifice dure civitas. Que vba
vident^r plana sz sunt valde de
ceptoria ¶ Et q hoc sit vere
fermentu videtur velle hermes in
1° qui dicit et nondum qd fer
mentu qfectione dealbat combus
tionem inhibet tinctura ingredi
facit tinctura continet ne fugiat
corpa levificat et se invice ingredi
facit et coniugit quod finis e opu
¶ Et morien9 fermentu auri auru
est sicut fermentu paste pasta est
¶ Et quibus omnibus liquide patet
q sol et luna sunt eiusde nature et
q luna precedit solem et ordinatur
albiditas i rubedo
ad ipm et quo sol est occultatus
i in humiditate et albedine q in magisterio precedit
i luna ¶ Et quomo de ventre lune
sol extrahit^r ¶ Ideo dicit senior q
sol est oriens in luna crescente.
¶ Et zeno in turba phor Sic tote
oms hui9 artis investigatores q
nisi dealbetis no potestis rubeu
facere ¶ Eo q due nature nichil
aliud sunt qz rubeu et albu et
de albo fit rubeu et albu ¶ Et
dardanus in eode Sciatis filii sie
q si parum auri in compone ponat^r
exiet tinctura patens candida.

i. alba

No. quomodo de lune sol ext[rahitur]

Et q sol e oriens in [luna] crescente

No. q de albo fit Rub[eum] et albu

No. dicit di[...] si parum auri [in] compone ponat[ur] exiet tinctura candida.

call himself an alchemist, for he would have no foundation in the principles and causes or the understanding of allegorical figures and discourses.[93] Nor is assiduous study and a well-stocked library enough; the would-be alchemist 'must naturally have the deepest and highest talent, so that he knows how to discern the figures and analogies of truths in their necessarily varied and manifold meanings'.[94] As much as the art requires contemplation, we are reminded that 'meditation without experience is worth nothing...so there is more experience to be sought than meditation'.[95] Like Rupescissa, Bonus displays a familiarity with astrology; in fact, he includes alchemy in a list of 'four noble sciences' together with astrology, natural philosophy and magic.[96] He informs his reader that a knowledge of planetary positions and astrological aspects is necessary when a new accidental form is infused in something, as in the art of images (i.e. the making of talismans), as well as in elective astrology, when one seeks the best time to construct a building or make a journey.[97] Such is not the case, however, with alchemy; so astrology is not necessary.[98]

In this chapter we have considered several of the most influential works on alchemy that appeared between the twelfth and fourteenth centuries in the Latin West, providing some sense of the variety of approaches to the subject from famous figures in the history of alchemy, discovering a mixture of matter-of-fact descriptions and explanations and enigmatic allegorical material, with a distinct sense in all authors of concealment of knowledge (sometimes in plain sight). The majority tended to focus on natural work in the laboratory, be that for *chrysopoeia* (gold-making) or *chymiatria* (chemical medicine), but some allude to other levels of experience, with additional goals or motivations. While the authors appear to have shared similar theoretical knowledge about their art, it is clear that they did not feel constrained to hand down one immutable set of laws, but instead added their own insights and perspectives to their transformative art.

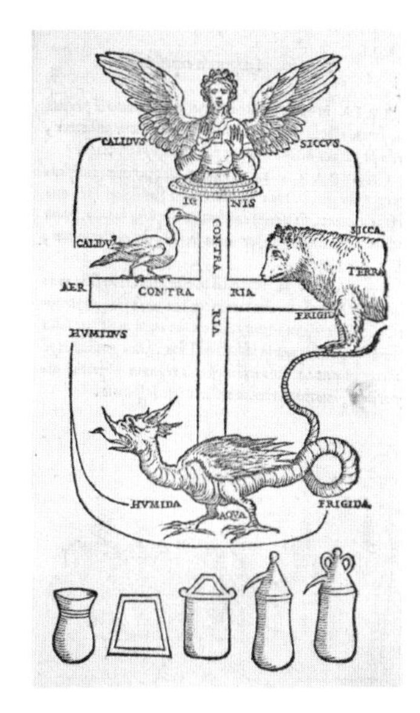

Woodcut from *Pretiosa margarita novella* by Pietro Antonio Boni of Ferrara. *C.* 1450–80.

Opposite:
Page with marginal notes from *Pretiosa margarita novella.*

THE RENAISSANCE, ESOTERICISM AND ALCHEMY

PETER J. FORSHAW

Darkness will appear upon the face of the deeps; Night, Saturn and Antimony of the Wise will appear. Blackness, and the Raven's Head of the Alchemists, and all the colours of the World will appear at the moment of conjunction. The Rainbow, too, God's messenger, and the Peacock's Tail...Finally, after the ashen-colour, whiteness and yellowness have been passed through, you will see the Philosophers' Stone, our King and Lord of Lords, come forth from the inner-chamber and throne of his glassy sepulchre, onto this worldly stage, in his glorified body, that is, Regenerated and Surpassingly Perfect, namely, a shining Carbuncle, most Temperate in its splendour, and whose most subtle and most purified parts, from the concordant peace of the mixture, are inseparably bound into One.[1]

Detail of the Treasure-House. In *Aurora Consurgens* ('The Rising Dawn'), 15th century.

This dramatic passage appears close to the end of the *Amphitheatre of Eternal Wisdom* (1595, 1609) of the German theosopher and alchemist Heinrich Khunrath (1560–1605), in the Isagoge or commentary to the third of the *Amphitheatre*'s four circular figures, that of the *Rebis* or Alchemical Hermaphrodite. It is a fine example of Northern Renaissance alchemy from the circle of alchemists centred around the Late Renaissance court of the Holy Roman emperor Rudolf II (1552–1612), who was a magnet for specialists in alchemy, magic and Kabbalah. Khunrath's vivid metaphorical description (and visual representation) of the alchemical colour changes preceding the appearance of the philosophers' stone in the glass vessel represents the culmination of processes that were already underway around the start of the Italian Renaissance in the early fifteenth century, with the appearance of high-quality illustrated alchemical manuscripts, the rediscovery of esoteric thought from classical antiquity and exploratory combinations of alchemy with other arts, such as Kabbalah.

Before we discuss Late Renaissance manifestations of alchemy and esotericism, however, let us travel back 200 years to the dawn of the Renaissance in the early fifteenth century, which saw the production of two of the greatest cycles of alchemical illustration in the history of visual alchemy, the *Buch der heiligen Dreifaltigkeit* (*Book of the Holy Trinity*)[2] and the *Aurora Consurgens* ('The Rising Dawn').[3]

The *Book of the Holy Trinity*, dating from 1410–19 is believed to be the oldest alchemical manuscript in German.[4] It has several levels of reading, combining alchemical, prophetic and political material, with its author, thought to be the Franciscan friar Ulmannus, declaring his intention to write a treatise that would be a 'true astronomy, alchemy, theology, [and] prophecy, infinitely profound, without beginning, without end',[5] to be made public at the time of the struggle with the Antichrist,[6] whose birth, defeat and death he announces, as well as the coming of the saviour emperor who will bring justice to Christendom. The fight takes place on a material and moral level, with the *Book* offering the secret of making gold and silver – 'true' alchemy being put to the service of righteousness – to produce the necessary financial and political resources to enable the emperor to carry out a reform of Church and Empire and to counteract the Antichrist's own infernal alchemy by which he seduces and recruits followers with false words and counterfeit gold.[7] Ulmannus propagates a mystique of the Trinity and above all the sufferings of Christ, his wounds and blood, as an allegory for laboratory processes, in particular mortification, calcination and purification.[8] At the same time, the author convincingly demonstrates his alchemical knowledge, in depictions of laboratory vessels and furnaces.[9]

In the image of the *Forma Speculi Trinitatis* ('Form of the Mirror of the Trinity'), Christ, bearing the five wounds of the cross, is fused with the body of a large black two-headed eagle.[10] His double nature as God and Man designates quicksilver, the source of all metals, from impure lead to pure gold. Several crowns represent the various metals to be transmuted, for this is a work of *chrysopoeia* or gold-making alchemy, and the author informs his reader that one must prepare all works according to this figure of the Holy Trinity.[11] A later image of Christ's resurrection from the tomb symbolises the successful production of the philosophers' stone (as alluded to in Khunrath's passage above).[12]

From the same period, its earliest version dating from the 1420s, the *Aurora Consurgens*, or *Rising Dawn*, is another theo-alchemical work, combining religious moralisation and alchemical symbolism,[13] but whereas the German work includes images of Christ and Mary along with more technical laboratory images within a fairly secular text, the Latin *Aurora* makes abundant use of a patchwork of biblical quotes (especially from the Song of Songs, Book of Wisdom, Psalms and Proverbs) in the form of seven parables (Of the Black Earth Wherein the Seven Planets Took Root, Of the Treasure-House Which Wisdom Built upon a Rock, etc),[14] but chooses to illustrate its message with figures from the alchemical bestiary, such as the sun and moon jousting, the former on a (fiery, sulphurous) lion, the latter on a (volatile, mercurial) griffin, with colour changes at different stages of the work represented by a menagerie of birds: a black raven for calcination and putrefaction, a white swan for solution or washing, a peacock for when many colours appear in the flask, eagles as volatile salts and a phoenix as a symbol of the perfect red philosophers' stone.[15]

Illustrations from *Buch der heiligen Dreifaltigkeit* by Ulmannus, in *Alchemisch-medizinisches Hausbuch des Pfarrers Valentinus aus Ottrau* ('Alchemical-medical household book of the priest Valentinus of Ottrau, Book of the Holy Trinity'), 1529.

I·N·R·I

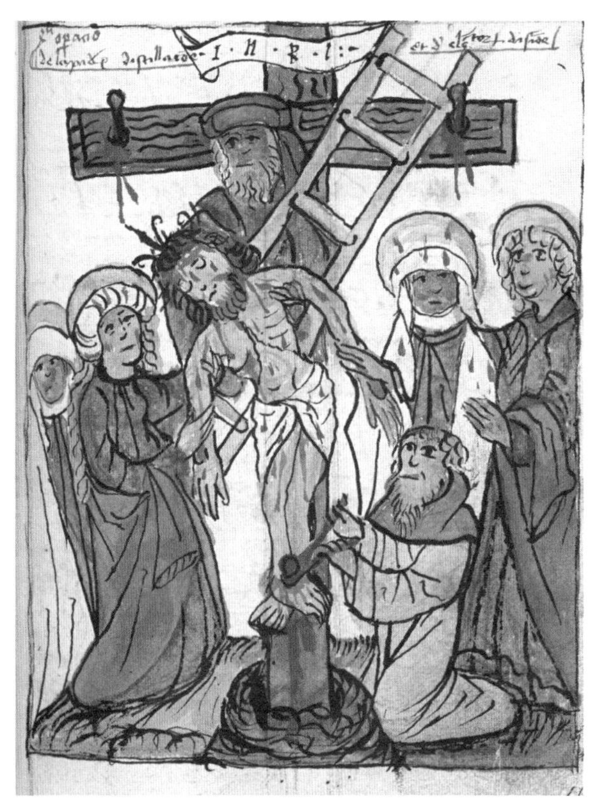
I·N·R·I

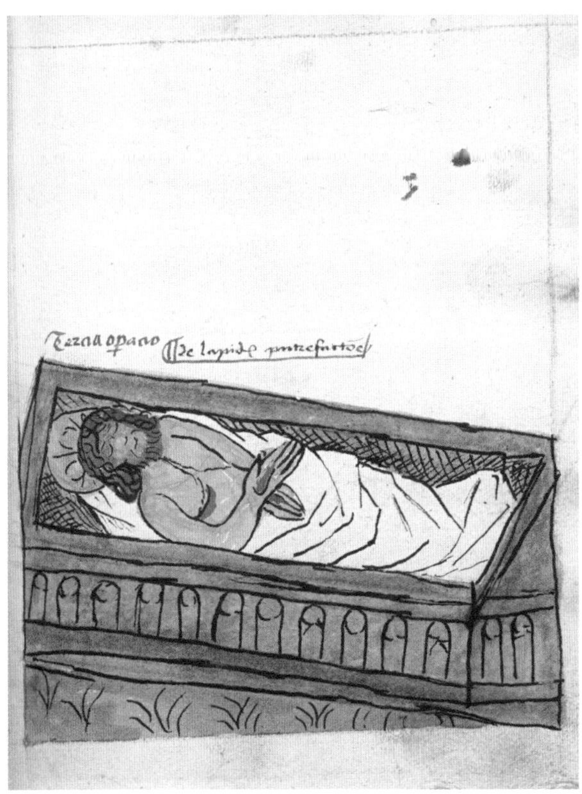

also schwer aqua fort bretz in borhet es ad oleu. von stund es im
gnet wann balb pulverisiert wie es ist distilliert auch dye ele
menten und truckten sy in de corpus, es wirt ein stein der hoher
tingiren ist weiss. de mochten in als vil solviren und congeliren
de tinctur als gross rot als er weiss thut. Saturnus mugt
de dess gleich werken als diss ist an putrefiren. Also kurtz
werkt er alle werk dy seind auff disen grat alle planeten
gleich sol stan. kein man anders kan finden. Anno domi
ni. cccc°. xvij. die veneris in hora martis solis angehaben
in die solis in hora solis veneris geendet, des dritten tags der
nechst diser vorgeschriben Endung beyde endung auff gegangen
sind gleich in den xvj tag in die mart vor ascensionis domini ein
zweyfalt summe wil dye endung sein mit acht flam flammen
zweifalt auff gegeben Als er ist auff dem grab gestigen.

Close connections are drawn between religion and alchemy. In the *Aurora*'s depiction of a treasure-house, we see an aged Hermes Trismegistus holding a tablet on which are depicted the secrets of the work.[16] 14 'cornerstones' of the house are discussed, each associated with a different virtue (1. health; 2. humility; 3. holiness, etc) and exemplified with quotes from scripture and alchemical authorities. For example, the seventh cornerstone is Faith, where we read,

> Faith saveth a man, and he who hath it not, cannot be saved. Faith is to understand what thou seest not. And the *Turba*: 'It is invisible like unto the soul in man's body.' And in the same it is said: 'Two things are seen, namely water and earth, but two are not seen, namely air and fire.' And Paul: 'Whosoever believeth therein shall not be confounded, for to them that believe not the stone is a stumbling stone and a rock of scandal.' And the Gospel: 'He that believeth not is already judged.'[17]

In this one passage, we have quotes from the Athanasian Creed,[18] St John's Gospel[19] and St Paul's Epistle to the Romans 9:33,[20] all linked with the Arabic alchemical treatise, the *Turba Philosophorum* ('Assembly of the Philosophers'),[21] which itself is citing ancient Greek authorities like Plato and Pythagoras, with the Christian virtue of faith being recommended for anyone taking up the challenge of preparing the philosophers' stone.

A fascination for relating the Christian religion to the mysteries of classical antiquity is, of course, one of the defining characteristics of many of the significant Renaissance thinkers associated with the history of esotericism. The Florentine philosopher, physician and priest Marsilio Ficino (1433–99) is famous for his concept of the *prisca theologia*, by which he meant representatives of 'ancient theology' such as Zoroaster and Hermes, Pythagoras and Orpheus, whom he believed anticipated the coming of Christ. As well as translating Plato from Greek into Latin, together with important Neoplatonist thinkers including Plotinus, Proclus and Iamblichus, Ficino also provided the Christian West with translations of the *Corpus Hermeticum*, the philosophical discourses linked with the legendary Egyptian sage Hermes Trismegistus. Although Ficino was also interested in what is known as the technical Hermetica, especially the arts of astrology and magic sometimes associated with Hermes, he is not known for any particular knowledge of alchemy, even if there is a pseudepigraphic *Liber de Arte Chemica* ('Book on the Chemical Art') in his name.[22] However, his influential astral-magical-medical treatise, *De Vita Libri Tres* ('Three Books on Life', 1489), was to become a work cited in later alchemical publications, partly due to shared interest in the medicinal virtues of potable gold, but especially for his reference to the elixir:

> Diligent natural philosophers, when they separate this sort of spirit from gold by sublimation over fire, will employ it on any of the metals and will make it gold. This spirit rightly drawn from gold or something else and preserved, the Arab astrologers call elixir.[23]

Illustrations from the *Book of the Holy Trinity*. Laboratory apparatus (top right). *Forma Speculi Trinitatis* ('Form of the Mirror of the Trinity') (bottom left). Christ risen from the tomb (bottom right), 1467–1500.

This singular statement constituted one of the major innovations in Renaissance alchemy, endowing it with 'a cosmic character that it had lacked in the Middle Ages'.[24] Ficino provoked a shift in how alchemy was perceived by identifying the Neoplatonic *spiritus mundi* (spirit of the world) with the alchemical *coelum* or quintessence: 'But let us return to the spirit of the world. The world generates everything through it...and we can call it both "the heavens" and "quintessence".'[25] In the same year, one of Ficino's friends, Ermolao Barbaro (1454–93), professor of natural philosophy at the University of Padua, wrote a famous account in his *Corollary on Dioscorides* of the discovery of a lamp in an ancient tomb near Este that was perpetually burning due to the powers of alchemical *aqua vitae* (water of life).[26] The value publishers placed on Ficino's *De Vita* can be seen in how its translation as *Das Buch des Lebens* in 1505, by the Strasbourg physician and humanist Johannes Adelphus Muling (1485–1523/55), was printed with two famous works on distillation: Hieronymus Braunschweig's *De Arte Distillandi* ('On the Art of Distilling', 1505) and Philipp Ulstad's *Coelum Philosophorum* ('The Philosophers' Heaven', 1526).[27]

Ficino, furthermore, corresponded with the humanist poet Giovanni Aurelio Augurelli (1441–1524), who was to become well known in the alchemical community as the author of a long poem on gold-making and the creation of the philosophers' stone, the *Chrysopoeiae libri tres* ('Three Books of Gold-Making'), published in Venice in 1515, in which he 'adapted Ficino's ideas to an alchemical context', being the first to refer to the passage from *De Vita* on the *spiritus mundi* 'to describe the artificial extraction of the spirit trapped within gold'.[28] Augurelli also published shorter poems, including one on the *Vellus aureum* ('Golden Fleece'), a favoured subject for writers interested in mythoalchemy – that is, interpretations of Egyptian, Greek and Roman mythology as concealing secrets of the alchemical art, a trend that would increase in popularity over the coming centuries.[29] In the prologue to *Chrysopoeia*, addressed to its dedicatee, the Medici pope Leo X, Augurelli entreats him to use alchemy 'to regenerate the soul of Italians and bring about a new golden age of peace and prosperity'.[30]

One of Ficino's younger contemporaries, Giovanni Pico della Mirandola (1463–94), began studying Kabbalistic works as part of his project to create a synthesis of Aristotelian and Platonic thought with esoteric doctrines gleaned from Ficino's *prisci theologi*. Pico is the first Christian-born author who is known to have studied a considerable amount of genuine Jewish Kabbalah.[31] This led to the publication in 1486 of 900 *Conclusiones Philosophicae Cabalisticae et Theologicae* ('Philosophical, Cabalistical and Theological Conclusions'),[32] in which he introduced the Christian West to his own Christian form of Cabala. His ideas had such an impact on the Christian West that he has been called the 'Father of Christian Cabala'.[33] Pico was particularly interested in novel exegetical techniques of the Jewish Kabbalists, ones which had no equivalent in Christian interpretation of scripture. The Christian reader found meaning but left the original text intact, but his Kabbalist counterpart could make use of methods that reshaped and transformed the written text, breaking it down into its constitutive elements, the Hebrew letters, and even reducing the letters themselves into their parts, discovering (or creating) an abundance of new meanings in the process.[34] Three main techniques were *Gematria*, arithmetical computations of the values of words, based on the alpha-numeric nature of the Hebrew alphabet; *Notarikon*, the manipulation of letters into acronyms and acrostics; and *Temura*, permutation, commutation or transposition of letters.[35]

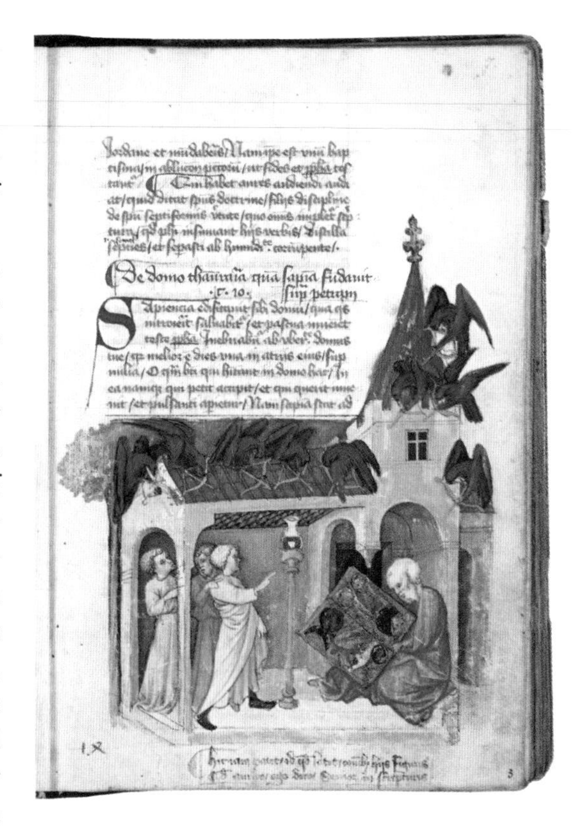

Treasure-House. *Aurora Consurgens,* 15th century.

Overleaf.
Joust of the Sun and the Moon. From *Aurora Consurgens.*

Such Kabbalistic approaches to texts gradually percolated into some alchemical works, which is hardly surprising: given the opacity of many of the texts, anything that might help in their interpretation was readily received. If anyone deserves the appellation Father of Cabalistic Alchemy or Chymical Cabala, it must surely be Pico's fellow Italian the Venetian priest Giovanni Agostino Pantheo (fl. 1517–35). Pantheo developed a hybrid 'Cabala of Metals' (Cabala Metallorum) in the *Ars Transmutationis Metallicae* ('Art of Metallic Transmutation', 1518), published with a *Commentarium Theoricae Artis Metallicae Transmutationis* ('Commentary on the Theoretical Art of Metallic Transmutation', 1519) and the *Voarchadumia contra Alchimiam* ('Voarchadumia against Alchemy', 1530).[36] There we discover that the most powerful Hebrew divine name יהוה, the *Tetragrammaton* YHVH, conceals alchemical secrets because each of its four letters represents one of the four elements (*Yod*: Air, *He*: Water, *Vav*: Fire, *He*: Earth).[37] As each Hebrew letter also has a numerical value, the values of the letters of the divine name YHVH were believed to indicate the relative proportions of the elements at different stages in the alchemical process.[38] Elsewhere Pantheo turns to the numerical analysis of words from different languages connected with alchemical substances.[39] The Greek *Thélima* and Hebrew *Reçón* both mean 'Will', but by the Kabbalistic exegetical technique of *Temura*, *Reçón* can be transformed anagrammatically into one of the Hebrew words for 'Earth', *Eretz* (as in Genesis 1:1 'God created the Heavens and the Earth').[40] *Thélima* and *Reçón* both appear in Pantheo's list of synonyms for 'Gold', and it seems likely that he had the equation of 'Will' with 'Earth' in mind as a kind of cover-name for gold.

Moving into the sixteenth century, Ficino's works had an influence on one of the most colourful figures in alchemy, the Swiss chemical, medical and social reformer Philippus Theophrastus Paracelsus (1493–1541),[41] although he showed little of Ficino's humanist 'renaissance' veneration of works from classical antiquity.

Although the Persian Abū Bakr al-Rāzī (*c.* 865–925/35) had written about medical applications of alchemy and John of Rupescissa about the medical virtue of the quintessences of animal and vegetable material in the fourteenth century,[42] Paracelsus is one of the first to redirect Western alchemy away from the medieval quest for gold to the preparation of medicine from minerals and metals. He is the first known writer of his period to call himself an iatrochemist, an adept combining medicine and chemistry,[43] with the assertion that human philosophy had reached its highest point in the spagyric or alchemical work, originating from careful obser-

vation and contemplation of the greater world. This art of separating and re-uniting required a profound knowledge of the relationships between primal, intermediate and ultimate matter; the transformations of the four elements, earth, water, air and fire; and particularly the interplay between the Tria Prima, the Paracelsian augmentation of the two classical agents of alchemy, sulphur and mercury, with the addition of a third principle, salt, denoting three modalities of matter.[44]

'Oratory-Laboratory' from *Amphitheatrum Sapientiae Aeternae* by Heinrich Khunrath, 1595.

Paracelsus emphasised the importance of iatrochemistry for the preparation of new medicines, including the highly refined homeopathic use of poisonous substances,[45] based on personal experiment, as a replacement for the book-learning of the university medical schools, who perpetuated what he considered to be the outdated theories of Aristotle (384–322 BCE) and Galen (*c.* 129–199 CE). Alchemy, he argued in *Das Buch Paragranum* ('The Book against the Grain', 1529–30), was one of the four pillars of medicine, together with philosophy, astronomy and *proprietas* or virtue.[46]

Paracelsus's work on longevity, *Libri quatuor De vita longa* ('Four Books on Long Life', 1560) shows evidence of familiarity with Ficino's *Three Books on Life*,[47] and many of the genuine and pseudepigraphic Paracelsian works reveal the influence of esoteric thought, in addition to an interest in alchemy. The influential pseudo-Paracelsian *De Tinctura Physicorum* ('On the Tincture of the Natural Philosophers', 1603), for example, affirms the value of Cabala for the alchemist:

> If you do not understand the use of the Cabalists and of the old astronomers, you are not born by God for the Spagyric art, or chosen by Nature for Vulcan's work, or created to open your mouth about the Alchemical Art.[48]

In the *Aurora Philosophorum* ('Aurora of the Philosophers', 1577), Paracelsus provides helpful insights into how he perceives the natures of magic and Cabala:

> Magic, indeed, is the art and faculty by which one attains the hidden properties, powers, and operations of elements, bodies, and their fruits. But Cabala itself, out of anagogy, seems to prepare a way for men to God: who may act with him, and prophesy from him. For Cabala is full of divine mysteries, just as Magic is full of natural secrets; indeed it teaches to speak from the nature of things to come and of things present. For its operation consists in knowing the inner nature of all creatures, both heavenly and earthly, what is hidden in them, what powers are concealed, for whom they were destined from the beginning, with what properties they are endowed. These and the like are the bond by which heavenly things are joined to earthly things, as can sometimes be perceived visually, and sensually, by their operations.[49]

The impact that such doctrines of magic and cabala had on Paracelsian forms of alchemy is evident in the entries that can be found in the Paracelsian dictionaries, Michael Toxite's *Onomastica II* (1574), Gerard Dorn's *Dictionarium Theophrasti*

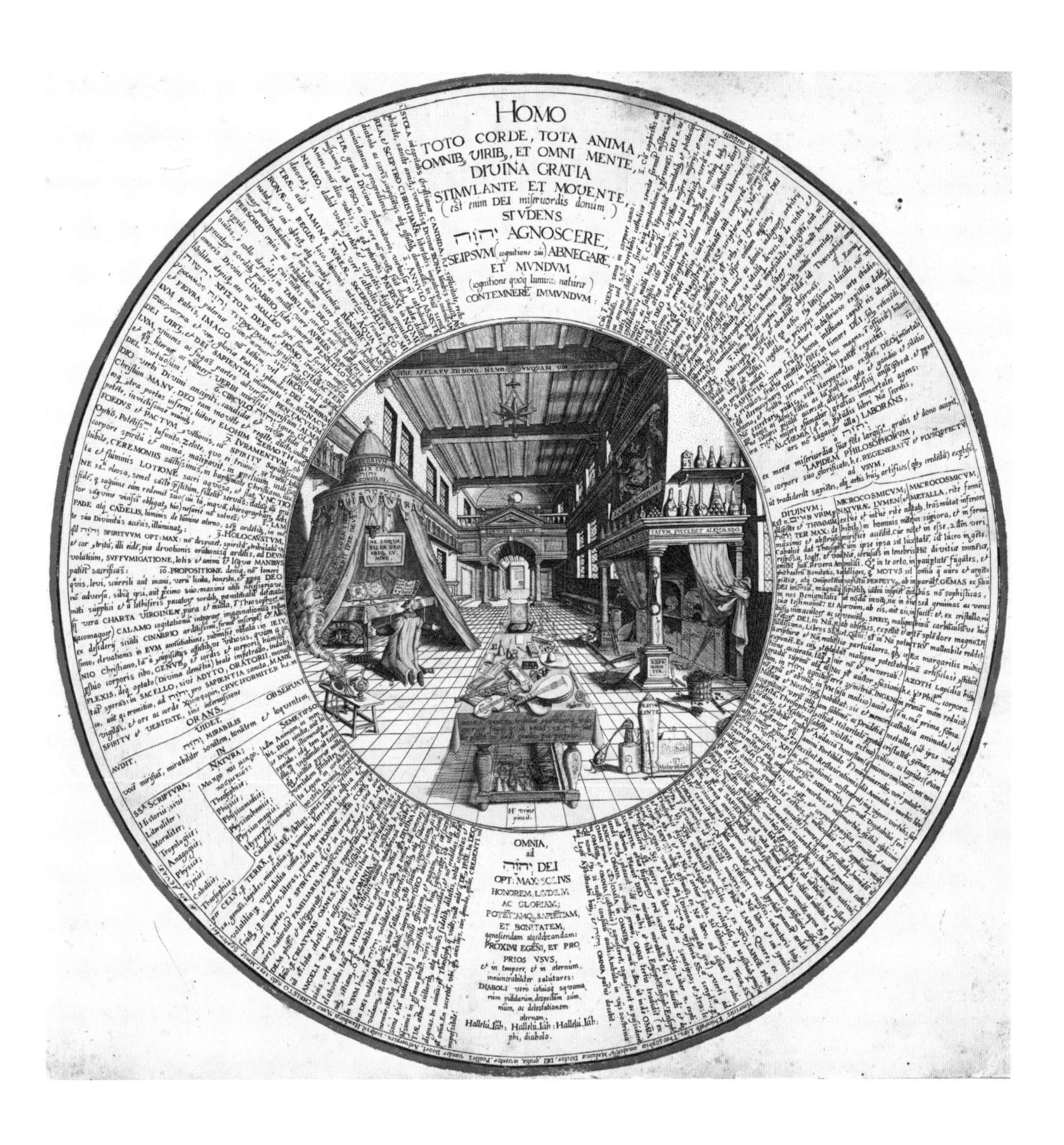
HOMO
TOTO CORDE, TOTA ANIMA
OMNIB, VIRIB, ET OMNI MENTE,
DIVINA GRATIA
STIMVLANTE ET MOVENTE,
(est enim DEI misericordis donum)
STVDENS
יהוה AGNOSCERE,
SEIPSVM (cognitione sui) ABNEGARE,
ET MVNDVM
(cognitione quoq luminis naturæ)
CONTEMNERE IMMVNDVM:
OMNIA,
ad
יהוה DEI
OPT: MAX: SOLIVS
HONOREM, LAVDEM
AC GLORIAM;
POTESTATEMQ, SAPIENTIAM,
ET BONITATEM,
agnoscendam atq. celebrandam:
PROXIMI EGENI, ET PRO
PRIOS VSVS,
et in tempore, et in æternum,
inenarrabiliter salutares:
DIABOLI vero istiusq squama
rum piscidarum, despectum sum
mum, ac detestationem
æternam.
Hallelu Iah: Hallelu Iah: Hallelu Iah:
phi, diabolo.

Paracelsi ('Dictionary of Theophrastus Paracelsus', 1583) and Martin Ruland's *Lexicon Alchemiae* ('Lexicon of Alchemy', 1612).[50]

And so we return to Khunrath, who is in many ways the heir of most of the works mentioned in this chapter and generally considered to be a follower of Paracelsus, espousing a broad-spectrum interest in various kinds of alchemical practice in the context of a wide-ranging esotericism. He has a proclivity for experimenting with hybrid forms of knowledge and practice, as well as a penchant for highly complex visual expressions of his occult philosophy or theosophy. This interest in mastering a variety of esoteric sciences and arts is clearly advertised in the full title of his magnum opus, *The Universal Ter-tri-une, Christian-Cabalist, Divinely Magical, and Physico-Chymical Amphitheatre of the Only True Eternal Wisdom* (1595 and 1609).[51] In another of his works, *De Igne Magorum Philosophorumque* ('On the Fire of the Mages and Sages', 1608), he makes one of his most idiosyncratic statements in the declaration that 'Kabala, Magic and Alchemy conjoined, should and must be used together with and alongside one another'.[52] This contributes to what the Kabbalah scholar Gershom Scholem considered to be Khunrath's 'definitive blending' of the three arts,[53] all of which he introduces as handmaids of the Eternal Wisdom in his book's title.[54]

The best-known visualisation of the dynamic relationships between Khunrath's religious and esoteric beliefs and his laboratory practice is the last of the four circular figures which first appeared in the rare 1595 edition of his *Amphitheatre*, the 'Theosophical Figure' of the 'Oratory-Laboratory', designed by himself, delineated by the mannerist artist Hans Vredeman de Vries (1527–1609) and engraved by Paullus van der Doort.[55] It is the culmination of a sequence of images that begins with a *Sigillum Dei* (Seal of God), featuring Christ at the centre of concentric circles of Hebrew divine names and the ten *sephiroth* associated with the Kabbalistic Tree of Life, and then continues with an engraving of Androgynous Adam and Eve, before their separation, before the Fall, surrounded by the four elements, with a description of their alchemical purification in terms heavily redolent of the spiritual regeneration of the adept – in the context of Grades of Cognition and a Ladder of Conjunction and Union – leading ultimately to the experience of deification. These two figures, Christ at the centre of macrocosmic creation and the microcosmic Androgyne, are followed by the *Rebis* or Hermaphrodite, representative of macrocosmic laboratory alchemy. All three are depicted in a two-dimensional manner, while the 'Oratory-Laboratory' is deeply three-dimensional and should be understood as the conjunction or compounding of all three preceding realms of practice and existence.

Khunrath's laboratory stands on the right side of the hall and it is obvious to the viewer that it is a practical workspace, with various furnaces, glass vessels and other implements used in physico-chemical experiments with gold-making, distillation and the preparation of spagyric medicine, where the diligent alchemist perseveres in his investigations of the secrets of nature and his attempts to transmute metals, discover the elixir and create the philosophers' stone. The laboratory's mantelshelf holds reagent bottles containing substances required in *chrysopoetic* and *iatrochemical* alchemy. High up on the second level of windows are additional bottles, some perhaps placed there to absorb the rays of the sun; on the ground a coal bucket encourages readers to get their hands dirty: 'Non pudeat te carbonum' (Do not be ashamed of coals), while the athanor advises patience: 'Festina lente' (Make haste slowly).

On the opposite side of the hall a bearded man in a long coat, probably Khunrath himself, kneels on a cushion, in earnest prayer before a pavilion labelled *Oratorium* (Oratory), calling to mind Moses's Tent of Meeting (Exodus 33:7) or the tent Jacob pitched at Bethel (Genesis 12:8).[56] This is where Khunrath focuses on communication with the divine through his practices of Christian Cabala and Divine Magic, seeking inspiration and illumination, indeed ultimately deification.

In the background is a portico, leading out of the main hall into a room with a bed and drapes on the left. As we, in our imagination, exit the room we pass beneath an architrave bearing the words 'Dormiens Vigila' (While Sleeping, Keep Watch). Khunrath was a firm believer in the value of dreams, familiar with both the biblical (and kabbalistic) revelations of Solomon, Jacob, Daniel and Joseph, as well as the alchemical visions of Zosimos of Panopolis and Arisleus; to which could be added the dreams of John Dastin and the dreams of John Dastin and Giovanni Battista Nazari. An important thing to bear in mind with Khunrath's alchemy is that he roundly condemns those who 'utterly un-Philosophically separate Oratory and Laboratory from each other'.[57] This novel sense of the connection between these two domains of practice becomes evident when he describes the properties of the philosophers' stone. On a macrocosmic level, it transmutes base metals into silver and gold, it turns pebbles into gemstones, makes gems and metals potable, cures animals and revives plants, and produces a perpetually burning water like that discussed by Ermolao Barbaro. It is also miraculously effective against sublunar spirits. On a microcosmic level, it offers many physico-medical benefits, being capable of routing all internal and external maladies of body, spirit or soul, purifying the body, preserving it from corruption and conferring long life. It stimulates innate genius, exalts the memory, expels evil spirits, cures melancholy and promotes cheerfulness. Last, but not least, on a divine level, it is the biblical Urim and Thummim by which God kabbalistically communicates with the devout theosopher about great and hidden things.[58]

This chapter has introduced some important works of Renaissance alchemy from the early fifteenth until the late sixteenth centuries. All of these texts, both manuscript and print, are concerned with laboratory alchemy, sometimes with transmutation, at other times with medicine, but they not only engage with the broader theme of Christianity but also represent experimental hybrids, combining material and approaches, for example, with Kabbalah, newly introduced to the Christian West, or associate with magic, natural and supernatural.

DAT ROSA MEL APIBUS.

ALCHEMY AND ROSICRUCIANISM

CHRISTOPHER MCINTOSH

Frontispiece of the *Summum Bonum* by Robert Fludd, 1629.

Over the many millennia of its history, alchemy has passed through different phases of development, one of the most significant of these being what we could call its Rosicrucian period, which is interwoven with the history of Rosicrucianism from about the early seventeenth century up to the present day. Before examining this interconnection in detail, it may be helpful to give a thumbnail account of Rosicrucianism and its basic chronology.

Rosicrucianism as a movement of ideas emerged in central Europe in the early seventeenth century, during a piod of great religious and political tension, especially in the collection of states that constituted the Holy Roman Empire of the German Nation. The Reformation had taken place a century earlier, dividing the Empire into rival religious camps but failing to bring the spiritual renewal that many had hoped for. A fragile settlement, the Peace of Augsburg, had been concluded in 1555 but was soon to collapse, and tensions were about to erupt into the Thirty Years' War (1618–48). In this uneasy atmosphere many people became preoccupied with millenarian prophecies, astrological portents and notions of an ancient esoteric worldview that offered the potential to heal the religious divisions and usher in a new age. This worldview was underpinned by various currents of thought including Gnosticism, Neoplatonism, the Hermetic tradition, Kabbalah, astrology and alchemy. It found expression in the writings of Cornelius Agrippa, Paracelsus, Johannes Reuchlin, Jacob Boehme, Heinrich Khunrath and others.

As the tensions mounted in central Europe a strange thing happened in the town of Kassel in the Duchy of Hesse, where in 1614 a curious text was published by the court printer Wilhelm Wessel. It was entitled *Fama Fraternitatis of the Praiseworthy Order of the Rosy Cross*, addressed 'to all the rulers, estates and learned of Europe', and was the first of three so-called Rosicrucian manifestos,[1] the other two being the *Confessio Fraternitatis* (*Confession of the Fraternity*) and the *Chymische Hochzeit Christiani Rosenkreuz* (*Chemical Wedding of Christian Rosenkreuz*). The author, or

Allgemeine vnd General

REFORMATION,

der gantzen weiten Welt.

Beneben der

FAMA FRA-
TERNITATIS,

Deß Löblichen Ordens des
Rosenkreutzes / an alle Gelehrte
vnd Häupter Europæ geschrieben:

Auch einer kurtzen RESPONSION,
von dem Herrn Haselmeyer gestellet / welcher
deßwegen von den Jesuitern ist gefänglich ein-
gezogen / vnd auff eine Galleren ge-
schmiedet:

Jtzo öffentlich in Druck verfertiget /
vnd allen trewen Hertzen communiciret
worden.

Gedruckt zu Cassel / durch Wilhelm Wessell /

ANNO M. DC. XIV.

co-author, was probably the Tübingen Protestant theologian Johann Valentin Andreae (1586–1654), along with his circle of friends in Tübingen. The text was partly an account of the tenets of the said order and partly an account of the life of its German founder, one Christian Rosenkreuz who, we are told, was placed in a monastery as a child and at the age of 16 embarked on a long journey which took him to Cyprus, Damascus and Egypt, along the coast of North Africa to Fez, across to Spain and thence back to Germany. On the journey he imbibed the wisdom of the wise men of the places he visited and on this basis established his order, initially with three other brethren from his former monastery. One can only speculate about the exact motives of Andreae and his friends in issuing this work, but possibly they intended it as a call in fictional form for a new dispensation in Europe that would heal religious divisions and usher in an era of peace and harmony.

The French scholar of Rosicrucianism Roland Edighoffer writes: 'The *Fama*, the *Confessio*, and the *Chemical Wedding* reflected a worldview based upon an analogical apprehension of God and of the world on the part of man, based upon earlier esoteric sources and traditions in Western esotericism. The rose in this context reveals symbolic attributes similar to those of the lotus in the Far East. It unfolds on the surface of the Earth-Mother, while the mystic rose is an attribute of the *Mater Dei* (Mother of God) and of spiritualized matter.'[2] Edighoffer emphasises the key influence on the Rosicrucian movement of the alchemist and physician Theophrastus Bombastus von Hohenheim, who wrote under the name of Paracelsus (1493–1541): 'Paracelsus...explains that man, regenerated by the Cross, thereafter receives a spiritual body whose glorification is symbolized by the rose; it is Christ, the Man-God, who transfigures us, just as the philosophers' stone transmutes metallic matter.'[3] Other key Paracelsian ideas that feature in the manifestos include the notion of the *Astrum* or world-soul; the concept of the *Liber Mundi* ('Book of the World'), which can be deciphered by someone with the requisite knowledge; and the belief in the elementary beings – that is to say, those that partake of the nature of the four classical elements of earth, air, fire and water.

Johann Valentin Andreae. Engraved frontispiece from his *Seleniana augustaliam*, 1649.

Opposite:
Title page from *Fama Fraternitatis* ('Report of the Fraternity, of the Praiseworthy Order of the Rosy Cross') probably by Johann Valentin Andreae, 1614.

In describing Rosenkreuz's homecoming the *Fama* includes a reference to alchemy, saying that Rosenkreuz 'could have boasted of his art, especially that of the transmutation of metals', but instead 'he set more store by heaven than by all splendour'.[4] To be fully accurate, the German text says 'heaven and its citizens'.[5] I found this use of the word 'heaven' somewhat puzzling until I came across a translation of a Hebrew treatise on alchemy, quoted in Raphael Patai's book *The Jewish Alchemists*, which says that 'the physicists called heaven by the name "fifth essence", because heaven is not subject to corruption'.[6] Another passage states: 'Just as the highest heaven does not on its own influence the existence of the world, but [does this] through the intermediacy of the luminaries and the stars...so for our Human Heaven it is proper that it should be aided by the things which are influenced by it, such as the medicines, or the herbs, which have virtues for the limb that we intend to cure.'[7] So what the passage may be saying is that the use of alchemy for healing purposes is infinitely more praiseworthy than the transmutation of metals. And the word 'citizens' perhaps refers to the luminaries and the stars. This interpretation is

supported by another passage in the *Fama*, which states: 'To the true philosophers [probably meaning alchemists] gold-making is a trivial matter and a side issue, in comparison with which they have a thousand better skills.'[8]

Thus, while alchemy is alluded to in the *Fama*, it is somewhat downplayed, as it is in the second Rosicrucian manifesto, the *Confessio Fraternitatis*, published in the Latin language in 1615. Very different in character is the third of the original Rosicrucian texts, the *Chemical Wedding of Christian Rosenkreuz*[9] – in fact so different that it is something of a mystery why it came to be grouped with the other two. The Christian Rosenkreuz of the title is not the founder of the Rosicrucian Brotherhood but rather an elderly hermit of humble piety who is invited as a guest to the wedding of a king and queen. The scenario takes place in a castle full of marvels and features many – clearly symbolic – figures: a lion, a unicorn, a dove, a phoenix, a Virgin, a Cupid, a Moor, a Lady Venus. Dramatic events unfold, such as the beheading of several 'royal personages' including the bride and groom, who are later brought back to life in an alchemical operation.

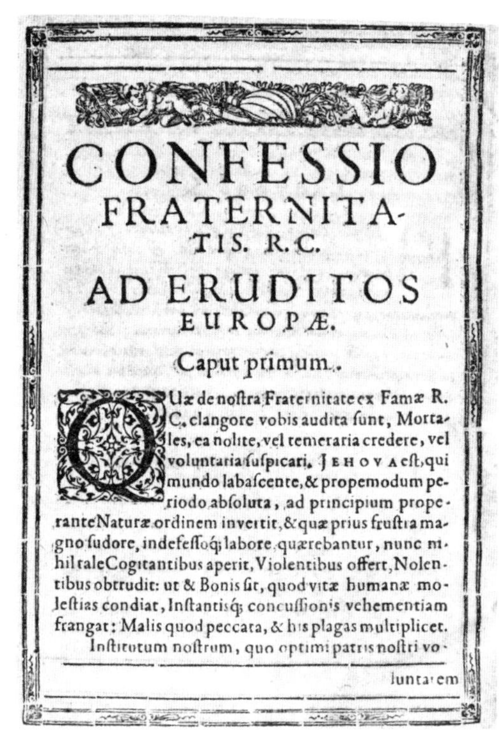

CONFESSIO
FRATERNITA-
TIS. R.C.
AD ERUDITOS
EUROPÆ.

Caput primum.

QUæ de noſtra Fraternitate ex Famæ R. C. clangore vobis audita ſunt, Mortales, ea nolite, vel temeraria credere, vel voluntaria ſuſpicari. JEHOVA eſt, qui mundo labaſcente, & propemodum periodo abſoluta, ad principium properante Naturæ ordinem invertit, & quæ prius fruſtra magno ſudore, indefeſſoq; labore quærebantur, nunc nihil tale Cogitantibus aperit, Violentibus offert, Nolentibus obtrudit: ut & Bonis ſit, quod vitæ humanæ moleſtias condiat, Inſtantisq; concuſſionis vehementiam frangat: Malis quod peccata, & his plagas multiplicet.
Inſtitutum noſtrum, quo optimi patris noſtri vo-

luntatem

First edition of the *Confessio Fraternitatis* ('Confession of the Fraternity'), contained within *Secretioris philosophiae consideratio brevis*.

Roland Edighoffer has offered a detailed interpretation of this text as an elaborate alchemical allegory. The story, he writes, 'describes the coming of a "perfectly pure" man...he who will complete the redemptive work of Christ by practising "deliverance" in the sense of the Greek word "maieutic"; that is to say, the birthing of nature which Paul says in the Epistle to the Romans "has been groaning in travail up to now"...This glorification of Nature...is interpreted as an alchemical *opus*: it is the Great Work, the effect of the double mystery of regeneration and of the *hieros gamos*, hence the sacred marriage.'[10]

The *Chemical Wedding* was written by Andreae in his early years and subsequently reworked for publication. Andreae's father, also called Johann, was an alchemist, and the young Johann would have been familiar with alchemical symbolism. Why he decided to publish the *Chemical Wedding* is unclear. Possibly he was dismayed by the way in which the story of Christian Rosenkreuz had been taken seriously by a large number of readers, and he may have thought that by publishing this third text he could emphasise that the whole thing was a fiction. If so, his strategy was not very successful, because the *Chemical Wedding* has only served to reinforce the association between Rosicrucianism and alchemy.

The *Fama* included an appeal to the learned of Europe to communicate with the Brotherhood. Although the latter's whereabouts were not revealed, this resulted in a flood of writings and publications that came to be known as the Rosicrucian furore. Some of these writings were requests to join the Rosicrucians, some were hostile, some purportedly came from members of the Brotherhood itself, some were from people who embraced the Rosicrucian vision and presented their own interpretation of it. Whether any received a reply is not recorded.

Opposite:
The exploits of the allegorical figure of Christian Rosenkreuz may have been inspired by earlier travelling monks. Shown here in the *Breviculum Codex* of Ramón Llull. Produced by Thomas Migerii in Northern France after 1321.

Overleaf:
Stage set for Mozart's *Magic Flute*, designed by Karl Friedrich Schinkel, 1816. The opera contains some elements of alchemical and Rosicrucian symbolism.

VERNUNFT

HEIT
NAJUR

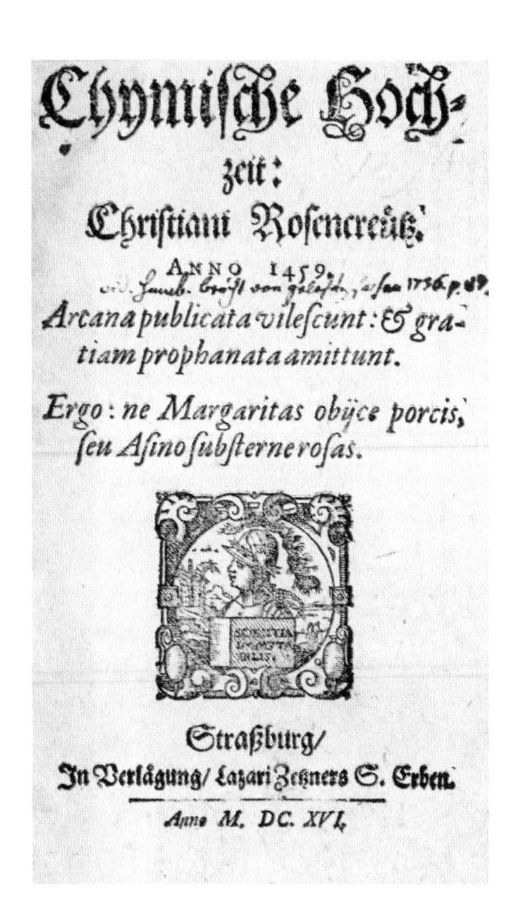

Chymische Hochzeit:
Christiani Rosencreütz.
Anno 1459.

Arcana publicata vilescunt: & gratiam prophanata amittunt.

Ergo: ne Margaritas obijce porcis, seu Asino substerne rosas.

Straßburg/
In Verlägung/ Lazari Zetzners S. Erben.
Anno M. DC. XVI.

Christiani Rosencreutz 95

schlaffen gnug hätte. Nach dem ich mich nun angezogen/ vnd meiner gewonheit nach die Stiegen hinab begeben/ war es noch zu frühe/ vnnd fande niemand anders in dem Saal/ bat deßwegen meinen Knaben mich ein wenig inn dem schloß vmbzuführen/ vnd etwas sonderlichs zuzeigen/ der war nun wie allweg willig/ führet mich auch als bald etliche Stiegen vnter die Erd/ zu einer grossen eysenen Thüren/ darauff waren nachfolgende wort von Kupfferen grossen Buchstaben angehefftet:

Diß hab ich also abgemahlt/ vnnd in mein Schreibtäfelein auffgezeichnet: Nach dem nun diese Thür eröffnet/ führet mich der Knab bey der Hand durch einen gantz finstern Gang/ biß wir wider zu einem kleinen Thürlein kamen/ daß war nur zugelenet: dann wie mich der Knab berichtet/ hatte man solches erst gestern eröffnet/ vnd die Sarch darauß genommen/ wäre also noch nit beschlossen worden/ wie wir nun hinein getretten/

Title page and page 95 of *Chymische Hochzeit Christiani Rosenkreuz* (*Chemical Wedding of Christian Rosenkreuz*), 1616.

Opposite:
A hand-coloured version of Michael Maier's alchemical poem *Atalanta fugiens* ('Atlanta Fleeing'), 1618.

By the 1620s the furore in Germany had largely dissolved in the chaos of the Thirty Years' War, but the impact of the manifestos continued to be felt elsewhere. Like a stone dropped into the collective mind of the age, it sent out ripples to various countries including England, France, Holland and Sweden. One person who helped to transmit the Rosicrucian current to England was the physician, alchemist and composer Michael Maier (1568–1622), author of an alchemical poem symbolically illustrating stages in the search for the philosophers' stone. It consists of 50 verses or epigrams, with illustrations by Matthias Merian and a series of fugues composed by Maier himself. The work was produced in a lavish edition by the celebrated printer Johann Theodor de Bry of Oppenheim. While in England Maier probably met the fellow Rosicrucian apologist Robert Fludd (1574–1637), also a physician and alchemist. Thus was the connection between Rosicrucianism and alchemy perpetuated.

Meanwhile, in the German lands little was heard of Rosicrucianism for some decades until it reappeared in the eighteenth century in a form rather different from the original movement. A seminal work in the emergence of this neo-Rosicrucianism was *Die wahrhaffte und volkommene Bereitung des philosophischen Steins der Brüderschafft aus dem Orden des Gülden und Rosenkreutzes* (*The True and Perfect Preparation of the Philosophical Stone of the Brotherhood within the Order of the Golden and*

ATALANTA
FUGIENS,
hoc est,
EMBLEMATA
NOVA
DE SECRETIS NATURÆ
CHYMICA,
Accommodata partim oculis & intellectui, figuris cupro incisis, adjectisque sententiis, Epigrammatis & notis, partim auribus & recreationi animi plus minus 50 Fugis Musicalibus trium Vocum, quarum duæ ad unam simplicem melodiam distichis canendis peraptam, correspondeant, non absq; singulari jucunditate videnda, legenda, meditanda,, intelligenda, dijudicanda, canenda & audienda:

Authore

MICHAELE MAJERO Imperial. Consistorii Comite, Med. D. Eq. ex. &c.

OPPENHEIMII

Ex typographia HIERONYMI GALLER[...]

Sumptibus JOH. THEODORI de BRY,

M DC XVIII.

Rosy Cross), published at Breslau (now Wrocław, Poland) in 1710 under the name of Sincerus Renatus, a pseudonym for Samuel Richter, a Protestant pastor from Silesia. In it he describes the rules and practices of a somewhat nebulous Rosicrucian order. Significantly, as the title of Renatus's book indicates, alchemy was a central preoccupation.

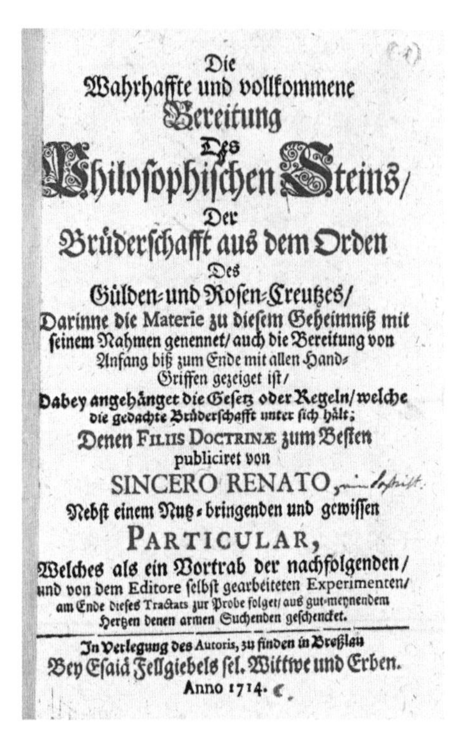

Die
Wahrhaffte und vollkommene
Bereitung
Des
Philosophischen Steins/
Der
Brüderschafft aus dem Orden
Des
Gülden- und Rosen-Creutzes/
Darinne die Materie zu diesem Geheimniß mit
seinem Nahmen genennet/ auch die Bereitung von
Anfang biß zum Ende mit allen Hand-
Griffen gezeiget ist/
Dabey angehänget die Gesetz oder Regeln/ welche
die gedachte Brüderschafft unter sich hält;
Denen FILIIS DOCTRINÆ zum Besten
publiciret von
SINCERO RENATO,
Nebst einem Nutz-bringenden und gewissen
PARTICULAR,
Welches als ein Vortrab der nachfolgenden/
und von dem Editore selbst gearbeiteten Experimenten/
am Ende dieses Tractats zur Probe folget/ aus gut-meynendem
Hertzen denen armen Suchenden geschencket.

In Verlegung des Autoris, zu finden in Breßlau
Bey Esaiä Fellgiebels sel. Wittwe und Erben.
Anno 1714.

Die wahrhaffte und volkommene Bereitung des philosophischen Steins der Brüderschafft aus dem Orden des Gülden und Rosenkreutzes (*The True and Perfect Preparation of the Philosophical Stone of the Brotherhood within the Order of the Golden and Rosy Cross*), by Sincerus Renatus (Samuel Richter), 1710.

Whether or not the order described by Renatus really existed at that time, by about the 1760s an order of the same name, Golden and Rosy Cross, had emerged as part of Freemasonry. To be admitted, one had to have passed through a regular Masonic lodge. And from then on, we find various Masonic systems of one kind or another invoking the Rosicrucian symbology. The Golden and Rosy Cross Order was grouped into circles of nine members each and had nine grades of initiation, each involving elaborate rituals. (This grade structure, slightly modified, was adopted by the English occult order the Golden Dawn and later by other Rosicrucian orders.)

Part of the order's curriculum involved reading recommended books and manuscripts on Christian theosophy, mysticism, Kabbalah and alchemy. In cases where a member had difficulty in understanding the material he could consult with his superior by letter. And if it was difficult for him to attend a circle he could even be received into the order by correspondence and initiate himself into the various grades. So in effect the Golden and Rosy Cross was operating a kind of mail order mystery school, possibly the very first of its kind.

Alchemy played a major part in the order's activities. Alchemical symbolism featured in the initiation ceremonies and members were supposed to have their own laboratories and work diligently at their furnaces, retorts and crucibles. As one progressed up the order one received more and more alchemical secrets and there survive today many alchemical books and manuscripts that circulated among the fraternity. These were practical recipe books, containing very detailed and precise instructions about alchemical processes, often with drawings of equipment. At the same time, the physical work in the laboratory was combined with a spiritual process in which the task of refining substances went hand in hand with inner refinement.

This approach to alchemy was closely bound up with a Gnostic worldview. In the Golden and Rosy Cross this comes out very clearly in the stated aim of the order, which was 'to make effective the hidden forces of nature, to release nature's light which has been deeply buried beneath the dross resulting from the curse, and thereby light within every brother a torch by whose light he will be able better to recognize the hidden God'.[11] This is typical of a strain of esoteric spirituality which was very strong in Germany at that time and has its antecedents in mystical writers like Meister Eckhart and Jacob Boehme.

By the early 1790s the Golden and Rosy Cross had petered out, but in 1785 there had appeared an influential Rosicrucian/alchemical work in three folio volumes, namely the *Geheime Figuren der Rosenkreuzer aus dem 16ten und 17ten Jahrhundert* (*Secret*

Opposite:
'The unseen incomprehensible chaos and visible, understandable chaos.' From *Geheime Figuren der Rosenkreuzer aus dem 16ten und 17ten Jahrhundert* (*Secret Symbols of the Rosicrucians from the 16th and the 17th Century*), 1919.

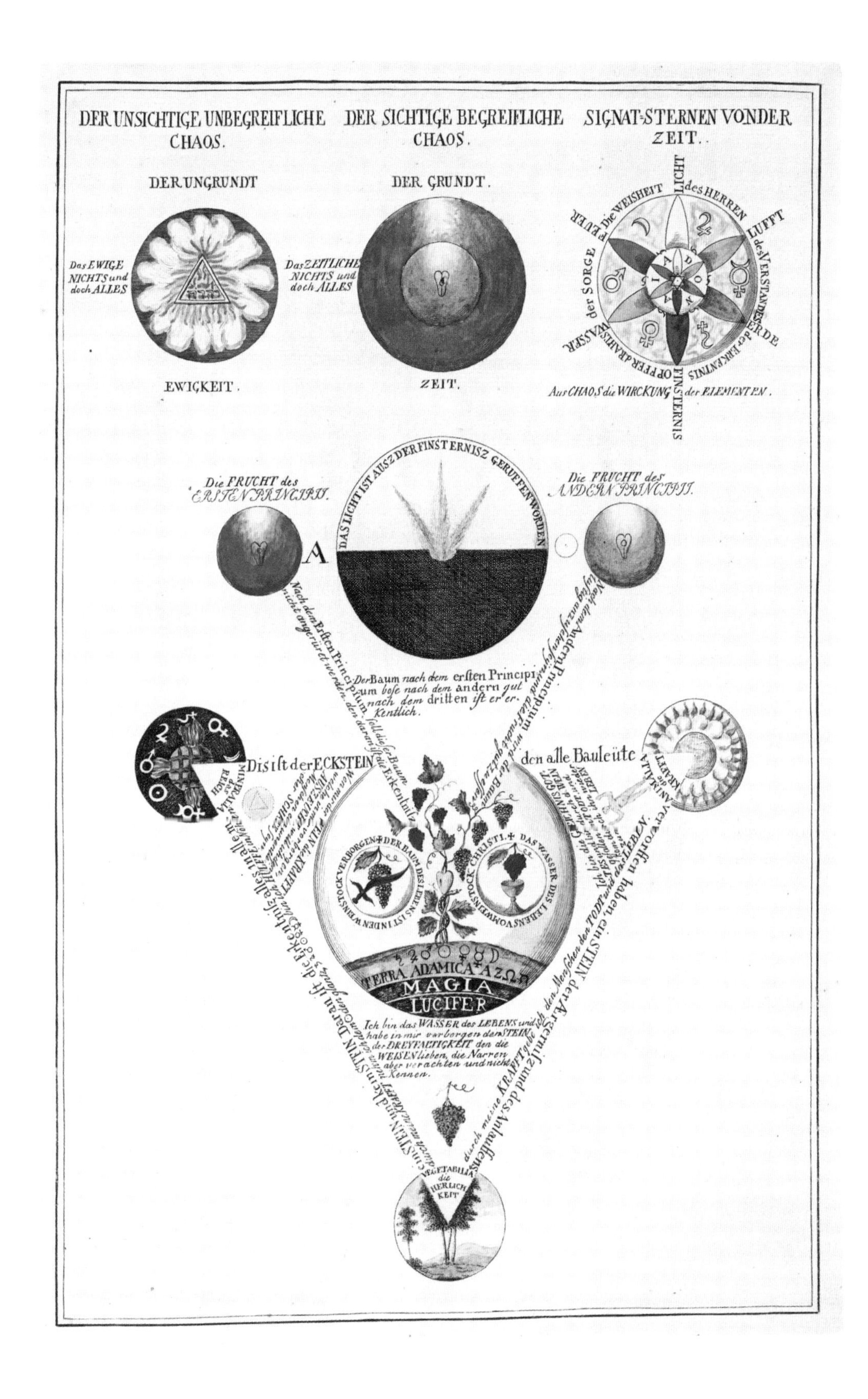
DER UNSICHTIGE UNBEGREIFLICHE CHAOS.
DER SICHTIGE BEGREIFLICHE CHAOS.
SIGNAT-STERNEN VON DER ZEIT.
DER UNGRUNDT
DER GRUNDT.
Das EWIGE NICHTS und doch ALLES
Das ZEITLICHE NICHTS und doch ALLES
EWIGKEIT.
ZEIT.
LICHT des HERREN
Die WEISHEIT
FEUER
LUFFT
des VERSTANDES
ERDE
der ERKENTNIS
WASSER
OPFFER
der SORGE
FINSTERNIS
Aus CHAOS die WIRCKUNG der ELEMENTEN.
Die FRUCHT des ERSTEN PRINCIPII.
Die FRUCHT des ANDERN PRINCIPII.
DAS LICHT IST AUSZ DER FINSTERNISZ GERUFFEN WORDEN
A
Der Baum nach dem ersten Principi um böse nach dem andern gut nach dem dritten ist er er. Kentlich.
Dis ist der ECKSTEIN
den alle Bauleüte
TERRA ADAMICA
MAGIA
LUCIFER
Ich bin das WASSER des LEBENS und habe in mir verborgen den STEIN der DREYFALTIGKEIT den die WEISEN lieben, die Narren aber verachten und nicht Kennen.
VEGETABILIA die HERLICHKEIT

Die Kunst ist gerecht, wahr und gewiß,
Dem, der gottsfürchtig, fleissig ist,
Und braucht sich der Naturen recht,
Macht ihn zum Herrn, nicht zum Knecht,
Eil nicht, bleib auf der rechten Bahn,
So wirst du Nutz und Freud viel han,

Und gönn't es Gott dir in dein'm Leben,
So thu reichlich den Armen geben,
Sey treu, und halt die Kunst im Still,
Denn das ist gewißlich Gottes Will,
Halt Treu und Glaub, denk mein dabey,
So bleibst du aller Nachred frey.

1604

MONS PHILOSOPHORUM.

Die Seel des Menschen überall
Verlohren ist durch einen Fall,
Durch einen Fall des Leibs Gesundheit
Verlohren und zerrüttet leid't.
Der Seel ein Heil wiederbracht ist,
Welches ist IEHOVA Jesus Christ.
Des Leibes Gesundheit wiederbringt
Von Angesicht ein schlechtes Ding,
Welches ist verborgen in diesem Gemähld,
Der höchste Schatz in dieser Welt,
In ihm ist die höchste Medicin
Auch der größte Theil der Reichthum,
Welchen uns der HERRE IEHOVA
In der Natur fürstellet da,
Pater Metallorum genannt,
Den Philosophis wohl bekannt,
Sitzend wol für des Berges Höhl,
Jedermann er sich darbeut feil,
Aber von Sophisten, so verblendt,
Am wenigsten er wird erkennt,
So an den Wänden herummer tappen,
Behängt mit sophistischen Lappen.
Zur Rechten wird gesehen da
Lepus, deut der Kunst Chymia,
Wunderbar'rweiß, und derselben Art
Erforscht wird durch des Feuers Grad,
Zur Linken denn find man auch frey,
Was der rechte Clavis artis sey;
Gleich wie ausbrüht die Henn das Huhn,
Zu subtil kann man ihm nicht thun.
Im Mittel des Berges vor der Thür
Steht der tapfre Löw mit grosser Zier,
Welchen der Drache Ungeheuer,
Vergeust sein edles Blut so theuer;
Wirft ihn wol in ein tiefes Grab,
Davon entspringt der schwarze Rab;
Welches denn Ianua artis heist,
Aquila alba davon entspreust;
Selbst der Crystall im Ofen sein,
Wird dir zeigen mit Augenschein,
Servum fugitivum geschwind,
Vielen Artisten ein Wunder-Kind.
Principium laboris ist
Der Mittler genannt zu aller Frist.
Dann auch im Faß zur rechten Hand
SOL LUNA des Firmaments Verstand.
Der Senior so pflanzen thut,
Rad. Rubeam & albam gut.
Nun fährst du fort mit Beständigkeit,
Arbor artis sich dir erzeigt.
Mit seiner Blüt verkündet er nun,
Lapidem Philosophorum.
Darob die Kron der Herrlichkeit,
Herrschend über alle Schätzeweit.
Sey fleißig, friedsam, beständig fromm,
Bitt daß dir GOTT zu Hülfe komm.
Erlangst du das, so laß dir sein
Die Armen stets befohlen seyn.
So wirst du mit der Engel Schaar
GOTT loben jetzt und immerdar.

Symbols of the Rosicrucians from the 16th and the 17th Century), published at Altona near Hamburg. This was a collection of material from various sources, put together by an anonymous compiler. In the tradition of the Golden and Rosy Cross the work contains detailed instructions for carrying out alchemical processes and also expounds a gnostically tinged religious teaching. The text is accompanied by a series of beautiful images, many of which have since appeared repeatedly in literature about Rosicrucianism and alchemy. This work has undoubtedly helped to keep the alchemical tradition alive and to perpetuate its association with Rosicrucianism.

The burial place of the fictional Christian Rosenkreuz in the *Philosophers' Mountain*. From *Geheime Figuren der Rosenkreuzer*. 18th century.

The nineteenth century saw a decline in alchemy, which by now had become generally divorced from chemistry. However, a few dedicated seekers, in Germany, France, England and elsewhere, continued to study alchemy, either as an activity in the laboratory or as a spiritual and symbolic language. The latter approach attracted an increasing following from around the middle of the century, stimulated by such works as Mary Anne Attwood's *A Suggestive Inquiry into the Hermetic Mystery*, published in 1850. Thenceforth there was an increasing number of apologists for the symbolic interpretation of alchemy, either alongside or as an alternative to the practical approach.

In the late nineteenth and early twentieth centuries the connection between alchemy and Rosicrucianism is apparent in the case of various members of the Hermetic Order of the Golden Dawn, which had been founded in London in 1888, which claimed Rosicrucian antecedents and had an inner order that was explicitly Rosicrucian. Richard Caron, in a searching article on the history of alchemy, mentions several Golden Dawn members in this connection. 'The Reverend William Alexander Ayton (1816–1908) joined the Order together with his wife in 1888...A passionate devotee of alchemy, which he had practiced since about the 1850s, he was continually pursuing alchemical books and manuscripts, some of which he translated. Ayton corresponded with Frederick Leigh Gardner (1857–c.1930), author of a *Catalogue raisonné of Rosicrucian books*',[12] who was a fellow Golden Dawn member. A. E. Waite, another member of the Golden Dawn and later founder of his own Rosicrucian order, also had a profound interest in alchemy. He was the author of *The Secret Tradition in Alchemy* (published in 1926) and translated and edited a number of ancient alchemical texts.[13]

In the German-speaking world someone who embodied both the Rosicrucian and the alchemical currents was Rudolf Steiner (1861–1925), founder of the Anthroposophical Society after breaking away from the Theosophical Society. Speaking in 1907, Steiner said that the Rosicrucian current had flowed underground for most of the nineteenth century but had recently re-emerged, making it 'possible again to make the Rosicrucian wisdom accessible and allow it to flow into the general culture'.[14] He also emphasised that 'Rosicrucian wisdom must not stream only into the head, nor only into the heart, but also into the hand, into our manual capacities, into our daily actions'.[15]

Commensurately, Steiner's approach to alchemy is both symbolic and practical. In a 2019 master's thesis Camila Rodríguez Galilea writes: 'Steiner's Anthroposophy its partly grounded on alchemical principles, which are contained in the different levels of the Anthroposophical narrative in a fractal projection. These interwoven narratives help to give birth to Anthroposophical practices such as Medicine, Biodynamic agriculture and Waldorf education, among others, which represents one of the most prolific contemporary uses of Western alchemy in esoteric practices.'[16]

In 1910 Steiner met a kindred spirit in the person of a young poet named Alexander von Bernus (1880–1965), who had already developed an interest in alchemy, and the two remained close friends until the end of Steiner's life. Whether Bernus would have actually called himself a Rosicrucian is open to question, but as an Anthroposophist and close associate of Steiner he would most likely have regarded himself as working in the Rosicrucian spirit. Bernus owned the former Benedictine abbey of Neuburg, near Heidelberg, where in 1921 he created the Soluna Laboratory to produce alchemical remedies. Later he moved the laboratory to Stuttgart, where it was bombed out in 1943. Subsequently he carried on the work in his mansion at Donauwörth in Bavaria. After the death of Bernus in 1965 Soluna was continued by his widow Isa until 1988 and since then has passed through various changes of management. Today it operates from a laboratory at Donauwörth. It is one of a number of enterprises in Germany that are keeping the alchemical tradition alive.

In France, alchemy and Rosicrucianism joined hands in an organisation called the Frères Aînés de la Rose Croix (Elder Brethen of the Rose Croix), led by Roger Caro (1911–92), which operated in the 1960s and 70s. Of this fraternity Richard Caron writes: 'Here Rosicrucianism and alchemy were closely associated with theology, notably in Kamala-Jnana's works: *Dictionnaire de philosophie alchimique* (*Dictionary of Alchemical Philosophy*, 1961), *Pléiade alchimique* (*Alchemical Pleiad*, 1967), and *Tout de Grand Oeuvre photographié* (*The Whole Great Work Photographed*, 1968).'[17]

In the United States, a major role in promoting alchemy in modern times has been played by the Ancient and Mystical Order Rosae Crucis (AMORC), founded by Harvey Spencer Lewis (1883–1939). Having made contact with various European initiatic orders, Lewis launched his organisation in New York in 1915 and declared himself Imperator. After a period in Florida he moved to San Jose, California, where AMORC finally settled. Lewis claimed to have mastered the art of alchemical transmutation, which he demonstrated to followers in New York in 1916. One account describes the event as follows:

> Fifteen of the twenty-seven members had received from the Imperator a card, telling what ingredients and objects they must bring for the operation. They promised to keep secret the names on the cards and to join the fifteen parts of the formula only three years after the death of the Imperator. After prayer and an address by the latter treating upon the laws of matter, a piece of zinc...was placed upon a little plate of china-porcelain, over the fire of a crucible; the various ingredients, among them the petals of a rose, were then presented by the fifteen brothers and sisters...to the Imperator who deposited them one after another upon the plate. After the sixteen minutes required, during which

> the operator concentrated a little known power of mind, the piece of zinc was transformed into gold, as was chemically established.[18]

Samuel Liddell MacGregor Mathers, founding member of the British branch of the Hermetic Order of the Golden Dawn. Painted by his wife and fellow occultist Moina Mathers, *c.* 1895.

In the early 1940s AMORC's librarian Orval Graves conducted a series of classes on practical alchemy and also published articles on the subject in the order's magazine, *The Rosicrucian Digest*. 'In those early classes, the techniques of Paracelsus were generally followed, artificial stones were created, and students would often take turns staying up throughout the night, to regulate the heat of the furnaces for the herbal work. A great sense of harmony prevailed.'[19]

One of the people who attended Graves's classes was a young German called Albert Riedel, who had immigrated to America in the early 1930s. Riedel went on to become an assiduous promoter of alchemy in his own right under the pseudonym Frater Albertus. Richard Caron writes of him: 'Born in Dresden, Riedel emigrated to the USA...and followed Graves's courses before developing his personal vision. In 1960 he published *The Alchemist's Handbook*, and simultaneously founded the Paracelsus Research Society. The activities of this group from 1960 to 1984 were enormous: the training of hundreds of pupils, most of whom were members of AMORC or the Golden Dawn; the preparation of tinctures, publishing of numerous bulletins reserved for members...and the publication by Albertus of several ancient and modern texts.'[20]

Looking back over the intertwined history of alchemy and Rosicrucianism one can identify, *inter alia*, the following salient interconnections:

1. The alchemical references in the Rosicrucian manifestos, especially the heavy alchemical symbolism in the *Chemical Wedding*.
2. The fact that, among the early Rosicrucian apologists, many were also alchemists.
3. The pursuit of alchemy by Rosicrucian orders such as the Golden and Rosy Cross and later AMORC.

In conclusion, it is fair to say that it is to a large extent thanks to Rosicrucianism that alchemy is still alive and well today.

TRANSITION TO MODERN SCIENCE AND ENLIGHTENMENT THOUGHT

TU NE CEDE MALIS, SED CONTRA AUDENTIOR ITO:
HENRICI KHUNRATH
ANNIHILEMUS
Blasphemias, Sit anathema maranatha.
DURABIT
Veritatem investientibus
SENECA Lib: de Morib.
LIBER Numinis et Luminis NATURÆ macrocosmicus, veridicus.
Sacrosanctus BIBLIORUM DEI codex, infallibilis.
Einer achts: der ander verachts und verlachts; der dritte betrachts; was machts? Ein jeder judiciret wie Eres entweder versteh et / oder aber
Es müste einer viel zuthun haben / der jeden und alle Thoren solte klug machen. Ja, wer kan solches? wil geschweigen / das Ich Doct. Henricus Khunrath es thue

ALCHEMY, CHEMISTRY AND MEDICINE IN THE EARLY MODERN ERA

GEORGIANA D HEDESAN

Illustration from *Amphitheatrum Sapientiae Aeternae* by Paracelsian physician Heinrich Khunrath, 1609.

Previous vignette: The Alchemist in Search of the Philosopher's Stone. Painting by Joseph Wright of Derby, depicting the discovery of phosphorus by Hennig Brand in 1669.

Traditionally, medieval Latin alchemy aimed at transforming matter for two chief purposes: transmutation of metals into gold and silver (henceforth referred to generically as *chrysopoeia*, 'the art of making gold'), or medicine.[1] Other purposes, like the making of pigments, glass, artificial gems or soil fertilisers, were also important but were often carried out more quietly, leaving less trace in manuscripts. This is probably due to the fact that they were part of the 'mechanical arts', usually undertaken in medieval guilds and as such covered by trade secrecy.

Chrysopoeian alchemy, which elicited interest among thirteenth-century Scholastics, came under intense scrutiny in the fourteenth century. Worried about attempts at counterfeiting gold and silver coin, Pope John XXII issued the 1317 bull *Spondent quas non exhibent*, effectively banning gold-making as a pursuit. In 1396, the Aragonese inquisitor Nicholas Eymerich (*c.* 1316–99) further condemned *chrysopoeia* in his *Tractatus contra alchimistas*.[2] The prohibition by religious authority was upheld by the secular one: Charles V of France (1380), Henry IV (1403/4) and Henry VI (1452) of England, and city authorities such as Venice (1488) and Nuremberg (1493). The secular proscription should not be understood to mean that all *chrysopoeia* was forbidden; instead, rulers sought to control the practice by, for instance, issuing special licences.[3]

The backlash notwithstanding, the fourteenth century was a time when alchemy flourished, producing the Pseudo-Lullian corpus (attributed, wrongly, to Catalan philosopher Ramon Llull, 1232–1315) and the work of the Franciscan friar Johannes de Rupescissa (*c.* 1310–*c.* 1362). What is remarkable about these works is that they make more room for medical purposes than had previously been the case. Rupescissa's work distinguished between *chrysopoeia* (in *The Book of Light – Liber lucis*) and medical alchemy (in *Book on the Consideration of the Quintessence of All Things – Liber de quintae essentiae omnium rerum*), but it is the latter that he is best known for.[4] His concept of the quintessence, a powerful substance extracted by means of distillation, would give medical alchemy the theoretical ground it had been missing. Distillation had proven its prowess as early as the twelfth century, when monks at the monastery of Salerno produced purified alcohol (usually termed *aqua ardens* – burning water – or *aqua vitae* – water of life). Under the imaginative pen of Rupescissa, alcohol was considered a lower form of the quintessence, and many others could be produced from various substances. There was in his view a scale of quintessences, increasingly powerful, with that produced from wine crowning them all, and being similar to the *quinta essentia*, the stuff that supposedly made up the heavens.

IL LEGNO DEL
LA VITA, DI MESSER GIO
uanni Braceſco, nel quale ſi dechiara qual
fuſſe la Medicina per laquale gli
primi Padri viueuano No
uecento Anni.

Il Legno della vita (*The Tree of Life*), by Giovanni Bracesco, 1542.

The Pseudo-Lullian corpus sought a different path from that of Rupescissa, but also of thirteenth-century alchemy: it conceptualised the notion of elixir, inherited from the Arabs, to propose a universal philosophers' stone. This stone had multiple purposes: it would equally better metals, living bodies and gemstones. This monistic approach sought to overcome the boundaries set between *chrysopoeia* and medical alchemy, providing a flexible tool that could, at least theoretically, unify alchemical pursuit. In practice, the difference seemed to survive in Pseudo-Lull's distinction between a mineral and a vegetable mercury (by which the initial solvent of the work was meant).[5]

In the early fifteenth century, the Pseudo-Lullian approach appeared dominant. In the late fourteenth century, Rupescissa's *Liber de quintae essentiae* was assimilated to the Pseudo-Lullian corpus under the name of *The Book of Natural Secrets, or of the Quintessence* (*Liber de secretis naturae, seu quinta essentia*). The notion of the philosophers' stone and the Pseudo-Lullian framework appealed to those who were interested in both the philosophical side of alchemy as well as its practice. The fifteenth century saw the rise of Pseudo-Lullian commentators, of whom the most important were George Ripley (*c.* 1415–*c.* 1490) in England and Christopher of Paris (mid-fifteenth century) in northern Italy.[6] There was a new revival in the sixteenth century through the medium of print, when Giovanni Bracesco (*c.* 1481–*c.* 1550) published a controversial treatise, *The Tree of Life* (*Il Legno della vita*, 1542), drawing on the reputation of the reputation of the Pseudo-Lullian corpus.

At the same time, the fifteenth century was a period when the alchemical technology of distillation became more sophisticated and increasingly widespread across Europe.[7] This coincided with the improvement of techniques for glassmaking and the availability of the raw materials needed for alchemical work. Furthermore, Rupescissa's view of quintessence provided theoretical explanation, cohesion and justification for distillatory practices. When Strasbourg surgeon

Hieronymus Brunschwig (*c.* 1450–*c.* 1530) produced his two books on distillation (the so-called 'small' and 'large' books of distillation, 1500 and 1512 respectively), he used the Rupescissan framework.[8] The same background – whether mediated or not through Pseudo-Lull – was present in subsequent publications: Philip Ulstadt's popular *Coelum philosophorum* (*The Heaven of Philosophers*, 1528), Conrad Gessner's *Thesaurus Euonymi Philiatri* (*The Treasure of the Medical Practitioner Euonymus*, 1554) and *Archidoxis* (1525) by Theophrastus von Hohenheim, known as Paracelsus.

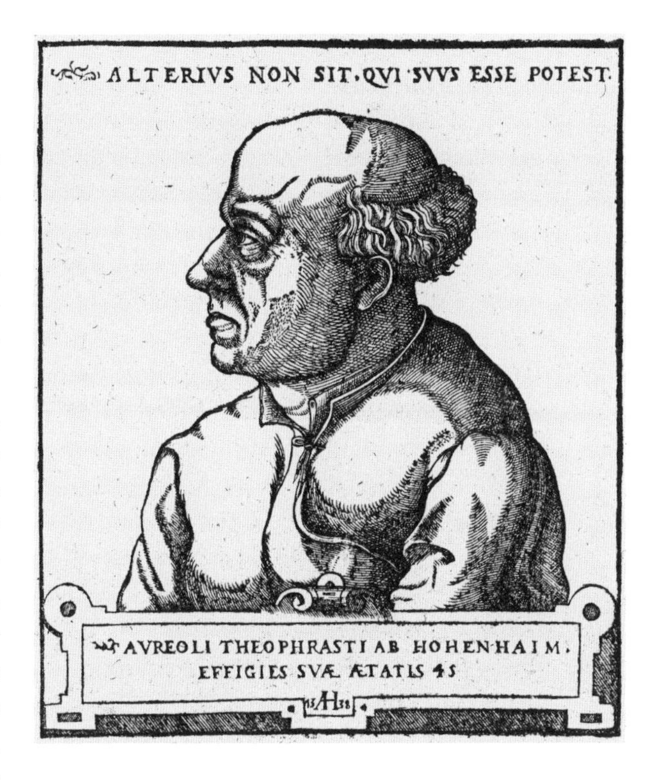

Paracelsus (Aureolus Theophrastus Bombastus von Hohenheim). Woodcut by A. Hirschvogel, 1538.

Distillation was mainly carried out for medical purposes, with the highly successful alcohol being used initially to treat burns and cuts (indeed, medical alcohol is still available to buy in some parts of the world). Distillation initially became popular among medical practitioners without university training – a motley group that would include friars, lower ecclesiastics, barber-surgeons, apothecaries, midwives, bone-setters, bathmasters, executioners and village wisemen and women. They were generally not expected to know Latin, and their interest lay chiefly in simple recipes. While know-how was presumably transmitted mainly in formal or informal groups, there was also a rising need for generic manuals and 'how-to' books written in vernacular, preferably containing images of apparatus. The newly developed technology of print seemed perfect for this purpose, as Brunschwig quickly found. His works on distillation were a runaway success. His 'small' book, containing basic instructions on distilling, went through six editions, plus seven more in a modified form.[9] The 'large' book had nine editions and continued to be published as late as 1614.[10] There were also small cheap versions and translations; the English version was published as early as 1527. It spawned an entire genre of books on distillation (*Distillierbücher*), which were written by common medical practitioners as well as trained physicians. Distillation can chiefly be done with plants, animals and some minerals. Plant distillation was by far the most popular, yielding medicinal 'waters'. Although distilling could be carried out in a wide variety of settings, including pharmacies and households, a popular location was in or near gardens. This proximity allowed for practitioners to control the quality of the plants being used and to cultivate more exotic kinds, which could then be explored in the still. Thus, when botanical gardens began to be established, first in Italy, then in other parts of Europe, distillation laboratories accompanied them. Such buildings could become institutions of their own, emblematic of a new way of acquiring knowledge. In Paris in 1640, a French alchemical physician, Guy de La Brosse (1586–1641), founded one of the oldest and most long-lasting scientific institutions, the Jardin du Roi (The Royal Garden), which survives even today under the name of the Jardin des Plantes. The garden featured a building where distillation was taught and practised. La Brosse's foundation was a triumph of the new medical learning that drew on the radical works of Paracelsus to combine alchemical knowledge and medicine.[11]

Distillation works with matter that is liquefied; however, some harder materials like metals were difficult to reduce to a solution. In the Middle Ages, metallurgists used corrosive acids (commonly termed 'strong waters', *aquae fortis*) to make such metals malleable. The strongest included spirit of vitriol (sulphuric acid), *aqua fortis* (hydrochloric acid) and *aqua regia* (nitrohydrochloric acid). The acid treatment could possibly work for the *chrysopoeia*, but such methods were less likely to be useful for medical purposes.

Hieronymus Brunschwig's book on distillation, *Liber de arte distillandi*, 1500.

In fact, the use of minerals or metals for medicine was controversial. The Greeks and Romans had mainly used plant and animal products, though some minerals and earths were approved by the authority of the two foremost physicians of late antiquity, Galen and Dioscorides. In the Middle Ages, however, the rise of alchemy led to a fascination with two metals in particular: gold and mercury. The two are quite different in nature: gold is hard, mercury is liquid; the first is yellow or yellowish-red, while the second is silvery (often described as white). Yet it was long known that the two were also compatible, creating what was called, by the Arab name, an 'amalgam': indeed, this was the traditional way of extracting gold from ore. It was also an initial method of rendering gold in a liquid or viscous form.

Gold was of course considered the 'king' of metals; it was not only attractive to the eye but also apparently incorruptible. Gold's qualities led some to believe that, if this virtue was somehow extracted from and transferred to the human body, it too could vanquish corruption – diseases and old age. The thirteenth-century Catalan physician Arnald of Villanova (*c.* 1240–1311) thought that gold was the natural remedy for human health, given for this purpose by God.[12]

The question of how gold's virtue could be transferred was not easily answered. It was thought by many that simply dipping gold coins or fine plates into water, wine or alcohol would make the liquid acquire powerful health properties. This was a cheap practice which was particularly popular in the fourteenth and fifteenth centuries but seems to have waned after that, or at least it is not as present in the textual record. In any case, noble persons could go a step further, ingesting gold in some shape or form. A popular method was eating or drinking gold leaf; this was done for health reasons and is still practised, though eating gold-leafed cake is nowadays a question of status rather than health. The futility of swallowing gold was already recognised in the early modern period; it is indigestible and is expelled from the body via faeces.

The medieval discovery of *aqua regia*, the first radical solvent of gold, seemed to bring a ingestible gold closer to reality. However, the solution was a corrosive, which if swallowed causes violent outcomes: one empirical treatise in 1602 praising gold dissolved in a 'mineral fire' (most likely a form of *aqua regia*) described it as causing fever, headache, stomach pain, inflammation of the kidneys, blood in the faeces and convulsions.[13]

Ironically, it was not evident that such symptoms were a bad thing; some argued that it was proof that the 'medicine' was working and that the impact of a treatment should match the virulence of the disease. The early modern period, especially before the eighteenth century, was racked by devastating contagions, such as the ill-defined plague or the newly discovered syphilis. Syphilis in particular arrived suddenly, being identified in northern Italy in 1494.[14] Present-day syphilis seems a milder form than that reported in the early modern age; this manifested itself as large and painful pustules and ulcers that covered the whole body, and skin, especially that of the nose, often wasted to the bone. Since syphilis manifested itself as a skin disease, ointments were often prescribed. The most powerful contained mercury, but the outcome was perhaps just as devastating as the disease. A German knight who contracted syphilis described in detail his symptoms, including ulcerations, bone waste, tooth loss and profuse sweating. He was in favour of a different remedy, originating from the New World: guaiacum. The opinion of physicians soon split between mercury and guaiacum, with the debate being fuelled by financial interests as well as religious and professional rivalry.

Among those who favoured mercury was Theophrastus von Hohenheim, called Paracelsus (1493–1541). His treatise on syphilis is one of the few works published during his lifetime. Paracelsus became city physician and a university lecturer in Basel in 1528, in the midst of the Lutheran Reformation; for his rebellious attitude

Alchemist physician Guy de La Brosse's Jardin du Roi, painted by Federic Scalberge in 1636.

Overleaf:
The chemical laboratory of the Jardin du Roi, in an etching by Sébastien Leclerc, *c.* 1676.

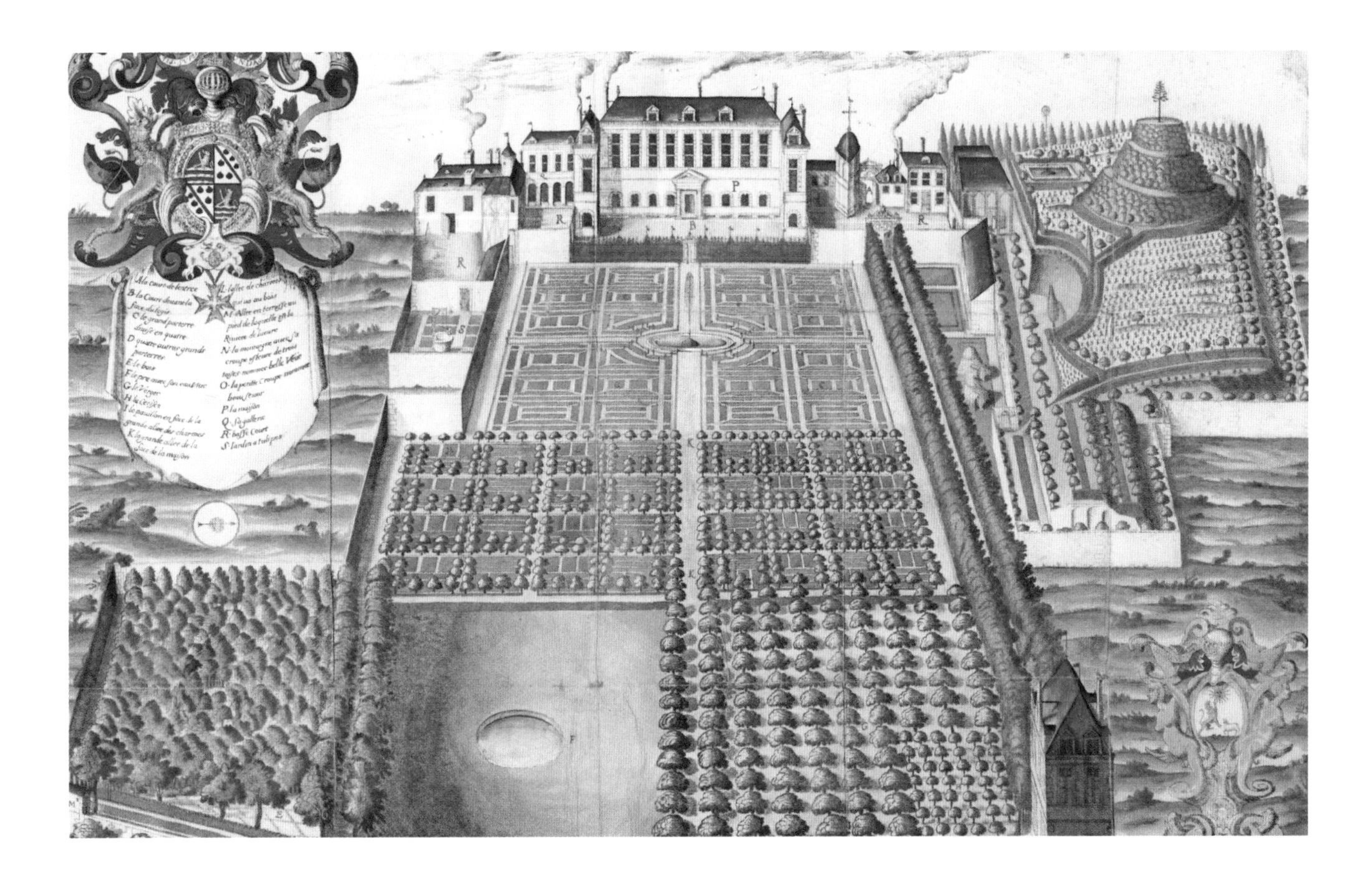

Se. le Clerc in et f.

he was in fact sometimes called 'the Luther of physicians'.[15] He was displeased with this title, considering himself superior to Luther. In a dramatic gesture befitting the period, he burned the traditional medical books in the public square. He proclaimed a new way of medicine, articulating this in one of his chief works, *Paragranum* (*c.* 1531). He incurred the wrath of some prominent Basel citizens and was thrown out of town and out of the academic world a year after his appointment. Thereafter he led an itinerant life, writing or dictating his new doctrines. He eventually settled in Salzburg. He died rather abruptly in 1541, leading to rumours that he had been poisoned. His followers saved his manuscripts, which started to be published in the 1560s, leading to a Paracelsian printing phenomenon. The demand for Paracelsus's writings led to less scrupulous followers forging many treatises published in his name. The work of sifting through the expansive corpus and distinguishing real Paracelsus works from these fakes started in the late sixteenth century and continues even today.[16]

In *Paragranum*, Paracelsus described alchemy as being one of the four pillars of medicine. He rejected the Galenic-Aristotelian tradition and argued that the new

Illustration from *Der Grossen Wundartzney* ('The Great Surgery'), by Paracelsus, published shortly after his death, 1562.

medicine had to rely on alchemical preparations. These included extracts from all realms of nature, including minerals and metals.

Furthermore, Paracelsus distinguished between medical and chrysopoeian alchemy, and rejected the latter outright, describing it as unethical. His attack on gold-making was well in line with late-medieval prohibitions, but it drew mainly on religious concerns for the pursuit of worldly wealth. Many Paracelsians subscribed to his view and wrote pamphlets against those alchemists who supposedly engaged in the practice only in the hope of becoming rich. These writings were part of a rising rhetoric against 'quackery', which newly formed societies like the Royal College of Physicians in London (founded in 1517), or older institutions such as the Italian *Protomedicato*, tried to crack down on.[17]

Empirics often favoured the use of minerals and metals like antimony and mercury and used Paracelsus's name to sell their products. As were those of the notorious and controversial Leonardo Fioravanti (1517–88) and Tommaso Bovio (1521–1609) in Italy. They were connected with the popular literature on 'secrets' and were often called 'professors of secrets'.[18] The most famous of these was Giambattista della Porta (*c.* 1535–1615), a Neapolitan who published a famous compendium called *Natural Magic*; his interests were eclectic, but he drew on Paracelsianism as well as the distillation vogue of the era. The Paracelsian link to magic was strong during this period, as Paracelsus himself had tried to bring all forms of knowledge together under the overarching framework of astronomy in *Astronomia magna* (1531), which comprised many kinds of magic as well as alchemy, astrology and mysticism.

ASPICIT ET INSPICIT

IO. BAPT. PORTAE
NEAPOLITANI
MAGIAE NATVRALIS
LIBRI XX.
Ab ipso authore expurgati, & superaucti, in quibus scientiarum Naturalium diuitiæ, & delitiæ demonstrantur

I De mirabilium rerum causis.
II De varijs animalibus gignendis.
III De nouis plantis producendis.
IIII De augenda supellectili.
V De metallorum transmutatione
VI De gemmarum adulterijs.
VII De miraculis magnetis.
VIII De portentosis medelis.
IX De mulierum cosmetice.
X De extrahendis rerum essentijs.
XI De myropœia.
XII De incendiarijs ignibus.
XIII De raris ferri temperaturis.
XIIII De miro conuiuiorum apparatu.
XV De capiendis manu feris.
XVI De inuisibilibus literarum notis.
XVII De catoptricis imaginibus.
XVIII De staticis experimentis.
XIX De pneumaticis.
XX Chaos.

CVM PRIVILEGIO.
NEAPOLI, Apud Horatium Saluianum. D. D. LXXXVIIII.

Title page from *Magiae naturalis* (*Natural Magic*) by the 'Professor of Secrets' (Giambattista della Porta), 1560.

If the empirics used Paracelsus's name as a form of legitimation, his better-educated supporters found their often-strident rhetoric inconvenient. The learned Paracelsians aspired to social acceptance and advancement, and found themselves increasingly pressed to distinguish themselves from reviled or prosecuted empirics. Attacking charlatanry was one way;[19] another was to reject the notion that Paracelsus's doctrine was a novelty. This was a path popular with those Paracelsian sympathisers who possessed or sought a medical degree. The best known was Petrus Severinus (1540–1602), a Danish physician who received his medical degree from the conservative University of Paris and then taught at the University of Copenhagen.[20] Severinus drew upon the contemporary fashion of tracing philosophical lineages (*prisca sapientia* or *prisca theologia*) to claim there was a Hermetic-alchemical tradition that went back to the legendary Hermes Trismegistus and included alchemists like Arnald of Villanova, Ramon Llull and, the last in the line, Paracelsus. He accompanied his claim with a philosophical understanding of Paracelsus that had a Neoplatonic and Hippocratic bent, in common with new tendencies in the medical thought of his time. His view was highly successful and was further articulated by physicians like Oswald Croll (1560–1608), Joseph Du Chesne (*c.* 1544–1609) and Jan Baptist Van Helmont (1579–1644).

Placing Paracelsus among a lineage of alchemists who came to be known as 'adepts'[21] gave a certain respectability to the movement. Yet it paradoxically undermined some of his doctrines, as it allowed sympathisers to look further afield in search of an elusive original philosophy. The presence of medieval manuscripts that were supposedly written by or referred to Plato, Democritus or Pythagoras seemed to support this search. The enigmatic *Emerald Tablet* of Hermes Trismegistus, apparently of Arabic origin, with its well-known adage 'as above so below', fascinated many of the philosophically minded. Connections were made between alchemy and Hippocrates, the founder of Greek medicine. The medieval *Turba philosophorum* – again of Arabic roots – pointed the way to pre-Socratic philosophy. The interest paid by Latin Geber to corpuscularian theories chimed with the belief that Democritus had been an alchemist, hence paving the way for the adoption of atomistic theories.[22]

The search for the source of alchemy led to an estrangement of alchemical philosophers from Paracelsus. The empirical appeal to him was already problematic, but Paracelsus himself became the target of accusations, none more influential than that laid forth by German theologian Thomas Erastus in 1570.[23] Erastus condemned Paracelsus for atheism, based on a treatise that we now know to be a forgery: *Philosophy to the Athenians* (*Philosophia ad Athenienses*). Claiming that all things sprang from an uncreated *mysterium magnum*, the work seemed to espouse the discredited pagan doctrine of the eternity of the world.[24] No Christian could uphold that openly, especially as many Paracelsian followers were keen to portray themselves as devout and interested in an alchemical exegesis of the Bible. This was compounded by reports of Paracelsus's unseemly behaviour and disorderly life, which horrified pious and conservative learned readers. The suspicion of 'atheism', an ill-defined term that could easily mean anything outside accepted theology, enveloped many Paracelsian supporters at the turn of the seventeenth century. In 1625, a work by the Paracelsian and formally Lutheran physician Heinrich Khunrath, *Amphitheatrum Sapientiae Aeternae* (1609), was condemned by the University of Paris for its support of the doctrine of Christ as the philosophers' stone.[25] This theme, which was not addressed by Paracelsus but was present in medieval alchemy, seemed to diminish the status of Christ and encourage materialistic views.

Appalled by *Philosophia ad Athenienses*, Khunrath's views and the radical Paracelsian currents of his time, Andreas Libavius (1555–1616) decided to separate alchemy from its Paracelsian connotations. In his *Alchemia* (1597), Libavius sought to clearly distinguish between traditional alchemy and Paracelsianism, which to him seemed a dangerous new doctrine, 'one of the devil's enterprises so that...he himself may rule at his own pleasure'.[26] Libavius's viewpoint seemed to be vindicated by the publication of the Rosicrucian manifestos in 1614 and 1615, which were not Paracelsian but drew on similar themes.[27] Libavius, an orthodox Lutheran schoolmaster, found the Rosicrucian calls for a new world order horrifying and tried to repudiate them. Medieval alchemy must have seemed a solace to him, in spite of the fact that the Rosy Cross brethren claimed medieval heritage and that the religious themes present in medieval alchemy were problematic for theologians.

Under attack from several quarters, Paracelsus's name was less and less evoked in alchemical literature and the term 'Paracelsian' acquired a negative meaning.[28] Those who stood by him in the mid-1650s were often alchemists who were not university-trained, like Johann Rudolf Glauber (1604–70) or Nicaise Lefevre (1615–69). Physicians were increasingly wary. La Brosse may have still put Paracelsus on the cover of his principal work in 1628, but he also took pains to distance himself from him. More dramatic was the change that can be detected in the writings of the Flemish physician Jan Baptist Van Helmont.[29] In 1608, a young Van Helmont had written a summary of Paracelsian doctrine (*Eisagoge*), describing Paracelsus as his guide to alchemical philosophy. Some 30 years later, he would fervently criticise Paracelsus, attacking his proficiency in healing and his philosophical tenets. What had happened? A simple answer would be the Spanish Inquisition; Van Helmont was placed under house arrest and prosecuted in the Spanish Netherlands for a treatise he published on the subject of natural magic. He was eventually released, but the encounter with the Counter-Reformation had long-lasting effects on his writing. Breaking free of Paracelsus may have seemed a good way to avoid further accusations of impiety or even heresy. At the same time, Van Helmont believed that he could write a better alchemical philosophy than Paracelsus, one that would not be suspected of atheism.

IDEA
MEDICINAE
PHILOSOPHI-
CÆ,
FUNDAMENTA CONTINENS TOTIUS DOCTRINÆ PARACELSICÆ, HIPPOCRATICÆ & Galenicæ.
AVTHORE
PETRO SEVERINO DANO
Philosopho & Medico.
16 16.
ERFURTI typis Martini Wittelii, Impensis Johannis Episcop. Bibliopolæ Erfurtens.

Title page of *Idea medicinae philosophicae*, by Petrus Severinus (Peder Sørensen), 1616.

Van Helmont's late views, as expounded in his posthumous *The Sunrise of Medicine* (*Ortus medicinae*, 1648), were coloured by the influence of medieval alchemy, as mediated by early seventeenth-century alchemical philosophers. His corpuscularian tenets drew on Geber, to which he adapted Paracelsus's own views.[30] His belief in a universal medicine had more to do with Pseudo-Lull than Paracelsus, though he stayed clear of claiming the acquisition of the philosophers' stone. Be that as it may, there are clear signs he believed in its existence. What he preferred to talk about was a universal solvent, which he usually termed the 'Alkahest', which he associated with Pseudo-Lull's philosophical mercury and saw as the key to the supreme panacea and the medicine of long life.

The Alkahest was a topic that preoccupied many alchemical philosophers who came after Van Helmont. Robert Boyle (1627–91) sought it; Glauber claimed to have it; Herman Boerhaave (1668–1738) still believed in its existence in the early eighteenth century. It seemed a more acceptable pursuit than the philosophers' stone. Yet, as manuscripts of medieval alchemy were increasingly published, it was clear that the *lapis* was the mark of special and rare alchemists, the adepts. Moreover, the *lapis* became associated with the acquisition of supreme knowledge. It was thought that those seeking adeptship should work secretly, in silence; those who achieved the *lapis* would not make their knowledge public, except to offer some guidance to aspiring alchemists, and then under pseudonyms or anonymously. One famous alchemist who was considered an adept was Michael Sendivogius (1566–1636), a Polish nobleman who published *The New Light of Alchemy* (*Novum lumen chymicum*, 1616) under a pseudonym. Sendivogius's work was highly influential in alchemical circles, both philosophically and as a presumed guide for achieving the philosophers' stone. In the mid- to late seventeenth century, would-be adepts seemed to be everywhere, but they could seldom be pinned down; they were often seen as itinerant, appearing in different locations to display their art. This usually involved a public or private demonstration of the philosophers' stone's efficacy in turning metals into gold. Such an adept is famously described by Johann Helvetius in his popular *Golden Calf* (*Vitulus aureus*, 1666). Yet these adepts were elusive; having proven their art, they often disappeared without a trace, not disclosing their secrets.

Robert Boyle was one who believed in adepts; he thought he had finally come into contact with a group of them and sought to gain membership and their secrets.[31] Boyle's group, the Asterism, was a secret society of the type that proliferated at the end of the seventeenth and into the eighteenth century. Such societies drew on the lore of the Rosy Cross and sometimes even claimed to be 'Rosicrucian'; the most famous of these was the late eighteenth-century Order of the Golden and Rosy Cross in Germany.[32] The groups were often aristocratic and even included princes and monarchs among their members. While some societies, like the one Boyle was in contact with, may have been meant as an association of adepts, most later ones were hierarchical, providing a gradual path to the level of adeptship. In this sense they seemed more closely aligned to medieval guilds than to the so-called Rosicrucian Order, and it is indeed telling that the most successful 'secret' society was that of Freemasonry, drawn at least theoretically from the guilds of masons.

The connection between these groups and the practice of the philosophers' stone became increasingly doubtful. There's no doubt that searching for the *lapis* continued, even into the late eighteenth century, but the understanding of the philosophers' stone became increasingly less physical. The emphasis fell on communicating the secret rather than attaining it by trial and error in the laboratory; in this sense, the case of Elias Ashmole (1617–92) is paradigmatic. He encountered a secretive adept called William Backhouse, who agreed to initiate him. Backhouse took on Ashmole as a disciple but, as Ashmole later claimed, did not communicate the secret of the philosophers' stone until he was on his deathbed. Ashmole strongly believed in alchemy but thought of it as an aristocratic pursuit that should involve initiation and instruction by a master. Just how much he laboured in the laboratory is unclear; his recorded activities were mainly antiquarian. He collected many volumes of manuscripts and published a famous collection of medieval English alchemical poems as *Theatrum Chemicum Britannicum* (1652).

With the likes of Ashmole, the achievement of the philosophers' stone seemed to hinge more on personal worthiness than on hard work. This emphasis on one's character and probity rather than empirical knowledge evinces a certain turn toward spiritual matters. Of course, the notion of the philosophers' stone as a 'gift of God' (*donum Dei*) was already present in medieval alchemy; Petrus Bonus of Ferrara (fl. 1323–30), discussed in Peter Forshaw's chapter 'Medieval Latin Alchemy' in this volume, also argued that there was a divine, supernatural element to alchemy. Bonus's view that the stone bestowed the gift of prophecy seems to have been shared by both Ashmole and Boyle, who believed that the philosophers' stone could allow the appearance of and communication with angels.[33]

Hippocrates, Dioscorides, Paracelsus and Theophrastus depicted on the title page of *De la nature, vertu, et utilité des plantes*, by Guy de La Brosse, 1628.

In the middle of the seventeenth century there was, however, a growing movement that drew on alchemy to affirm purely spiritual goals. Founded by the German mystic Jacob Boehme (1575–1624) and based on earlier antecedents, it used alchemical and Paracelsian themes to think about spiritual rebirth during one's own life.[34] Of course, the pursuit of self-transformation and laboratory alchemy were not mutually exclusive; Boehme supporters could be practical alchemists, as the first expounder of Behmenist doctrine in England, Theodore Graw (fl. 1600–61), or Helmontian follower Joachim Poleman (fl. 1662) were.[35] Yet the belief that the pursuit of the philosophers' stone and that of spiritual rebirth were one and the same was less common. We know that Dionysius Andreas Freher (1649–1728) believed in the identity of the pursuits, though he may not have been a practising alchemist.[36]

Isaac Newton (1643–1727) preferred the laboratory path, but envisaged it as solitary work.[37] Newton did not seem to seek the communication of the *lapis*, as Ashmole and apparently Boyle did; he did, however, share their views of its secrecy. He worked assiduously in his Cambridge laboratory on abstruse alchemical processes. As his attention shifted to working for the Royal Mint, and his energy faded, he apparently stopped, but he was still willing to invest in others' labour in the laboratory.

It has long been thought that the pursuit of the philosophers' stone ceased in the early eighteenth century, scorned by increasingly sceptical circles of chemists. It is clear that, just as the *lapis* became increasingly confined to solitary practitioners or esoteric groups, mainstream alchemists – often calling themselves chymists or chemists – sought to distinguish themselves from these. The older rhetoric, condemning the pursuit of gold-making as impious, foolish or fraudulent, was now revived, this time by those who sought to legitimise the field in academia and the learned societies. The fraudster narrative was particularly advanced by Etienne-François Geoffroy in his famous 'Some Deceptions regarding the Philosophers' Stone' ('Des supercheries concernant la pierre philosophale'), a 1722 address to the Royal Academy in France. Yet, as Lawrence Principe points out, his intention was not to deny metallic transmutation but to warn against quackery.[38] What is particularly remarkable about Geoffroy is that he made no distinction between gold-making and the philosophers' stone, a view that seemed to have prevailed during the period. The threefold purpose of the *lapis* in Pseudo-Lull was by then denied, a process that Van Helmont's notion of universal medicine as different from the philosophers' stone may have had a lot to do with.[39]

ORIENS
VIDEAMINI
1604
COLLEGIVM FRATERNITATIS
16
18
S S
FA MA
VIDEAMINI
adfratres
ad I.D.
Fratri
Nostro T.S.
SEPTENTRIO
IVVA PATER
CAVETE!
IESVS NOBIS OMNIA
AGNOSCO
C.R.F.
NOTA
VENITE DIGNI
MOVE AMVR
PVTEVS
OPINIONVM
OCCIDENS

The impossibility of metallic transmutation was not, however, established by either Geoffroy or other French academicians. In fact, many of the chemists at the Royal Academy continued to pursue the goal of transmutation in private.[40] Such activities may prove to be the norm rather than the exception in other countries and settings as well, though much more research is needed in this direction. Certainly, the chemist Carl F. Wenzel (1740–93), a believer in metallic transmutation, was awarded the medal of the Royal Danish Academy of Sciences in 1776, at the recommendation of Christian Gottlieb Krantzenstein (1723–95), Professor of Experimental Physics and Medicine at the University of Copenhagen, and another believer.[41] Yet the Lavoisierian 'Chemical Revolution' put such beliefs and pursuits beyond the pale of mainstream chemistry, as Lavoisier taught that elements were unique and immutable.

During the nineteenth century, chemical transmutation was an unacceptable doctrine and was frequently derided in the histories of chemistry written in the period. It was only at the turn of the twentieth century that the pioneers of atomic science Sir William Ramsay (1852–1916) and Frederick Soddy (1877–1956), under the influence of Glasgow Professor of Chemistry John Ferguson (1837–1916), would revive the framework of transmutation to explain radioactive decay and change.

By then, however, alchemy had become co-opted in the occultist framework of the late nineteenth century, which saw it as part of a larger group dominated by magic. This view conceived of alchemy as a form of mysticism, closely linked to religion and spiritual beliefs. Just as transmutation of matter seemed an impossibility, transmutation of the soul appeared to some as the real goal of alchemy. Yet the fate of this thesis will be discussed in more detail in the chapter on alchemy and academia.

The supposed temple of the 'Rosy Cross' secret society. Illustrated in *Speculum sophicum rhodo-stauroticum* ('The Mirror of the Wisdom of the Rosy Cross'), by Daniel Mögling, 1618.

ALCHEMY AND CHEMISTRY

HJALMAR FORS

The Medical Alchemist. Oil painting by Franz Christoph Janneck, 1800s.

Ever since the Middle Ages and up until the beginning of the eighteenth century, alchemy and chemistry were essentially synonymous. Perceived to designate the same thing, the terms were used almost interchangeably.[1] Then, in the 50 years leading up to about the mid-eighteenth century, something happened. A group of practitioners began to shed the use of the Arabic definite article *al-*. In publications, orations, pamphlets, textbooks, etc, they made it clear that this decision represented something new. They were chemists now and were not to be confused with those other practitioners, the misguided and confused alchemists. Thus, the 'chemists' shed their association both with the past and with a certain practice which had started to have a strong whiff of fraud about it: *chrysopoeia* – that is, the making of gold through artificial means.

This process has often been described as a journey from darkness to light, from confusion to clarity. Only recently has it been observed that this narrative is a rhetorical construct carefully crafted during the eighteenth century. Its purpose: to rebrand a particular branch of practical laboratory alchemy as chemistry to improve its public standing and reputation. But just as the jazz musician instructed to *fake it 'til you make it* finally *does it*, this act of renaming accelerated and facilitated a process which would deeply transform chemistry over the course of a century. In the present chapter, I look at some important aspects of the historical process through which alchemy and chemistry went their separate ways. As we will see, the separation was a complex affair. To understand the break-up of this troublesome relationship, it is necessary to consider both similarities and dissimilarities, or to put it differently, both continuities and disruptions. Only in this way can we arrive at a balanced, albeit preliminary, view of how alchemy and chemistry became two distinct knowledge areas, pursued by, for the most part, distinctly separate communities of practice.

A historical field rife with misunderstandings

Let us begin by looking at what alchemists and chemists *do*. At the very deepest level it is possible to identify a continuous, foundational research aim which unites both traditions. This is the quest to make visible the imperceptible inner structures and forces of matter and to understand their relationship both to each other and to the phenomenal world, the world of the senses. The origins of this chemico-alchemical quest to comprehend that which underlies the surface of things can be traced to late antiquity. There are also other common traits which show up in the earliest of records and continue to exist in an unbroken methodological tradition to this day. Both alchemists and chemists engage in practical hands-on work with the transformation of solids, fluids and vapours. They conduct it in a designated workspace, *the laboratory*, and use equipment such as glass and clay or porcelain containers of many different kinds, shapes and sizes; equipment for separation including sieves and filters, stills and modern-day ultracentrifuges; as well as cooling and heating devices such as water baths, coal-fired furnaces and electric hotplates. That there is a long, and strong, continuity which unites alchemy and chemistry is obvious when one considers methods, equipment, materials and the organisation of the workspace. I have already mentioned their shared theoretical assumption: that it is possible to

The Physician Herman Boerhaave, Professor at the University of Leiden. Oil on canvas by Cornelis Troost, 1735.

unveil nature's secrets by manipulating matter in the controlled setting of the laboratory.[2]

It is important to emphasise these similarities and this continuity, because as soon as we move on to discuss historically situated chemical and alchemical practice, the diversity is overwhelming. History presents us with a vast number of chemico-alchemical projects. *Chrysopoeia* is just one of them. Others include the production of powerful medical remedies; the improvement of mining, manufacture and industry through innovation; to obtain personal wealth; to bring about revolution by overwhelming the world economy with alchemical gold; to learn about God's creation to praise Him better; to discover previously unknown substances; and so on. Many such activities, for example the improvement of mining works and the production of novel medicines, were part of the alchemical enterprise in the seventeenth century and were simply brought wholesale into chemistry in the eighteenth.[3] Alchemy and chemistry's shared heritage, premises and methodology were in fact no secret. And as we will see, eighteenth-century chemists possessed neither theories nor methods, which clearly set them apart, as a *community of practice*, from the numerous alchemists who were still around throughout the period. From the laboratory point of view, the eighteenth-century switchover from alchemy to chemistry largely entailed two things: (1) the abandonment of a specific research programme, that is, *chrysopoeia*. (2) The gradual abandonment of certain theories about matter, and the theoretical choice not to pursue a search for grand synthetic chemical theories for a period of about 20 years (*c.* 1760–80), during which time it was also considered improper to present such theories to the public.

What, then, prompted chemists to present this as a radical change? Answers provided by adherents of traditional historiography and by social historians and sociologists of science differ much on this point. A summary of my view, as a social historian of science, may be as follows: chemists were not only driven by scientific goals but also sought to enhance their prestige and obtain stable sources of funding. Their primary way to achieve these and other similar goals was through institutionalisation; that is, to join up with or connect to large organisations such as early modern states, universities and scientific academies. This was a long and arduous process: it is not whimsical to propose that a chemist is an alchemist with a university degree and secure job opportunities at universities or in industry. In the eighteenth century, chairs of chemistry would be established at major universities, and chemistry would become integrated into curricula. Chemists would to an increasing degree gain positions as specialists in state administrations. They would also stake out a place for chemistry in the public sphere, presenting their science as an activity capable of improving innovation and industrial production. Thus, chemistry sought to reinvent itself as *nothing but* a rational and useful science concerned with practical improvement. This made it possible for chemists to embrace and contribute to the utilitarian and Enlightenment ideals which were prevalent throughout the century.[4]

The importance of chemistry's successful institutionalisation must be considered in the light of the fact that alchemy, since the earliest of times, had a weak or non-existent formal institutional foundation. Often supported by royalty and other affluent

patrons, it flourished mainly outside academies, universities and similar places. Transmitted as it was in a similar way to artisanal knowledge, there were no formal guilds with apprenticeships, degrees or diplomas. Consequently, anyone could proclaim himself, *or herself*, an alchemist. A problem with this state of affairs was that alchemy became a venue of charlatanry. But on the positive side, from the perspective of many practitioners, this permitted alchemy to become a means of social advancement for enterprising and able individuals.[5] Equally important: up until very recently, formal institutions almost always were male, homosocial environments which excluded women. This was never the case for alchemy. Although the great majority of practitioners were male, there have always been women alchemical practitioners, from the mythical Maria the Jewess to the present day. When alchemy became chemistry, it was gendered as a male enterprise. As with other sciences and early modern knowledges, as soon as alchemy entered universities and state administrations women practitioners were excluded.

Bernard de Fontenelle. Oil on canvas by Louis Galloche, 1723.

Opposite: Allegorical engraving by Matthaeus Merian in *Atalanta Fugiens* by Michael Maier, 1617. The woman is engaging in alchemical 'cooking' with two fish symbolising mercury and sulphur.

Furthermore, towards the end of the seventeenth century, alchemy's reputation had begun to deteriorate. For this reason, chemists striving to occupy a new and more prominent social position gained several benefits from slashing connections with the past. But they went further: they denied all the continuities outlined above and initiated a process of 'othering' most historic predecessors in the field. In this way alchemy was turned into a useful foil. It was portrayed as the *other* against which victorious chemistry could define itself as useful, *Enlightened* and progressive. This was, however, to a large part based on active concealment of facts. The influential Bernard de Fontenelle (1657–1757) is a good example. A long-time secretary of the French Academy of Sciences, Fontenelle went to great pains to hide the existence of 'alchemical' research at the academy, going so far as to present outright lies in his eulogies of deceased members.[6]

Let us now try to produce a summary of the position of an imagined adherent of the traditional historiography of eighteenth-century chemistry. The story would go something like this. By the turn of the century, alchemy (that is, *chrysopoeia*) had exhausted itself. Already considered fraudulent by many in the seventeenth century, it did not survive Enlightenment scepticism. In short: most scientifically minded people regarded it as a fake, or at least deeply flawed, science. And rightly so, for chemistry's continuous advances proved that the basic premise of alchemy – the belief that it was possible to transmute one metal into another – was incorrect. Then followed a long period of theoretical confusion and polyphony, which was eventually resolved by Antoine Laurent Lavoisier's (1743-94) revolutionary innovations.[7]

The traditional view is problematic in numerous ways but nevertheless contains an important insight. Although alchemy continued to thrive throughout the eighteenth century, the chrysopoetic research project was increasingly shut down, or went underground, in most well-funded, high-status settings. This was not exclusively due to a social process of shaming, scepticism or withdrawal of funding. Many among the younger generation were hesitant to take it up, and other projects beckoned. This makes a certain sense from an experimentalist's point of view. The final decades of the seventeenth century had been an alchemical golden age. Practitioners included such luminaries as Robert Boyle, Isaac Newton and Wilhelm Homberg. These men were brilliant and relentless thinkers and experimentalists. Yet they ultimately failed at gold-making, despite access both to excellent resources and networks, and hard and continuous laboratory work, often lasting a lifetime. Even those who were deeply sympathetic to the chrysopoetic project took notice. Urban Hiärne was by far Sweden's most prominent chrysopoetic alchemist in the decades around 1700. In 1757 one of his sons, the mining official Erland Fredrik Hiärne, summarised his father's lifelong work in this way:

> If it had been possible to win the stone of the philosophers through labour, it would never have escaped my father, may he rest in peace, in his younger years; for the sharpness and power of his mind was far reaching, and his knowledge of nature deep.[8]

Thus, we should not be surprised that many able chemical experimentalists turned their laboratory efforts to other projects. But we should also not take their explanations as to why they did so at face value. And even in the eighteenth century, there were some chemists who wanted to dampen the highly strung Enlightenment rhetoric. Johan Gottschalk Wallerius, a prominent chemist who nowadays is mostly known for proposing the distinction between *pure* and *applied* chemistry, is one example. Admonishing Torbern Bergman, his younger colleague and successor as chair of chemistry in Uppsala, he stated:

> Though alchemical trials have failed many, who are less knowledgeable about the properties of metals, chemistry should, nonetheless, mostly thank the alchemists for their significant discoveries. For this reason, these [discoveries] should not be called fancies, without any exceptions. It would be better to distinguish the speculations of sensible alchemists from the guesswork of the ignorant, and to [properly] distinguish alchemists from frauds.[9]

Bergman's viewpoint – that alchemy was a futile enterprise pursued mostly by ignorants and frauds – would be dominant during the remainder of the eighteenth century, and throughout the nineteenth and most of the twentieth. But Wallerius's position is in fact rather similar to that held by contemporary historians of alchemy. The reason for this is mainly due to the development of what is called the *new historiography of alchemy*.

The new historiography of alchemy has established an important and highly pertinent fact also observed by Wallerius: *approached as an experimental laboratory practice, quite a lot of alchemy works*. It is possible to reproduce some of the processes described in alchemical literature. For generations, an overwhelming number of historians and chemists proposed that alchemical illustrations, descriptions of procedures and notebook annotations were mostly unconnected to laboratory reality. Instead, they were said to be the product of dreams, or of over-imaginative or unstable minds. We now know that this is not true. Although it requires much skill and patience, it is possible to follow alchemical texts and laboratory journals to establish which substances were used and how they were used. It is, furthermore, possible to conduct replication experiments and obtain at least some of the described results. From this follows that some, although certainly not all, early modern alchemists possessed a complex and precise understanding of the transformations of matter, articulated this knowledge theoretically and tested their theories using highly specialised skills and methods. That is, they possessed a skillset which we today would describe as chemical. Historians' realisation of this is a rather recent development. Beginning in the early 1990s, it has been powerfully argued by, in particular, two scholars: William R. Newman and Lawrence M. Principe.[10]

Thanks to scholarship conducted during the last three decades or so, historians

Urban Hjärne. Ink and wash on paper by Elias Brenner, *c.* 1700.
Hjärne was Sweden's most prominent chrysopoetic alchemist.

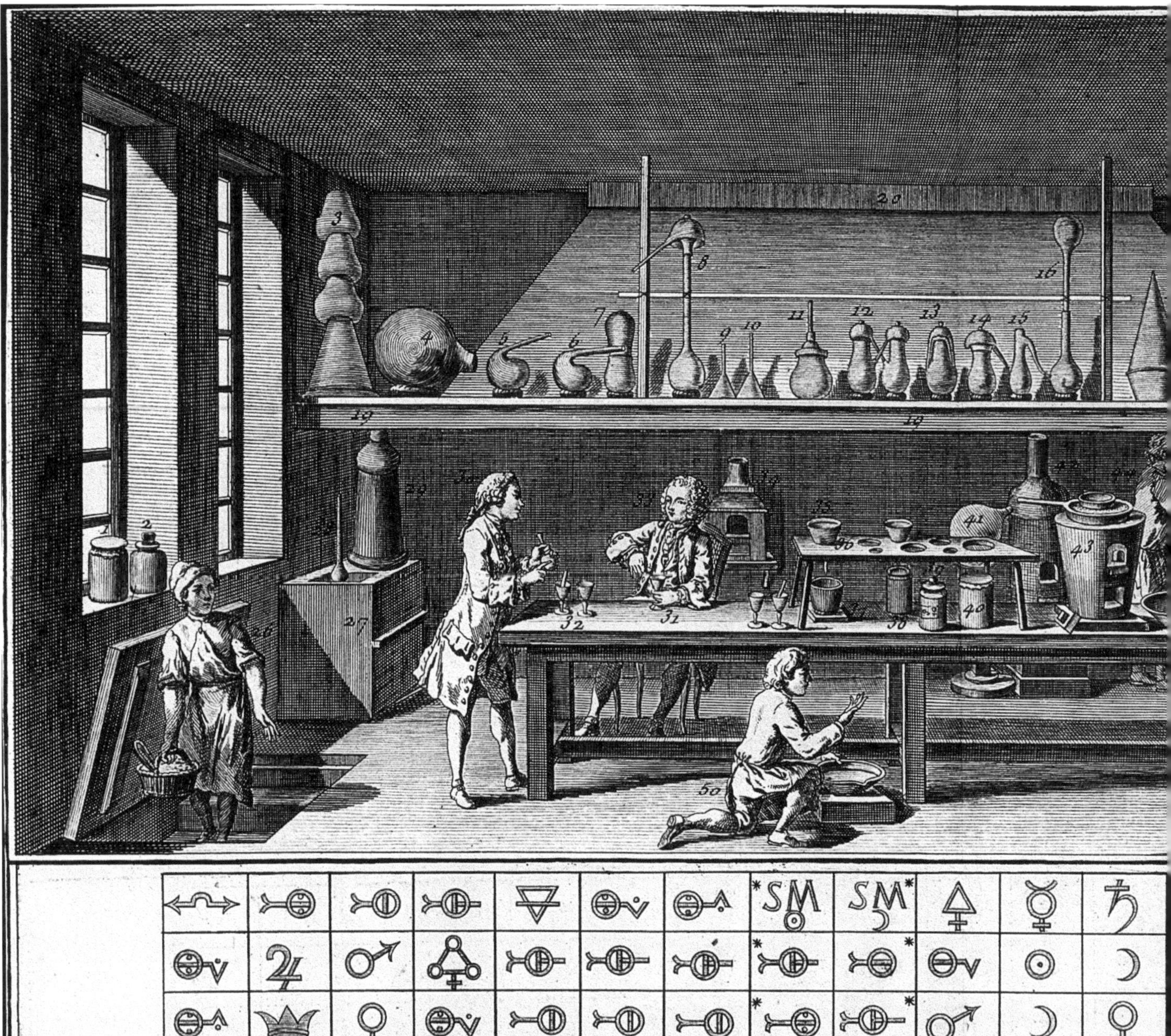

3
4
5
6
7
8
9
10
11
12
13
14
15
16
20
19
19
1
2
27
31
32
33
35
36
40
41
43
50
SM
SM
SM

Engraving of an 18th-century French chemical laboratory. Beneath the laboratory is a table showing elective affinities, or displacement reactions of different chemical substances. By Prévost after L. J. Goussier, 1763.

have vastly expanded our knowledge about early modern alchemy and its relationship to chemistry, as well as other sciences. The range of studies is impressive. Just to mention four, published roughly ten years apart: Betty Jo Teeter Dobbs's *The Janus Faces of Genius* (1991) established beyond doubt Isaac Newton's deep preoccupation with alchemy. Newman and Principe's *Alchemy Tried in the Fire* (2002) revealed the deep influence of Jan Baptist Van Helmont on subsequent generations of chemists and alchemists. John C. Powers's *Inventing Chemistry* (2012) was a careful investigation of the work of Herman Boerhaave. Jennifer M. Rampling's *The Experimental Fire* (2022) comprises an overview of English alchemy from medieval times to the beginning of the eighteenth century.[11] However, it should be noted that historians' appreciation of alchemy has not progressed in a linear fashion. As discussed in Georgiana Hedesan's contribution to this volume, the first half of the 20th century saw a nascent historiographical tradition concerned with understanding and explaining the pursuit of alchemy in its historical context. The journal *Ambix* was, and still is, the most important venue for such scholarship. In this context should also be mentioned Sten Lindroth's *Paracelsismen i Sverige: till 1600-talets mitt* ('Paracelsianism in Sweden: Until the Middle of the Seventeenth Century'). Published in 1943, it applied to alchemy the hermeneutic understanding foundational to Swedish *idéhistoria*. Lindroth's book was a remarkably astute attempt to comprehend alchemy as a historical phenomenon. Nevertheless, it remains an unsung classic in the genre, as it is neither available in English nor primarily focused on the usual canon of authors and historical figures. International scholarship of the 1950s and early 1960s would not be influenced by it. Snide remarks by influential historians of science such as George Sarton and Herbert Butterfield ensured that the historical study of alchemy fell into disrepute. Indeed, presumably inspired by Anglo-American scholarship, Lindroth too would show much less sensitivity to, and understanding of, alchemy in his later works.[12] Only later in the 1960s and in the 1970s would Frances Yates and Betty Jo Teeter Dobbs champion hermeneutic methods, by seeking to understand, rather than denounce and pass judgement on, the actors and practices they studied.[13]

The picture is far from complete. But with several caveats, it would seem that laboratory alchemy and laboratory chemistry should be regarded as a continuous enterprise: a single tradition. From this point of view, the much-touted 'scientific revolution in chemistry' was not a revolution at all. Rather, it was a gradual shift of research programme and research interests, which played out over a period of 100 years or so.

Comprehending the material world

I have emphasised how strongly laboratory practice joins alchemy and chemistry together. But theory was also necessary. This was true whether an early modern or eighteenth-century alchemist or chemist attempted to manufacture the Philosophers' Stone or tried to discover new medicines, dyes, alloys, metals, solvents, methods to preserve the hulls of ships from shipworm, etc. The practical work of both alchemists and chemists needed to be steered by concepts such as a theory about the nature of matter. In the following I will give a very brief overview of historic theories of matter current in the Western world, from antiquity onwards. My purpose here is not to give a full account of how these theories functioned as tools to comprehend the transformations of matter. Rather, it is to point to what eighteenth-century chemists felt that they needed to move on from. Thus, this framing provides the perspective needed to comprehend what it was about eighteenth-century chemistry that *really* can be regarded as a radical break with earlier traditions, and which formed the humble beginnings of modern chemistry.

The key historical process can be summarised as follows: at the beginning of the eighteenth century, chemists proceeded from theories of matter which assumed that a small number of elements, principles or other simple, homogeneous substrates made up all things. As the century progressed, these were increasingly regarded as a priori constructs; that is, unfounded assumptions. At the end of the century, chemists had abandoned them and had begun to develop the modern notion of chemical elements. Exactly which steps in this long process were important, and how, has long been a matter of debate. I will return to that debate towards the end of the chapter.[14]

Ever since antiquity and well into the eighteenth century, the most well-known and well-used scheme to comprehend the transformation of matter had been suggested by Empedocles and developed by Aristotle. It was a theory which made intuitively good sense. The objects which were present in our world were divided into four categories: solids, fluids, vapours (gaseous matter) and, finally, subtle and penetrating matter, such as the flames of a fire. This was of course the theory of earth, water, air and fire, the four elements from which all the 'stuff' – that is, individual objects – in the world were made. With this theory came an adjunct theory which stated that each element was characterised by two primary qualities. Hence *earth* was cold and dry, *water* was wet and cold, *air* was hot and wet and *fire* was hot and dry. Matter was transformed when these four elements and their qualities interacted and mixed with each other. Today we would say that what Empedocles and Aristotle described were states of matter and not the 'stuff' objects are made from. We have a different notion of what the term *composition* means. It is important to grasp this distinction. Without it, it is not possible to understand the history of chemistry.

When late moderns speak about what material objects consist of, we usually take this to mean their *chemical composition*. Thus an iron nail consists of iron; a steel nail consists of iron and carbon, while a stainless steel nail contains chromium as well. Most late moderns (leaving the few of us versed in particle physics aside) would say that chemical composition – the proportion and arrangement of chemical elements – confers the perceptible properties of an object. This basic framework for thinking about material objects was developed by eighteenth-century chemists. It began in a reconceptualisation of the metals.

A prominent – many would say defining – strain of research within alchemy was concerned with the transformation of metals. The Aristotelian theory stipulated that a metal, just like every other substance on earth, was composed from the four elements. Hence, the possibility of transmutation, i.e. that one metal could be changed into another, followed from this theoretical framework. The classical theory, however, was not particularly well suited to describing the transformative processes taking place at mines, smelting works and metallurgical laboratories. For this reason, alchemists – whom we here may consider metallurgical specialists – developed an adjunct theory. This was drawn from a specific passage in Aristotle but elaborated in the corpus of texts attributed to the early medieval Arab alchemist Jābir ibn Hayyān. The *mercury-sulphur theory of the metals* stated that metals were not directly produced from the four elements but by two intermediate substances composed from the four. These intermediate substances were sulfur and mercury. They were *absolutely not* identical to the chemical elements we now designate by these names. Rather, they were considered exhalations which emerged from the bowels of the earth. Mercury was steamy and wet, sulfur smoky and dry. Metals were created when these two combined. The type of metal produced was defined by the relative proportions of mercury and sulfur. It is important to emphasise that this theory had a solid foundation both in observation and experiment. It was also remarkably long-lived, insofar as it undergirded alchemical experimentation from the eighth century to at least the eighteenth.[15]

Jābir ibn Hayyān's mercury sulphur theory depicted in *Viridarium Chymicum* ('The Chemical Garden'), by Daniel Stoltzius von Stoltzenberg, 1624.

There were, however, other theories. One such theory was conceived by Theophrastus Bombastus von Hohenheim, aka *Paracelsus*. According to Paracelsus – ever the iconoclast – the four classical elements were useless. Instead, he proposed that three chemical principles made up all material objects. These were the Tria Prima: *sulfur*, a fiery and vaporous principle; *mercury*, a fluid principle; and *salt*, a principle designating solids. The change from the classical elements may look slight, and at a cursory glance one may think that Paracelsus simply combined *air* and *fire* into a single principle and claimed credit as a revolutionary innovator. In fact, the Tria Prima were something rather different from the classical elements. They were introduced to describe laboratory *products* – that is, the products obtained when decomposing natural objects in the laboratory. The theory was strongly connected to the widespread adoption of a new type of laboratory apparatus: the still. Introduced into Christian Europe from the Arab world in the Middle Ages, it was a new and powerful instrument. By the time of Paracelsus, European alchemists and other specialist distillers had begun to distil almost everything which came into their hands. Stills permitted the decomposition of a wide range of matter, whether they were sourced from the animal, plant or mineral realm. The Tria Prima were used to describe the products of distillation. The vapours which emerged, perceptible but lost, were called *Sulfur*. The fluid part, that is, the main product of distillation, was called *Mercury*. The dross, which remained at the bottom of the still when the operation was completed was called *Salt*.

But Paracelsus's Tria Prima was not the last great theory of this type. At the start of the seventeenth century Jan Baptista Van Helmont formulated an essentially monistic (in some later schemes, dualistic) model. According to this theory, matter consisted ultimately of water (or of water and earth). Through various processes, these combined to form *mixtures* of even greater and greater complexity. These mixtures were given names such as secondary mixts, supercompounds, moleculae, etc.

As an example, the influential seventeenth-century chemist Johann Joachim Becher held that all matter was ultimately composed from earth and water, while the Paracelsian principles, salt, sulfur and mercury, were composite bodies.[16]

Using theories and methods such as these, sixteenth- and seventeenth-century investigators unveiled many secrets of nature. They accumulated a precise and impressive knowledge about matter and its transformations and learned to refine or produce substances such as phosphorus, bismuth, antimony and true porcelain. But in hindsight, it is also possible to see that they proceeded from premises that eighteenth-century chemists would work hard to replace. In particular: the premise that there existed a small and finite set of underlying elements or principles, from which all material objects were created. Some laboratory workers in the late seventeenth century and an increasing number in the eighteenth began to perceive a disjunction between inherited theories of matter and laboratory practice. Thus it was that Robert Boyle famously suggested that some, albeit certainly not all, elements and principles were a priori constructs, the existence of which were not possible to prove through laboratory labours. Older histories of chemistry take the publication of Boyle's *The Sceptical Chymist* (1661) to be a watershed moment.[17] They conveniently omit to mention that, well into the eighteenth century, influential chemists such as Georg Ernst Stahl and Herman Boerhaave postulated underlying, imperceptible principles.[18] In fact, one of the chemical principles would survive to the very end of the eighteenth century. It was the sulphuric – fire – principle, reconceptualised as *phlogiston* by Stahl.[19] Phlogiston is still occasionally derided in introductory lectures in chemistry. Taken out of context, it seems ridiculous when described as a material object with negative weight, which was imagined to fulfil the functions of oxygen in combustion. But let us recall the Aristotelian element of fire, with its inherent tendency to rise upwards. Phlogiston is not so ridiculous when perceived in the light of an older theory which did not assign weight to all types of matter. And the phlogiston concept was not stupid at all. To the contrary, it remained in widespread use precisely because it could explain what happened during combustion, albeit in a way which we now consider to be incorrect. This said, by the middle of the eighteenth century phlogiston was beginning to be perceived as an anomaly. This was not because alternative explanations for combustion existed but because the basic conceptual structure provided by a small group of principles was not conductive to the search for new substances. As F.L. Holmes has observed: 'The conceptual structure within which Stahl worked and wrote was less conductive to a search for new acids, alkalis, metals, and earths, or their combinations, than to a search for the underlying composition of those already known.'[20]

As we will shortly see, Holmes's observation almost paraphrases certain mid-century chemists. And Boyle's questions, once posed, continued to nag: if neither principles nor the classical elements existed, exactly what was below the surface of perceptible phenomena? Eventually, around the middle of the century, the issue was temporarily resolved by simply refusing to talk about it. In an article which has received too little attention, Lissa Roberts proposed three factors which together constituted the identity of a chemist:

Carl Linnaeus's Mineral Realm. Illustrated in a later edition of his *Systema Naturae*, 1767–70.

Tab. I.
Linnæi Syst. Nat. T. III.

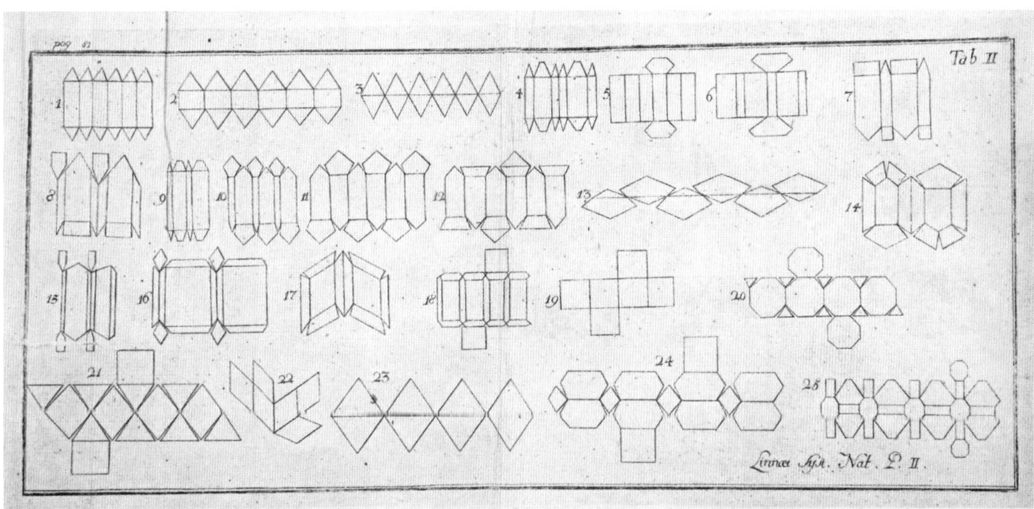
Tab II
Linnæi Syst. Nat. T. II.

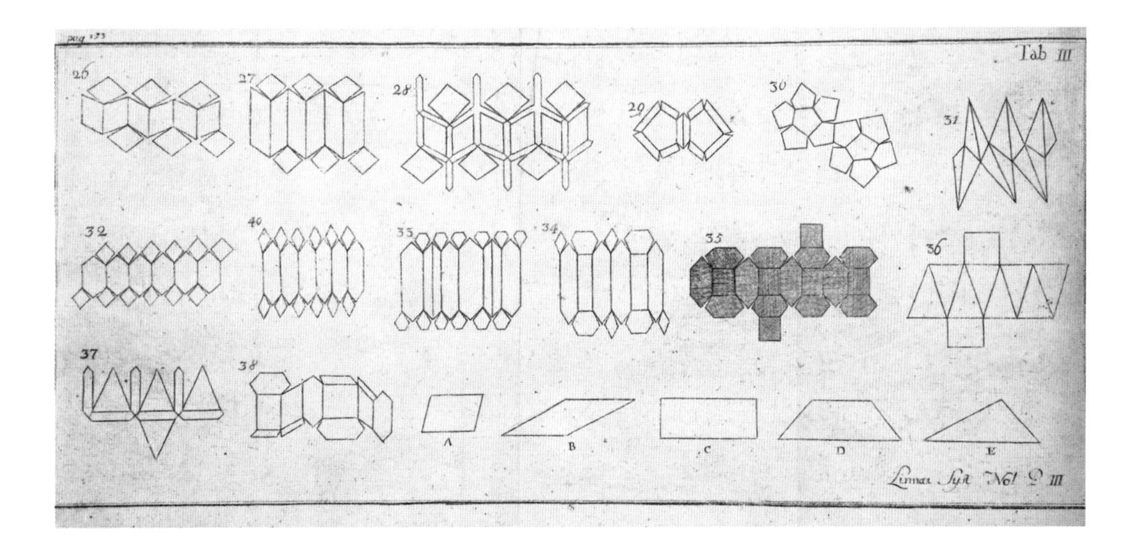
Tab III
A
B
C
D
E
Linnæi Syst. Nat. T. III

> Perceived manipulative abilities and technical acumen in laboratory settings; acceptance and use of polite, theoretically neutral discourse (that is, in the sense of not asserting an overall system) for communication in general and experimental reporting in particular and success in situating oneself in a recognized network of active participants.[21]

For lack of a general theory of matter which they did not share with alchemists, chemists simply chose to abstain from theoretical discussion for a few decades. It was an elegant solution, as it identified anyone wading into a discussion quoting Paracelsus or Jābir as impolite, overly invested in theory and not part of the scene.

Metals, earths and airs as species

In the beginning, God created heaven and earth. He also created animals and plants, each according to its kind. Now, what if He also created each metal and earth according to its kind? I know of no eighteenth-century chemist who drew this inference. But it is likely that Carl Linnaeus did. Styled by many as the 'second Adam', Linnaeus's reform of natural history took Europe by storm, beginning with the first edition of his *Systema Naturae* (1735). There, clearly delineated and in full parallel with the Animal Realm and the Plant Realm, we find the Mineral Realm, complete with families, species and varieties. Linnaean mineralogy never gained influence comparable to that of his work in zoology and botany. Nevertheless, his systematic methods deeply influenced both chemistry and mineralogy. In the following, I will give a brief overview of how a succession of eighteenth-century chemists discovered and sought to define a group of new metals. The story that follows is not found in standard textbooks on the history of chemistry, for I make a novel case: that early eighteenth-century chemical investigation resulted in the conception of metals as species. This constituted the foundation for an enormous change in perception of the material world, which preceded what we usually consider the 'chemical revolution' by several decades. A fully articulated conception of metals as species had been articulated by mineralogists and chemists in Sweden and Germany since the 1730s.

Aristotle, and Jābirian alchemy, held that metals were created through the combining of subterranean exhalations. In the sixteenth century, authors such as Vannoccio Biringuccio, Paracelsus, Georgius Agricola and Johannes Mathesius began turning the knowledge of miners and smelters into a topic of scholarly investigation. Judging from the information we can garner from these authors, it seems that early modern European miners and smelters held that metals were generated through a process of generation in which a fluid called *ghur*, or *gur*, was the key component. When discussed in a Jābirian framework, ghur was considered to be created from sulfur and mercury. It was found in close proximity to metalliferous ores and was the agent which transmuted metals.[22] Although this theoretical framework sounds straightforward, the ores found in a mine had to be thoroughly worked before they could be considered finished products. Many of these processes, in particular roasting and smelting, transformed the ore. Today, we consider this process refinement: a small amount of metal is disassociated from a matrix of less valuable minerals and rocks. To early modern miners, smelters, assayers and alchemists/chemists, however, it was an open question as to whether these processes constituted refinement or a perfection of nature's work, i.e. transmutation. Who was to say that metals were not created through these processes? Especially in central Europe, alchemists would be hired to improve practical processes in mining works. As the eighteenth century progressed, the role of these alchemists would seamlessly be taken over by mining officials who were proficient in chemistry, and sceptical of alchemy, as outlined above.[23]

These chemist-mineralogists, or mineralogical chemists, were highly proficient in the identification of minerals. For them, metal intended for the market was the end product of the chemical/metallurgical process of transformation. As we have seen, earlier theories had argued the existence of increasingly complex corpuscles built from simpler elements. This mode of reasoning built on the theoretical framework established by Aristotle, which also opened the possibility of metallic transmutation.[24] From the 1730s onwards, mining officials and chemists Georg Brandt (1694–1768) and Axel Fredrik Cronstedt (1722–65) and their followers abandoned this mode of reasoning entirely. Thus, they regarded as unimportant, or at least stayed clear of, a topic which had engaged chemical theorists such as Robert Boyle, Johann Joachim Becher, Georg Ernst Stahl and many others. The search for the underlying principles of metals was not in their interest. Working as assayers, their task was to investigate minute differences between metalliferous rocks by means of chemical analysis. They chose Linnaean natural history as their guiding theory. With its concept of stable species presented in tabular form, and emphasis on careful attention to minute differences between species, it worked very well as a theoretical framework for assaying and mineral analysis.[25]

But the concept of metals as stable species also had a long pre-history. As established by William Newman, it was already known by certain medieval alchemists that some metals, in particular gold, were virtually indestructible. They could always be recovered intact by the scrupulous laboratory worker, regardless of how strenuously he subjected them to attempts at chemical decomposition.[26] Thus, although the chemist-mineralogists did not yet perceive metals as chemical elements in the modern sense, they were methodologically inclined to accept that metals were the end product of chemical investigation, just as they were the end product of the mining endeavour. This line of reasoning was especially prevalent at the Swedish Bureau of Mines (*Bergskollegium*). This institution, which also was a significant scientific environment, would see a remarkable series of chemists, who spearheaded the discoveries of new metals. The first to make such a claim was the head of the Swedish Bureau of Mines chemical laboratory: Georg Brandt. His paper 'Dissertatio de semi-metallis' (1733) is generally considered to present the discovery of cobalt. It should be noted, however, that cobalt minerals of various types were well known to several authors prior to Brandt. His innovation was primarily conceptual. He argued that he could prove beyond doubt that cobalt ore contained a hitherto unknown pure substance, which he regarded as a metal indestructible through chemical means. Thus, by clearly and unambiguously laying claim to the

ACTA LITERARIA ET SCIENTIARUM SVECIÆ, Anni MDCCXXXV.

I.

GEORGII BRANDT,

Med. D. Cenſoris rei Metallicæ, & Directoris Laboratorii Chemici Stockholmenſis

Diſſertatio De

Semi - Metallis.

PEr ſemi - metallum intelligitur corpus quodcunque, quod formam metalli habet, & colore ac pondere ad metalla proxime accedit; ſub malleo vero perfecte extendi non poteſt.

Forma metalli eſt figura reguli, ſcilicet in ſuperiori parte, gibba & globoſa, qua ejusmodi corpora igne fuſa, ab omnibus aliis rebus igne liquefactis manifeſte diſtingvuntur.

Ut metalla hactenus cognita ſex tantum numero ſunt, ita huc usque non plura novi ſemi - metalla, nempe argentum vivum, regulum antimonii, bismuthum vel marcaſitam, regulum cobalti, regulum ar-

A ſenici

discovery of a new metal, Brandt made a 'chemical discovery' in a very modern sense. The significance of this has rarely been noted. Proceeding by way of careful chemical analysis, Brandt decomposed the minerals he investigated into their component parts, using the laboratory means at his disposal. Having done this, he arrived at a previously unknown solid, i.e. unblended metal. Finally, he gave that metal a name, *cobalt*, and issued a challenge to all who presumed to not accept his claim: decompose cobalt further and publish your results, or accept cobalt as a new metal. With the publication of his paper, Brandt outlined an investigative trajectory which is followed by chemists to this day. From Brandt's time onwards, any chemist who laid claim to the discovery of a new substance would need to argue as he did. One should pay special heed to the influence of natural history and of Linnaeus on Brandt and his followers. Linnaean natural history, too, proceeded through careful distinction and observation of minute differences. And description and naming were important paths to gaining systematic overview.

'Dissertation on Semi-Metals': Georg Brandt's account of his discovery of Cobalt, in *Acta literaria et scientiarum Sveciae* ('Swedish literary and scientific journals'), 1735.

One of Brandt's most prominent students was Axel Fredrik Cronstedt. The discoverer of nickel (1751), Cronstedt created the first comprehensive mineralogical system which used chemical composition as its guiding principle, and advocated a programme of discovery of new metals. The following passage from Cronstedt's work illustrates clearly that this was fully articulated by 1758:

> There is no danger attending the increasing [of] the number of metals. Astrological influences are now in no repute among the learned, and we have already more metals than planets within our solar system. It would perhaps be more useful to discover more of these metals, than idly to lose our time in repeating the numberless experiments which have been made, in order to discover the constituent parts of the metals already known. In this persuasion, I have avoided to mention any hypotheses about the principles of the metals, the processes of mercurification, and other things of the like nature, with which, to tell the truth, I have never troubled myself.[27]

Looking at Cronstedt's statement, it may even be interpreted as a rebuttal of the Stahlian concept of phlogiston. It is significant that the term and the theory of phlogiston had no place in his thinking or oeuvre. Furthermore, Antoine Laurent Lavoisier was 15 years old when Cronstedt published these words. The English translation used here was published in 1770. That alone should give pause for thought to anyone inclined to argue that Lavoisier's work set in motion a chain of events which revolutionised chemistry by refuting the phlogiston theory and by putting chemistry on a new foundation by making older 'alchemical' views obsolete.[28]

What about Lavoisier's revolution?

In our efforts to understand the relationship between alchemy and eighteenth-century chemistry, we have now arrived at a persistent myth which has plagued scholarship. It is that of the 'scientific revolution in chemistry'. For a very long time many historians claimed that one major event and one major actor changed everything. This was the *chemical revolution* of Antoine Laurent Lavoisier between 1772 and 1789. The idea can be traced to the late nineteenth-century French historian Marcellin Berthelot. It is tinged with the worship of scientific heroes current in that era, proceeds from a selective reading of Lavoisier and his supporters, and has been subject to careful scrutiny and scathing criticism over and over again during the last 30 years. Admittedly, it is based on statements made by Lavoisier and other late eighteenth-century chemists.[29] Lavoisier did not appreciate history but viewed it, rather, as a cumbersome ballast, an impediment to progress. It was an attitude which facilitated the portrayal of his own innovations, and other recent developments in his discipline, as an innovative break with the past. This attitude comes through in his theoretical writings and in his proposal for a new chemical nomenclature. As noted by Marco Beretta, in Lavoisier's *Méthode de nomenclature chimique* (1787),

> two thirds of the 1055 new names as well as their related substances were rooted in the recent discoveries made in pneumatic chemistry and in most cases that had no counterpart in the old chemical nomenclature. All of a sudden the longstanding dictionary of alchemy and early modern chemistry was cancelled and substituted with a new dictionary which, being a result of an exceedingly rapid reform project, had no history.[30]

The myth of the scientific revolution in chemistry is a difficult legacy, insofar as it warps understanding not only of alchemy but of eighteenth-century chemistry up until the early 1770s. This said, the work of Lavoisier certainly was important. Debating Lavoisier's contribution, one must remember that no one denies that he, and his associates, made an important contribution to chemistry. The disagreement among scholars concerns to what extent the group's contribution constituted a qualitative leap of knowledge which deserves the designation *revolution*, and exactly which innovations should be assigned to Lavoisier personally. It may therefore be of use to present Lavoisier's own views on the matter. He was often accused of claiming the 'discoveries' of others as his own, and as an answer to his critics he made an account of his individual contribution to chemistry in 1793:

> the theory of oxidation and combustion; the analysis and decomposition of air by metals and combustible bodies; the theory of acidification; the exact nature of a large number of acids, particularly vegetable acids; the first ideas of the composition of vegetable and animal substances; the theory of respiration, to which Seguin contributed with me.[31]

How then did Lavoisier's achievements, including those that are more visible with hindsight, fit into the story outlined in this chapter? My answer is that Lavoisier and his associates created a theory which joined together two major research strands pursued by eighteenth-century chemists. The first of these was the programme of finding out which earths and metals could not be reduced to other constituent parts and defining them as stable species. The second I have omitted to discuss. The reason is that its connection to alchemy is much less obvious. It was the work of Joseph Priestley, Carl Wilhelm Scheele, Lavoisier and others on the composition of air. These investigators brought the methodology and mindset established through the investigation of minerals to the realm of air and succeeded in reducing air itself to a set of well-defined component parts: *species of air* which could not be decomposed further. The theoretical synthesis proposed by Lavoisier and associates consisted of a theory which accounted for how the components of air, in particular oxygen, interacted with earths and metals. Or to put it in Lavoisier's own words, quoted above, 'the theory of oxidation and combustion; the analysis and decomposition of air by metals and combustible bodies'. This was a crucial contribution, but it did not comprise the 'foundation' of modern chemistry. Chemistry was built on the solid foundations of almost two millennia of laboratory investigations. And as for developments in the eighteenth century: the innovations of Lavoisier and his collaborators relied heavily on theoretical advances which had begun in mineralogical chemistry in the 1730s. This included the reconceptualisation of matter itself as consistent, not of the elements of antiquity or variations thereof, but of the discrete, homogeneous parts of objects which resisted decomposition in the laboratory: that is, gold, zinc, cobalt, nickel, platinum, chlorine, oxygen, etc. As for the notion that Lavoisier personally introduced the modern concept of the chemical element, it is misleading. Rather, Lavoisier *and his associates* renamed a set of substances as elements. These substances were already established to be stable chemical species; i.e. modern chemical elements in all but name. The innovation, as with much of Lavoisier's work, was based on novel theoretical and experimental insight but was in significant part a linguistic reframing of facts already known. It should also be noted that chemistry's period of explosive innovation did not end with the eighteenth century.[32] John Dalton's atomic theory and Jacob Berzelius's nomenclature reform and investigations of atomic weights were no mere mopping-up exercises. They were important contributions in their own right. And they were followed by the establishment of organic chemistry, with the discovery that living tissue – both plant and animal – is composed almost entirely of carbon and hydrogen. Undoubtedly, a few decades into the nineteenth century the school of alchemists who had begun to call themselves chemists almost precisely a century before had inaugurated a new era. But it had been a long process to get there.

When does change begin and when does it end? There was no specific point in history when modern chemistry was born. Most contemporary scholarship agrees that there was no singular chemical revolution but rather a series of transformative steps, reconceptualisations, or perhaps smaller revolutions, which took place not only during the eighteenth century but also extended into the centuries which preceded and followed.[33] Above, I have presented some select, and in my view important, aspects of this long process. I have mentioned that chemistry became institutionalised, gaining position in academies and universities. I have touched upon chemistry's theoretical development, i.e. the gradual abandonment of

classical elements and Paracelsian principles. I have described the emergence of a systematic programme of metal discoveries which began in the 1730s, and I have mentioned the chemists' reinvention of their science as a useful Enlightenment enterprise. I have also pointed to the fact that the latter processes were already well underway by the late 1750s.

Although much work remains to be done, I believe that the story presented here works well to fill the gap between the new historiography of alchemy and the developments which took place in chemistry in the decades preceding and following the year 1800. Much of this interpretation is based on my own research. But I should point out that mine is not a singular view. Its first outline was traced in a path-breaking essay by Robert Siegfried and Betty Jo Teeter Dobbs, 'Composition: A Neglected Aspect of the Chemical Revolution', in 1968. A crucial piece of the puzzle is provided by William Newman in his important work on the so-called negative-empirical definition of an element. The interpretation and general framing also rely heavily on the work of Holmes and Principe.

In this essay I have suggested that there was never a break between alchemy and chemistry. I suggest that contemporary chemists may claim the title of alchemists, should they want to. Seen from this point of view, chrysopoetic alchemy – gold-making – is but an abandoned research programme within a joint and continuous alchemical and chemical tradition.

Antoine Lavoisier and His Wife Marie-Anne. Oil on canvas by Jacques Louis David, 1781.

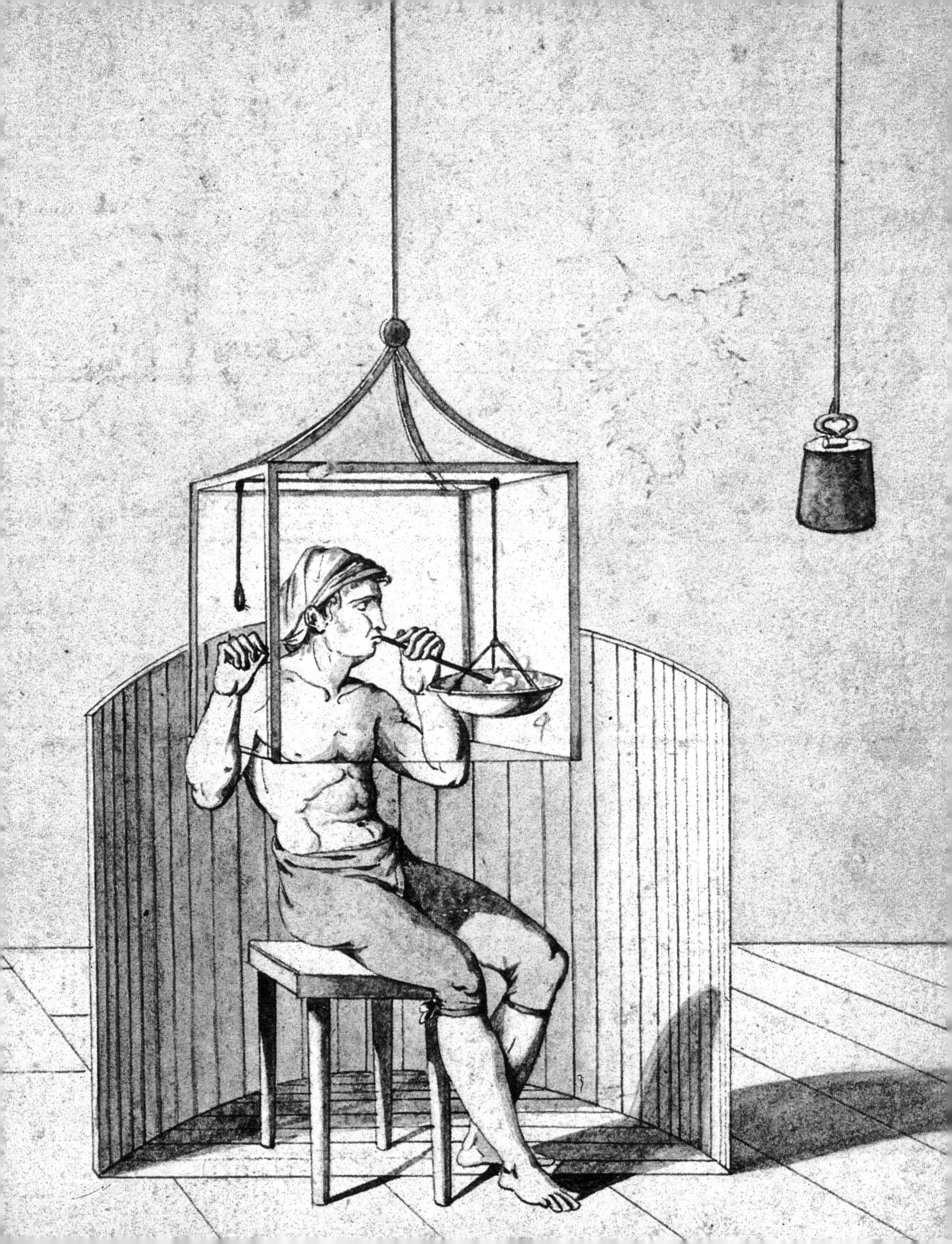

ALCHEMY IN THE ACADEMY

From Rejection to Reconstruction

GEORGIANA D HEDESAN

'A man with his head in a glass container.' Detail from a drawing by Marie-Anne Lavoisier, recording Antoine Lavoisier's experiment to monitor the effects of respiration on the human body.

In the nineteenth century, alchemy had little place in the academy. In the aftermath of the 'chemical revolution' of Antoine Lavoisier, the theory of transmutation of metals, and of matter in general, had been discredited: Lavoisier believed in a nature made of elements, each unique to itself and unchangeable. With this theory, he ushered in the new science of chemistry, purged of its taints of the old matter theory. Just as chemists pondered and marvelled at the achievement of Lavoisier, they felt scornful or embarrassed by the transmutational pursuits of their forebears. Certainly, their science was named differently – chemistry, not alchemy – but the common roots were not easily brushed aside.

The first to seriously commit to a study of the history of alchemy and chemistry was Johann Friedrich Gmelin (1748–1804), professor of medicine, chemistry, botany and mineralogy at the University of Göttingen. Thoroughly imbued with the spirit of the Enlightenment, Gmelin published a huge three-volume *History of Chemistry* (1797–99) that embraced Lavoisier's insights and sought to set 'rational' chemistry in opposition to 'irrational' or superstitious alchemy, which had been superseded.[1]

Gmelin's tone was repeated in other histories of chemistry, such as Thomas Thomson's two-volume *History of Chemistry* (1830) and Hermann Kopp's four-volume *History of Chemistry* (1843–47).[2] Yet the dismissive attitude began to change at the end of the nineteenth century. In later life, Kopp produced a better-documented and more nuanced *Alchemy in Ancient and Modern Times: A Contribution to the History of Culture* (1886).[3] An important role in the revival of interest was played by the polymath French chemist Marcellin Berthelot (1827–1907), who wrote and published a reputable translation of Greek alchemists, *Collection of Ancient Greek Alchemists* (four volumes, 1887–88), which is still in academic use today.[4] His work on Greek alchemy was taken up by the chemist Edmund Oscar von Lippmann (1857–1940),[5] while Julius Ruska (1867–1949) and his pupil Paul Kraus (1904–44) focused on Arabic alchemy.

In Great Britain, John Ferguson (1837–1916), professor of chemistry at the University of Glasgow, became known for his interest in the history of Latin

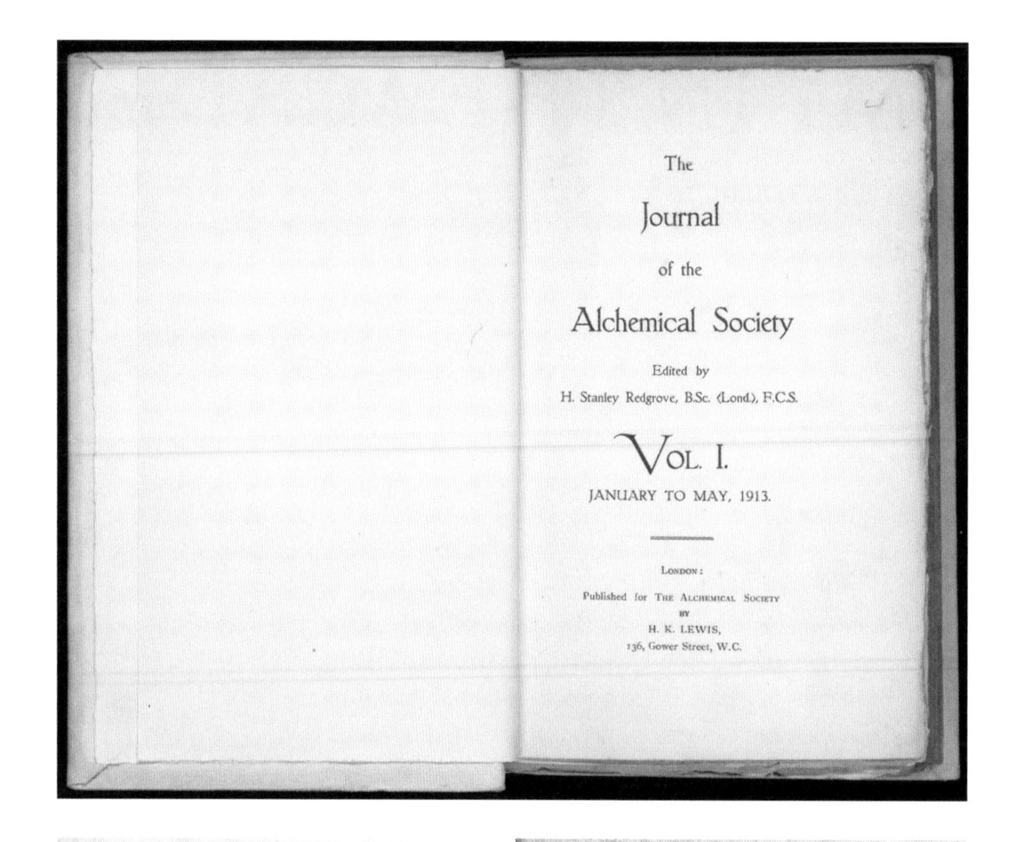
The
Journal
of the
Alchemical Society
Edited by
H. Stanley Redgrove, B.Sc. (Lond.), F.C.S.
VOL. I.
JANUARY TO MAY, 1913.

LONDON:
Published for THE ALCHEMICAL SOCIETY
BY
H. K. LEWIS,
136, Gower Street, W.C.

Geschichte
der
Künste und Wissenschaften
seit der Wiederherstellung derselben bis an das Ende
des achtzehnten Jahrhunderts.

Von
einer Gesellschaft gelehrter Männer
ausgearbeitet.

Achte Abtheilung.
Geschichte der Naturwissenschaften.
II. Geschichte der Chemie

von
Joh. Friedr. Gmelin.

Erster Band.

Göttingen,
bey Johann Georg Rosenbusch.
1797.

Some English
Alchemical Books

Being an Address
delivered to
The Alchemical Society
on
Friday, October 10th, 1913,
by
Professor John Ferguson, LL.D.,
Honorary President.

(Reprinted from Volume II. of
THE JOURNAL OF THE ALCHEMICAL SOCIETY,
Edited by H. Stanley Redgrove, B.Sc.)

LONDON:
Published for The Alchemical Society
BY
H. K. LEWIS,
136, Gower Street, W.C.
1913.

alchemy.[6] At Glasgow, he taught Sir William Ramsay (1852–1916) and hosted Frederick Soddy (1877–1956), both chemists who were known for their contribution to the discovery of radioactivity. Ramsay and Soddy concurred that radioactivity proved the validity of the theory of alchemical transmutation. Ramsay's claim of effecting artificial transmutation in 1907 (later proven wrong) and Soddy's conceptualisation of the new theory in terms of alchemy in 1909 (*The Interpretation of Radium*) opened up the possibility of a new type of chemistry, not as divorced from alchemy as it used to be.[7]

In this heyday of atomic science, it seemed possible for alchemy to be discussed openly by scientists. In 1912, chemist H. Stanley Redgrove (1887–1943) founded the Alchemical Society in London, with Ferguson as president. The society brought into dialogue scientists like Ferguson and Redgrove, and occultists such as Edward Arthur Waite (1857–1942) and Isabelle de Steiger (1836–1927). The society, which was active until the First World War broke it apart in 1915, sought to reconcile alchemical transmutation and atomic science.[8]

Stipple engraving of Johann Friedrich Gmelin, professor of medicine at Göttingen. Artist and date unknown.

Opposite:
Journal of the Alchemical Society, founded in London in 1912 (top).

Johann Friedrich Gmelin's three-volume *Geschichte der Chemie* ('History of Chemistry') (bottom left).

Founded in 1912, the Alchemical Society enjoyed lectures on topics such as this address by Professor John Ferguson on books of interest, 10 October, 1913. Reprinted in the second volume of the Society's journal, London (bottom right).

In 1919, Sir Ernest Rutherford (1871–1937) achieved the first artificial transmutation using radioactive materials. Rutherford's physics background made him hesitant to embrace the alchemy framework of Ramsay and Soddy. He seemed to finally succumb to their views in his last published book, *The Newer Alchemy* (1937). Yet atomic science in the 1920s and 1930s moved away from the realm of chemistry to that of physics. When the link between alchemy and atomic science made another appearance, it was in the popular book of Louis Pauwels and Jacques Bergier, *The Morning of the Magicians* (*Le Matin des magiciens*, 1960).[9]

The interwar period was marked by a rising interest in the study of alchemy. In 1935, two academics at Queen Mary College in London, chemists F. Sherwood Taylor (1897–1956) and James Riddick Partington (1886–1965), as well as Douglas McKie (1896–1967), a science historian at University College, London, became founding members of a new organisation, the Society for the Study of Alchemy and Early Chemistry (SSAEC, later the Society for the History of Alchemy and Chemistry, SHAC), which still survives today.[10] Like the Alchemical Society, SSAEC-SHAC comprised both chemists and individuals interested in the occult, including psychotherapist Gerard Heym (1888–1972). SSAEC-SHAC centred on the publication of an academic journal, *Ambix*, which barely survived the war and its aftermath, but it is now a thriving and important publication in the history of alchemy and chemistry. In the immediate post-war period, alchemy became an unfashionable topic, as the newly established academic discipline of the history of science gave rise to a new form of positivism. Practical alchemy was pursued only by individuals outside the scientific establishment, such as Eugene Canseliet (1899–1982, the pupil of a mysterious adept called Fulcanelli), Archibald Cockren

(1880–1949), Baron Alexander von Bernus (1880–1965), Armand Barbault (1906–74) and Frater Albertus – Albert Richard Riedel (1911–84).

The Golden Castle, a mandala painted by C. G. Jung in *The Red Book*, 1928.

The man who established the history of science as an independent subject was George Sarton (1884–1956), a Belgian positivist. Sarton, who became the chair of the new history of science department at Harvard University, had little patience with alchemy, placing it in the category of 'pseudo-science'. In a 1950 article, he dismissed alchemists as 'fools or knaves, or more often a combination of both in various proportions'.[11] This definition disregarded more than 70 years of historical research in alchemy, instead reverting to Enlightenment opinions of the topic.

Sarton's take was upheld by Herbert Butterfield (1900–79) in *The Origins of Modern Science, 1300–1800* (1952). This influential work, often used as a textbook in the early departments for the history of science, did not attack only alchemy but historians of alchemy as well:

> Concerning alchemy it is more difficult to discover the actual state of things, in that the historians who specialise in this field seem sometimes to be under the wrath of God themselves; for...they seem to become tinctured with the kind of lunacy they set out to describe.[12]

The reasons why this new field held alchemy and its historians in such low esteem are complex. According to historians Lawrence M. Principe (b. 1962) and William R. Newman (b. 1955), one of the chief reasons was the proliferation of a 'spiritual' perspective of alchemy. By this they mean a view that 'held (and holds) that the operations recorded in alchemical texts corresponded only tangentially or not at all to physical processes'.[13]

Principe and Newman attributed this view to Mary Anne Atwood (1817–1910), an occultist writer who published *A Suggestive Inquiry into the Hermetic Mystery* in 1850.[14] Atwood thought that alchemy was primarily a spiritual change in the practitioner of the art, with transmutation of metals being an unimportant sideshow. As she claimed, 'Self-Knowledge [is] at the root of all Alchemical tradition.'[15] Her views chimed with many thinkers involved in the occultist revival of the late nineteenth and early twentieth centuries, which mainly took place in France and England.

The views of 'spiritual alchemy' promoted by Atwood and others influenced Carl Gustav Jung (1875–1961), the Swiss psychologist who began as a follower of Sigmund Freud before breaking from him. Jung became interested in alchemy from the point of view of psychology and formulated a perspective of it that is often referred to as the Jungian interpretation of alchemy. He agreed with both positivists and spiritual occultists that the alchemists' laboratory work was not significant.[16] This dim view of practical alchemy was accompanied by a fervent interest in the symbols and images that alchemists used. Jung argued that what alchemists saw in the material changes taking place in the laboratory were actually the contents of the collective unconscious. In other words, alchemists transcended the personal conscious and the personal unconscious, reaching to a kind of prime matter of psychology, where archetypes – embodied in myths and symbols – were formed.

Yet the Jungian view would only become significant after the rise of counter-culture in the late 1960s and in the 1970s. This shift is important to recount, as it deeply affected the way academia looked at alchemy as well as other subjects that had previously been sidelined or rejected.

The 1950s and 1960s were still dominated by a positivistic view of science that derided alchemy and sought to drive a wedge between 'alchemy' and 'chemistry'. The 'spiritual interpretation of alchemy' seemed to give plenty of fodder to those historians of science who used the terms 'mystical' or 'mysticism' disparagingly; they employed it in order to separate the wheat (chemistry) from the chaff (alchemy).[17] The strident tone of criticism which we have encountered in Butterfield suggests a raising of the stakes in the rhetorical battle between chemistry and alchemy. The reason for this needs to be teased out: my view is that it had a lot to do with the rise of the paradigm of the 'scientific revolution' in the first half of the twentieth century.

The term was coined by the historian of science and neo-Kantian Alexandre Koyré (1892–1964), who immigrated to France after the Russian Revolution of 1917. He sought to formulate a different and superior type of change that took place not in politics but in science. Just as the French Revolution separated the early modern world from the modern, he thought that the Scientific Revolution marked the transition of a pre-scientific world to the new, scientific one, ushering in progress and reason.[18]

The Scientific Revolution, in Koyré's vision, took place almost exclusively in the seventeenth century. It was, he believed, a revolution at the level of ideas; it introduced mathematics and physics as the fundamental level of reality, prefiguring the findings of the modern age (particularly those of Albert Einstein). Koyré's heroes were Copernicus, Galileo, Descartes and Newton. Notably, he had little interest in Francis Bacon or Robert Boyle, as Koyré excluded experimentation and chemistry from the Scientific Revolution.

This canonising of the 'great heroes of the Scientific Revolution' held sway until well after the Second World War, as Koyré's ideas were disseminated by Sarton and Butterfield. Perhaps because the heyday of the history of science was played out in Anglo-Saxon lands, some of the greatest historians focused on two important figures in English science, Boyle and Newton. This interest produced one of the most fertile Atlantic alliances in the field: that of A. Rupert Hall (1920–2009), a Newton scholar, and his wife, Marie Boas Hall (1919–2009), a Boyle scholar.

Both of the Halls had a disparaging view of alchemy, because they thought that this practice had been, if not destroyed in the Scientific Revolution, then at least rejected by their heroes Boyle and Newton. In 1962, Rupert Hall described alchemy as 'the greatest obstacle to the development of rational chemistry'.[19] In turn, Boas Hall dedicated herself to proving that chemistry, of the type introduced by Lavoisier, had already appeared in the seventeenth century in the singular figure of Robert Boyle.[20] In her idol-building exercise, Boas Hall had no room for any alchemical interests that Boyle may have held, since these were just 'mystical' (by which she meant irrational and not to be dignified by interest).

In the 1960s, however, new perspectives were emerging. In 1966, French philosopher Michel Foucault (1926–84) affirmed that each era was characterised by an 'episteme' – a worldview – which defined it.[21] This, he added, was incommensurable to the one that came after or before: the Renaissance episteme was fundamentally different from the modern one. Thus, all forms of culture that existed in it – astrology, magic, alchemy, occult knowledge – formed a coherent whole that was later cast down and destroyed by the modern worldview. Foucault seemed to regret the loss of that Renaissance episteme, but other readers, steeped in the positivistic paradigm, thought its demise was a progressive development.

One of the most famous scholars who combined positivism with a sensitivity for the Foucauldian approach was Brian Vickers (b. 1937), a literary scholar of Francis Bacon. Vickers preferred to talk about 'mentalities' rather than episteme, defining the 'occult mentality' belonging to such 'occult sciences' as alchemy, astrology, natural magic, numerology, etc. These did not fit the rising mentality of the science. Vickers criticised the 'error...that the occult sciences in the Renaissance were productive of ideas, theories, and techniques in the new sciences'.[22]

The focus of his attack was the Warburg scholar Frances Yates (1899–1981), who dared to be different. In her landmark book *Giordano Bruno and the Hermetic Tradition* (1964), Yates claimed that Renaissance magic was a factor in the advent of modern science, an idea that came to be known as the Yates Thesis.[23] That magic, Christian Cabala or the rather ill-defined 'Hermeticism' could somehow lead to the Scientific Revolution shocked the traditional historians of science, still vigorously upholding the pristine ideal of science originating, as it were, *ex nihilo* in the minds of the likes of Galileo and Descartes. The result was an attempt to discredit Yates's ideas as being ridiculous and unfounded.[24] An outraged Vickers wrote a 29-page review article criticising her 'wish to rewrite the history of science' in the early modern period, concluding that it 'is an edifice built not on rock nor on sand but on air'.[25] His dismissal of the Yates Thesis did not end with the death of the author, as Vickers continued to attack her views, which he claimed to be 'wholly unfounded'.[26]

Yates herself had little interest in alchemy. Her focus was on the Hermetic corpus, *Corpus Hermeticum*, not on what had often been known as the 'Hermetic art', alchemy.[27] Yates's Hermetic tradition was chiefly learned magic. As such, the subject of alchemy remained either isolated from the Yates debate or a vague part of it, in the sense that alchemy could be seen as a secondary ingredient of the general soup of 'occult sciences'.

Yet there was one subject in which alchemy was dealt with separately: that of Isaac Newton (1643–1727). The generation that followed Newton struggled with both the knowledge that he was involved in alchemy and the hefty manuscripts on the topic he left behind. Most scholars preferred to shove the subject under the carpet, until it could not be done so anymore. In 1936, Newton's theological and alchemical papers were sold at a Sotheby's auction; quite a few of the alchemical ones were bought by the economist John Maynard Keynes (1883–1946). A perhaps over overexcited Keynes proclaimed Newton 'the last of the magicians'.[28]

In the aftermath of Yates's momentous book, historian Richard Westfall

(1924–96) claimed that Newton had been influenced by the Hermetic tradition, which he seemed to understand in a rather amorphous fashion, quite similar to Vickers's 'occult sciences'.[29] This was carried further by Betty Jo Teeter Dobbs (1930–94), who published *The Foundations of Newton's Alchemy: The Hunting of the Greene Lyon* in 1975. Dobbs introduced the views of Jung into the history of science, claiming that:

> In recent years, however, the insights of twentieth-century analytical psychology as applied to alchemy by C. G. Jung have come to provide a really promising approach to the problem, allowing as they do for an understanding of the many factors in alchemy which are not only obscure but patently irrational.[30]

This acceptance of the 'irrationality' of alchemy seemed to connect Dobbs with preceding positivists like the Halls.

Yet, beyond the limitations of the theoretical framework, *The Foundations* was also the first work to tackle the alchemical manuscripts of Newton in a serious fashion. Dobbs did not, however, stop here. In 1992, she followed her study with another monograph, *The Janus Faces of Genius*, which went far beyond *The Foundations*. Dobbs argued that all of Newton's ventures, including alchemy, experiment, mathematics and prophecy, were connected by his belief in 'the unity of Truth and its ultimate source in the divine'.[31] Her analysis further maintained that Newton had a 'religious interpretation of alchemy'.[32] Interestingly, *The Janus Faces of Genius* marks a tacit renouncement of the Jungian views Dobbs introduced in *The Foundations*: Jung only appears in a footnote in this book, without further commentary.[33]

Dobbs's new take on Newtonian alchemy provided a perspective into Newton's motivations and beliefs but almost none into what he was actually doing in the laboratory. It inspired scholars interested in the religious facets of early modern scientific culture. However, William R. Newman thought that Newton's alchemical papers had more to yield in terms of practical alchemy and sought to uncover them in a systematic fashion.

In the 1980s, the field of the history of science was rapidly changing. At the core of the transformation lay the rising framework of the sociology of scientific knowledge (SSK). SSK emerged against the background of increasing questions as to the progressive nature of science. History of science, as described by Sarton and Butterfield, was meant to look at science only within the boundaries of what it defined science to be: a rational and progressive phenomenon, separate from social, political or economic influences. Science was supposed to exist outside society, in a kind of glorious ivory tower. As such, it was always an intellectual construct kept pure by the policing of established historians and philosophers of science.

In the 1970s, sociologists like David Bloor (b. 1942) and Barry Barnes (b. 1943) encouraged investigations into how scientific truth claims were constructed. This approach appeared the most amenable to studies of laboratory science, leading to several groundbreaking analyses of scientific practices.[34] These were initially focused on contemporary studies, but in 1985 Steven Shapin (b. 1943) and Simon Schaffer (b. 1955) produced a foundational work on seventeenth-century science, focusing on the disagreement between Robert Boyle and Thomas Hobbes about the nature of scientific knowledge.[35]

Shapin and Schaffer's work contributed to a shift of interest in early modern studies, already evident at the start of the 1980s. There was a new emphasis on experiment, practices and practitioners. Charles Webster's *The Great Instauration* (1975) had already drawn attention to the experimental origins of modern science.[36] Although Webster (b. 1936) emphasised the influence of Francis Bacon on the Scientific Revolution, he also examined the importance of Paracelsianism and Helmontianism in the period. In 1977, Allen Debus (1926–2009) brought together

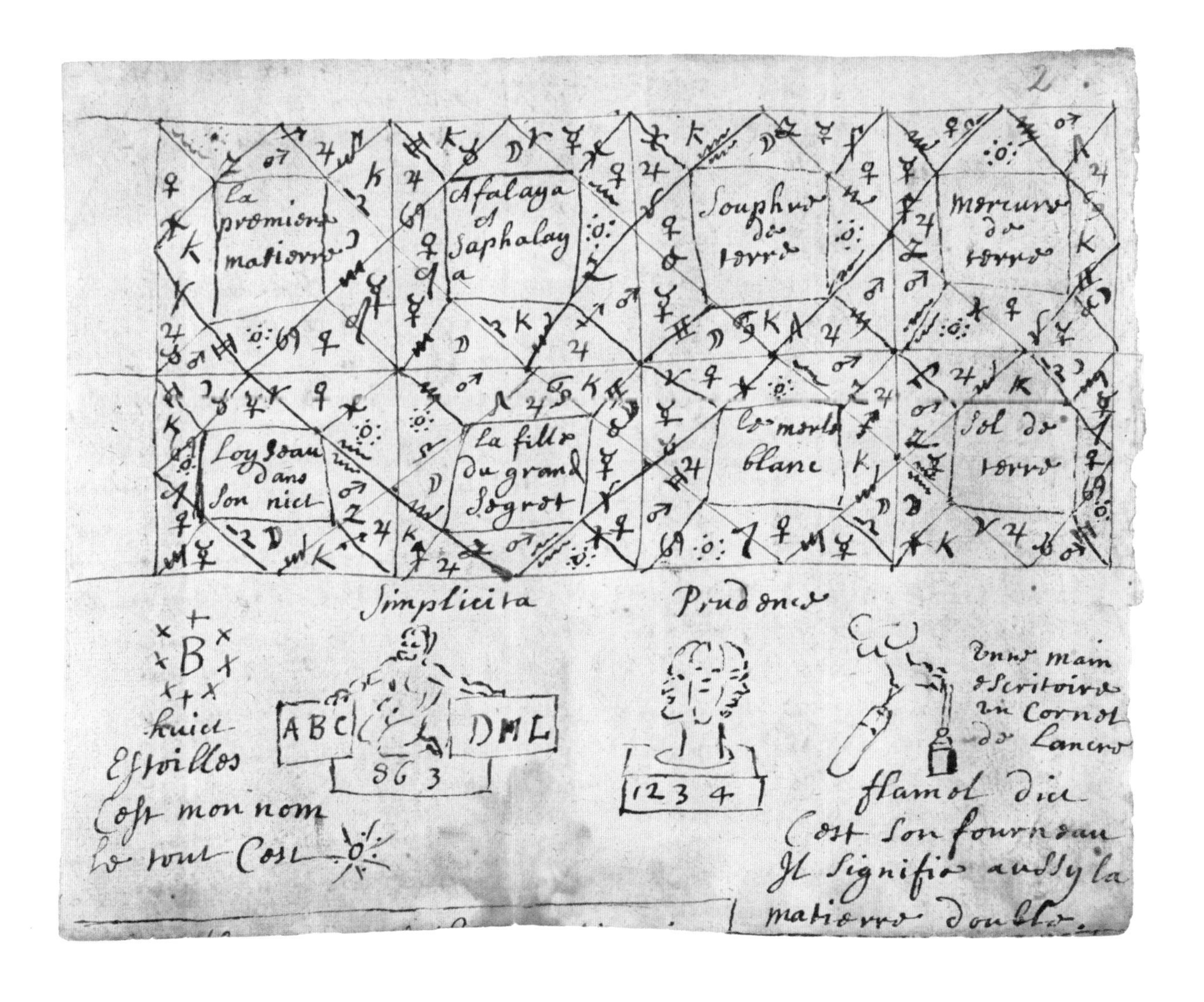

One of several alchemical and astrological notes and sketches in a small undated manuscript by Sir Isaac Newton.

these movements under the name 'chemical philosophy'.[37] Notably, Debus drew a distinction between 'chemical philosophers' and alchemists.

The new emphasis on practices and the laboratory led to a reconsideration of alchemy as a valid subject of inquiry in the history of science. The movement has come to be known as the 'new historiography of alchemy', and it was spearheaded by William R. Newman and Lawrence M. Principe. Their opening salvo was the article 'Alchemy vs Chemistry: The Etymological Roots of a Historiographic Mistake' (1998).[38] In it, the two scholars argued that there was no difference between alchemy and chemistry before the eighteenth century and any attempts at separating them in that period were not only presentist but futile. Their aim was to render obsolete attempts, such as that of Boas Hall, to separate Boyle's rational 'chemistry' from his irrational 'alchemy'. By the time they published 'Alchemy vs Chemistry', Newman and Principe had devoted a lot of attention to Boyle's period and drew very similar conclusions. Newman studied a little-known American alchemist, George Starkey (1628–65), who was both a highly proficient chemist who taught Robert Boyle the secrets of the art and a seeker of the philosophers' stone.[39] Principe, in turn, focused on Boyle's interest in alchemy; ultimately, Starkey's and Boyle's views converged, in being involved with both chemistry and alchemy. Newman and Principe argued that it was anachronistic to separate the two fields. They proposed to treat the pre-eighteenth century combined field of alchemy and chemistry as a hybrid form called 'chymistry'.

'Alchemy vs Chemistry' was followed up by another article two years later, this time aimed mainly at occultist views of alchemy.[40] Principe and Newman's bête noire was the Jungian interpretation, which as we saw was introduced by Dobbs. Yet they also attacked other widespread views of alchemy, such as that it was inherently vitalistic and spiritual.

Indeed, the existence of corpuscularianism in medieval and early modern alchemy was revealed by Newman and constitutes one of his chief contributions to the field.[41] Newman and Principe also showed that Starkey's extravagant imagery, written under his alchemist persona, could be decoded in the laboratory as being of a practical, not spiritual, nature.[42] Similarly, an alchemist writing under the adept name of Basil Valentine, Principe has pointed out, embedded practical recipes in obscure language and images.[43] This led the two scholars to argue that alchemists deliberately and consciously coded their laboratory findings in language and imagery that appear 'mystical' or 'spiritual' but are in fact nothing of the kind. To prove this, Principe and Newman turned to laboratory work themselves.

The framework of reconstructing experiments to enlighten the history of science was almost non-existent before the 1980s and the advent of SSK. Nowadays it is a hot field called 'experimental history of science', which is part of the wider domain known as RRR (reconstruction, replication, re-enactment, reproduction and reworking).[44] This sudden growth overlaps the rising interest in experiment and experimental science in general.

Experimental history of science had some antecedents in the 1950s and 1960s, in the reproduction of Galileo's and Newton's experiments or pharmacological recipes.[45] In the 1990s, H. Otto Sibum (b. 1956), the acknowledged creator of the field of experimental history of science, reconstructed scientific instruments and used them to recreate experiments.[46]

To these endeavours two more traditions should be added: firstly, that of art restoration and conservation, and secondly, 'experimental archaeology', which was established in 1979 by John Coles (1930–2020). All of these fields converged in the 1990s and 2000s in a development sometimes called history of science's 'material turn'.[47]

RRR is fraught with complexity and difficulty. There are acknowledged limitations: reliving the past is impossible, and absolute truth-claims cannot be made by any reconstruction. Recreating recipes, for instance, is limited by present conditions: materials, practices, instruments, laboratory conditions, mental conditioning. There is also the challenge of unifying the different realms of science and history, creating a close relationship between the library and the laboratory.[48]

The results so far have been encouraging. Principe has been able to prove some of the stranger texts and images of alchemy can in fact translate to laboratory processes.[49] Newman has been able to decode and reproduce some of Newton's experiments, showing the complexity of the alchemical knowledge that he accumulated. His preliminary findings were published in his 2018 book *Newton the Alchemist*, which in itself shows the importance of taking Newton's practical alchemy seriously. Various projects have been unfolding, including the reconstruction of experiments and practice in a French manuscript of the sixteenth century, the reproduction of alchemist Antonio Neri's *rosichiero* glass and the recreation of ancient techniques of producing mercury.[50]

The importance and consequences of such reproductions cannot be understated. It is clear that experimental history of science and RRR can illuminate the history of practical alchemy in a way that has perhaps not been done before. Here we are no longer dealing with analogies between nuclear chemistry and alchemical transmutation, or with textual witnessing of certain alchemical demonstrations. Could we then assume that most of the alchemical texts, even the most abstruse ones, hide a laboratory process? From the limited proof we have seen, it is possible that many of them might. We should not, however, let enthusiasm take the place of evidence. This evidence, in turn, must be collected with rigour and care. Conclusions must not be preconceived but drawn from empirical practice. We must also allow for a sizeable degree of error and disappointment, particularly when decoding complex texts. We already know that alchemists, especially in respect to their 'higher' secrets, were master concealers. We should also not forget that contemporaries, much more familiar with the ingredients and techniques of the period, were often struggling to understand and reproduce some previous alchemists' claims. Even where we have recipes, we must account for historical deterioration. Quackery cannot be excluded: it is possible that some would-be alchemists simply made up fantastic-sounding texts that could never be decoded because there's nothing there. Finally, there is the complex problem of spiritual alchemy, which uses alchemical terms and themes for spiritual practices.[51]

This highlights the importance of textual availability and accuracy before a

reproduction is even tried. There are hundreds of publications on the topic of alchemy dating from the early modern period; to these should be added hundreds if not thousands of unedited alchemical manuscripts scattered all over the world. These texts and images need to be extracted, correlated and understood in a historical manner. We must properly understand the genealogies of text, its changes and its practices; patterns can be drawn out to connect both writing and images.

The size of the alchemical textual wealth is so big and alchemy has been so little studied that it is my belief that the task requires the intervention of computational humanities. If we were to transcribe the thousands of publications and manuscripts in a computer-ready format, we could begin the task of correlating terms, practices and texts.[52] With the advances of IT, machine learning and AI, and assuming the availability of funds, a new computer-facilitated insight into alchemy becomes possible.

Old-fashioned research and methods of looking at alchemical text and images remain equally important. Close traditional reading can be carried out in parallel or in conjunction with computer-based data analysis (often called distant reading) and laboratory experimentation. The hope is to create a virtuous cycle whereby all these elements enforce each other, and more knowledge will be accumulated. Such projects could and should be interdisciplinary, bringing together historians, scientists and artisans, and other professions that could shed light on a bygone era.

Explanations will no doubt continue to be debated. There should be room for alternative views of alchemy and for considering the human body and spirit as a valid 'laboratory', as Chinese *Neidan* did.[53] We should further be wary of tendencies to overly 'materialise' or 'scientify' alchemy by projecting our own era backwards. However, all caveats considered, I like to believe that, in the light of the 'new historiography', the positivistic dismissal of alchemy, of the alchemist's persona as a searcher into the secrets of nature, and of the value of alchemy in itself, will soon become a discredited claim.

'Pendant cross (back). Gold, partly enamelled. Northern European, 16th Century. The red enamels bear similar characteristics to the rosichiero glasses described in Kunckel's *Ars Vitraria Experimentalis.*

Macrocosmus.
Micro cosmus.
Pituita
Sanguis
Cholera

ALCHEMY IN CONTEMPORARY SPIRITUAL, INTELLECTUAL AND ARTISTIC MOVEMENTS

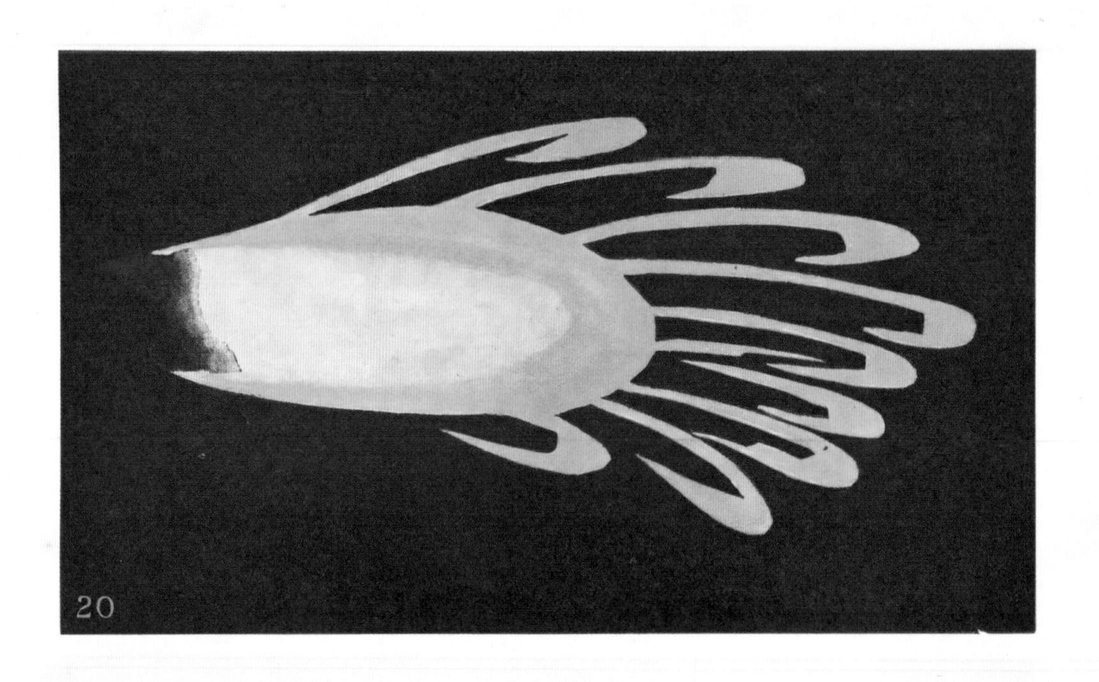

THEOSOPHICAL SCIENCE AND MODERN ALCHEMY

MARK S. MORRISSON

In the 15 November, 1895 issue of *Lucifer, A Theosophical Monthly*, sandwiched between 'Musings of a Neophyte' and an anonymous recounting of 'An Astral Experience', was a remarkable article entitled 'Occult Chemistry'. It not only encapsulated Theosophy's synthesis of spiritual and scientific enquiry, as its title implies, but it also signalled major developments in Theosophy's engagement with modern science – which was then on the cusp of the age of nuclear physics. Summarising an experiment by Annie Besant and C. W. Leadbeater, the nine-page piece imitated the style of scientific journal articles, incorporating mathematics and tables and even referring to itself in scientific parlance as a 'paper'.

Besant (1847–1933), the Theosophical luminary, was the editor of *Lucifer* and leader of the Theosophical Society; Leadbeater (1854–1934) was a major occult writer, spirit medium and key figure in the society. Though most chemists in the nineteenth century understood atoms to be fundamental particles, Besant and Leadbeater claimed to have viewed something revolutionary: a *subatomic* structure. Their experiment was not performed in a university physics lab, however. Rather, the two researchers lay in the grass in the quiet tranquillity of Box Hill, Surrey. Besant and Leadbeater continued these subatomic astral sessions across most of the rest of their long lives, and their further results were published in illustrated books in 1908, 1919 and (posthumously) 1951.

While it may be all too easy to pass over this unorthodox research as pseudoscientific fantasy – Peter Washington dismissively noted that 'the authors made several of their chemical discoveries while sitting on a bench in the Finchley Road'[1] – the article and the research behind it indicated a significant new form of Theosophical engagement with contemporary physical science. It couldn't have come at a more opportune time. At the beginning of what would become the nuclear age an invisible material world increasingly beckoned to both physics and Theosophy, bringing multiple domains of culture and knowledge into generative interaction.

High Ambition (figure 20) and *Selfish Ambition* (figure 21) from *Thought-Forms*, by Annie Besant and C. W. Leadbeater, 1905.

Previous vignette: 'Vitruvian man.' Engraving (detail) by Johann Theodor de Bry on the title page of *Utriusque cosmi maioris scilicet et minoris metaphysica* (*History of the Macrocosm and Microcosm*), by Robert Fludd, 1617–21.

By the nineteenth century, the widespread scientific acceptance of the atomic theory of Daltonian chemistry demonstrated that most scientists had rejected alchemical conceptions of the elements and their transmutation. Still, the spiritual dimensions of alchemy and Hermeticism had not been abandoned in Western esotericism (or even as a matter of historical or religious interest to some scientists). The ideas of Western esotericism coalesced in the seventeenth century, bringing together several alternatives to Judaeo-Christian religious orthodoxies and incorporating aspects of far older traditions. Pioneering scholar Antoine Faivre identifies Western esotericism's four major beliefs: belief in 'symbolic and real correspondences...among all parts of the universe, both seen and unseen'; in a 'Living Nature', animated by life energy or divinity; in the power of a religious creative imagination to explore unknown realms between the material world and the divine; and in the 'experience of transmutation' of the inner person, who is connected with the divine.[2] One can see these ideas throughout much of the occult revival in Britain, Europe and the United States in the later nineteenth century. They were manifested in Theosophical thought and in societies such as the Hermetic Order of the Golden Dawn (a secret organisation in its early years but now one of the best known, because its members included such famous figures as W. B. Yeats, Aleister Crowley, Florence Farr and MacGregor Mathers). But a major paradigm shift in physics and chemistry would strongly reconnect that concept of inner spiritual transmutation to the material world, reinvigorating the older understanding of

Annie Besant and C. W. Leadbeater, London, 1901.

correspondences found in sources such as the so-called *Emerald Tablet* and expressed in the loosely translated phrase 'as above, so below'.[3]

At the turn of the century, researchers in the fields of radio-chemistry and what became nuclear physics discovered that atoms were not fundamental particles, that they could transmute into other elements through radioactive decay, and that they could even be artificially induced to do so. This 'modern alchemy', as it was frequently called by scientists and journalists, introduced a scientific, material understanding of an invisible world to occult or esoteric thinking, and it caught the attention of Theosophists. The nature of matter itself, down to the subatomic level – rather than just the chemical interactions among atoms – was foregrounded as a key research programme for science, and Theosophists and other occultists could increasingly see the experimental probing of the nature of matter as an index of spiritual meaning in the universe. Indeed, for both esotericism and physics, this focus on an invisible world of matter and energy conveys ideas of meanings that cannot easily be accessed directly.

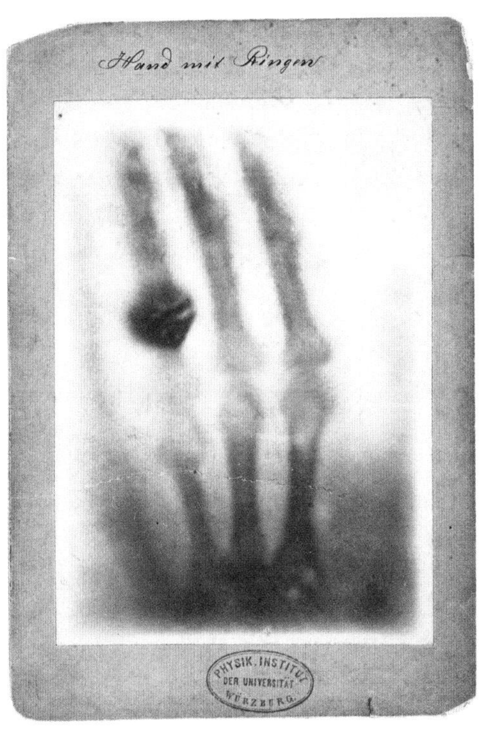

The discovery of X-ray. Wilhelm Röntgen's image of his wife Anna Bertha's hand, made in his lab at Würzburg, 22 December, 1895.

Though born of a soon-to-be-abandoned understanding of ether physics from the previous century, the timing of Besant and Leadbeater's 'occult chemistry' experiment could not have been more auspicious. Claiming a foothold in the field of experimental particle physics would allow Theosophy to declare a relationship to a new scientific paradigm while asserting a spiritual meaning for the nature of matter, a meaning modern science itself could not supply. To update Theosophy's claims to truth and meaning, Besant and Leadbeater began moving it beyond its founder's oracular practice of basing scientific claims on the wisdom one acquired by revelation from hidden Masters. A new Theosophical science would embrace, not reject, experimental methods.

Before unpacking the complex relations 'modern alchemy' mediated, a brief synopsis of scientific developments is in order. In his lab in Würzburg in November 1895, Wilhelm Röntgen discovered X-rays while sending an electric current through a cathode ray tube shielded with black cardboard. His discovery would help shift physics onto an astonishing new path. Indeed, the period from 1895, when the 'Occult Chemistry' article was published, to 1919, the year of the last *Occult Chemistry* volume brought out during the authors' lifetimes, witnessed several major experimental discoveries and theoretical breakthroughs that set the stage for the dominance of particle physics and the mathematical formalisms that describe quantum mechanics. In 1896, Henri Becquerel accidentally discovered rays emitted by uranium salts – a phenomenon that Marie Curie would call 'radioactivity'. Soon Curie and her husband, Pierre Curie, would discover the radioactivity of several elements, including a new one in 1898: the highly radioactive radium. The year before,

J. J. Thomson discovered the electron, the first subatomic particle to be identified in a laboratory, and several other researchers would pave the way for Ernest Rutherford's nuclear model of the atom and its 1913 refinement by Niels Bohr. In 1919, in a key act of modern alchemy, Rutherford transmuted nitrogen into an oxygen isotope and a hydrogen nucleus by bombarding it with naturally occurring alpha particles.[4] In so doing, he discovered the proton (a hydrogen nucleus). The race to build particle accelerators to further probe the nuclei of heavier elements would soon follow.

I don't use the word 'transmutation' fancifully; it was the term many chemists and nuclear physicists themselves used. That one atom could transmute into another, naturally or artificially, refuted Daltonian chemistry's foundational certitude that an atom was immutable, fundamental and indivisible, as the very word 'atom' signified. Indeed, the idea of 'alchemy' was there from the start, as recounted in an (oft-repeated) origin story for nuclear physics. In a physics lab at Canada's McGill University in 1901, chemist Frederick Soddy (1877–1956) and physicist Ernest Rutherford (1871–1937) discovered that radioactive thorium was transforming into an inert gas. As Spencer Weart tells it, 'Soddy recalled, "I was overwhelmed with something greater than joy – I cannot very well express it – a kind of exaltation." He blurted out, "Rutherford, this is transmutation!" "For Mike's sake, Soddy," his companion shot back, "don't call it *transmutation*. They'll have our heads off as alchemists."'[5] Despite Rutherford's alarmed reaction to the term, this phenomenon of radioactive elements transforming into other elements during radioactive decay was frequently figured as alchemical transmutation. Some in the press and elsewhere even imagined that radium could be a modern philosophers' stone – and that radiation might be what medieval alchemists called the elixir of life.

38 Popular Science Monthly

Will Radium Restore Youth?

Power of Mysterious Energy to Renew Worn Body Cells Is Recognized by Physicians

LONGER life.
New hair on bald heads.
A third set of natural teeth.

Renewed vigor of youth.

These are a few of the promises that the more optimistic radium experts are making for the use of radium emanation or radioactive waters in the treatment of disease and old age. And while recent glowing accounts of miraculous cures by radium therapy—as the new treatment is called—have not been accepted generally by the medical profession, there is a growing belief among physicians that radium emanation does have a definite place in medical treatment.

The "Spark of Life"?

Is it possible, as one physician has suggested, that in radium may be found the "spark of life"—the mysterious electronic force that drives the life of the world?

Constantly releasing mysterious energies with which the scientific world was not familiar until recently, radium, when introduced into the human system, produces a "kick" which, experts tell us, bombards our tired life cells into renewed activity.

"The fact is," says Dr. C. Everett Field, director of the Radium Institute of New York and one of the more conservative radium therapists, "that radium does not do the healing. But, for that matter, neither does any other form of healing. The healing exists within the organism. And radium, I am convinced, in some cases, is the most efficient medicine to give needed stimulus to the healing apparatus of diseased organisms."

Even now, Doctor Field believes, radioactive treatment may prolong life at least 15 years.

For internal treatment two methods are used—either doses of radioactive water, or the administration of extremely minute quantities of radium itself. In the first case either waters from springs examined and found to contain traces of radium are used, or radium is used to make ordinary water radioactive. The difficulty with spring waters is that they lose their radioactive power when bottled and transported and so lose their value when not consumed at their source.

Radioactive Drinking Water

Because of this fact, a group of physicians interested in the use of radium as a curative stimulant have invented an ingenious device for imparting radioactive properties to ordinary water. As designed for use in the home, this instrument consists of a case containing an arrangement of glass tubes and vessels in which emanations from radium salts in solution are imparted to air, which is then mixed with the water.

A much simpler apparatus, available for office use, somewhat resembles a hypodermic syringe, containing special capsules of radium salts. Pushing a plunger forces air through the radium capsules and into a glass of water and is said to make the water radioactive. The doses of radium in each case are constant, because radium emanates at a constant rate and only a certain amount can be dissolved in water, no matter how many times a day the apparatus is brought into use.

These constant emanations have been one of the great marvels of science ever since Madame Curie isolated radium from uranium ore in 1898. One gram of radium emits 136,000,000,000 atoms a second. As each atom disrupts, what is known as an *alpha* ray is emitted. This resultant atom disrupts in turn, emitting another *alpha* ray. And so the process continues. In the meantime, the mass of radium deteriorates into radium emanations. Radium thus loses half its strength in 1700 years and none of the original is left after 20,000 years of radiation.

Radium, a salt rather than a metal, is not a drug. It does not combine chemically with any known substance in the body. When taken in soluble form, 25 or 50 per cent of it remains in the body for four or five days. After that the rate of excretion is about only one per cent a day. Wherever it is located, it carries on a constant bombardment in releasing its energy, imparting strength to the tissues, cells, and protoplasm of the body. And when once these begin to function actively, they begin to rebuild themselves.

Radium Is a Stimulant

In external treatment by radium itself, emanations from a certain quantity of radium are allowed to focus on parts of the body over the diseased organs. In this way the curative functions of the diseased portion are stimulated to activity. The atrophying of diseased tonsils has been the most successful use of this form of treatment. The cure of cancer by radium, while claimed by some physicians, is generally discredited at the present time.

Whether radium treatment will prove able to restore youth to old age, grow new sets of teeth and perform other marvels that its more ardent supporters predict for it, only time will tell.

If radium treatment proves to facilitate the process of cell elimination, it will have gone a long way toward delivering the world from its enemies of disease.

"Saratoga Springs" in the home. This apparatus is said to make ordinary drinking water radioactive

For office use, a handy syringe forces air through capsules of radium salts and into a glass of water, imparting radioactivity, its sponsors claim, to the drinking water

Close-up rear view of new household apparatus for manufacturing radioactive water, showing arrangement of glass tubes for charging water with radium emanations

Adverts in the 1920s claimed that a daily glassful of radium-infused water would restore youth. Published in *Popular Science Monthly*, 1923.

During the early years of the 'modern alchemy' of nuclear physics, Soddy frequently explored alchemy in his writings. He speculated that the transmutation of gold from a heavier element might release energy that would be more valuable by far than the gold itself, and he contemplated all that might be accomplished for humanity with abundant energy supplies and the ability to transmute matter.[6] Even the alchemy-resistant Rutherford named his last book *The Newer Alchemy* (1937). He used the word 'transmutation' several times in the text, even when referring to nuclear processes rather than to medieval alchemy.

The term 'modern alchemy' led to far-reaching science fiction, occult writings and popular fiction involving various ways in which nuclear physics might transmute the elements, either imperilling or saving the world. Concerns about the gold standard in an environment in which artificially produced gold might flood the market were especially intense, becoming common themes in pulp magazines, such as Hugo Gernsback's *Amazing Stories*, *Amazing Stories Quarterly*, *Amazing Studies Annual*, *Astounding Stories*, *Wonder Stories* and others, as well as serious discussions of monetary theory and policy. This trope in fiction and in popular science writing helped portray scientists as (sometimes nefarious) magicians with secret, almost miraculous powers to destroy or protect an economy. The moral significance of the new science was, then, a major concern.

Even before pulp fiction magazines picked up the theme, in his 1914 novel *The World Set Free* H. G. Wells built on what he had heard in Soddy's 1908 lectures on 'The Interpretation of Radium', portraying the artificial creation of gold as part of a series of economic and social calamities that eventually lead to nuclear warfare. As I have argued elsewhere, fascination with the new science as a modern alchemy was not a one-way line of transmission in which occultists simply reinterpreted the science in alchemical terms. It also involved scientists, especially chemists, reading alchemical texts republished by occultists and reinterpreting alchemy as a significant early chapter in the history of chemistry, not a discredited, embarrassing pseudoscience. A revival of interest in alchemy was widespread, and the circulation of alchemical tropes and concepts was multidirectional and vibrant.[7]

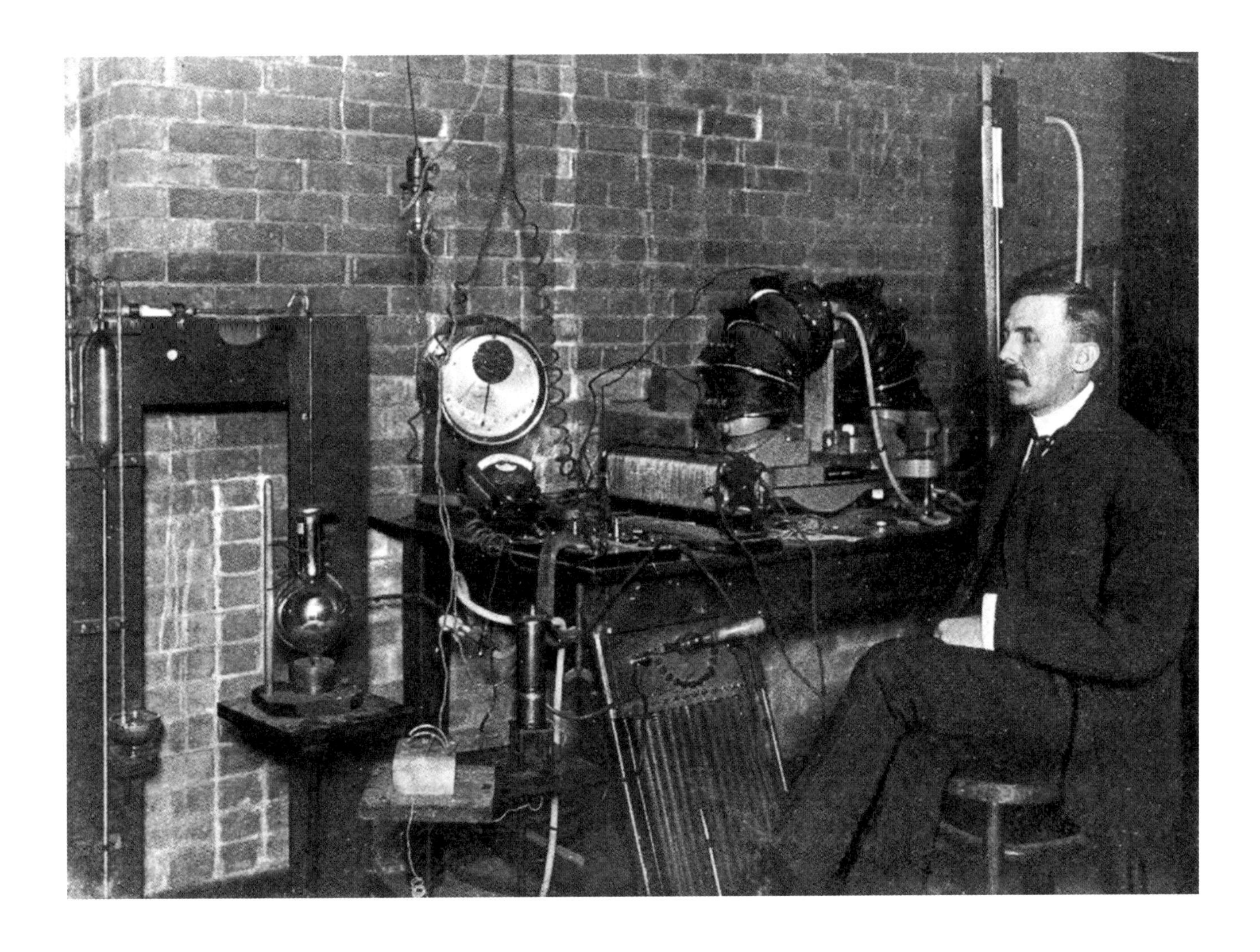

Ernest Rutherford with transmutational lab equipment at McGill University, 1905.

Modern alchemy drew together cutting-edge science and centuries-old writings. The confluence of the modern and the medieval, of science and esoteric spirituality, reveals an approach to the nature of the material and spiritual worlds that we might call Theosophical science. It is common for thumbnail sketches of the history of the emergence of a scientific discipline to tell that story as a series of discrete discoveries and theoretical breakthroughs. But science always occurs in social and cultural contexts, and just as alchemical tropes used by scientists and journalists derived in part from a burgeoning cultural engagement with occultism in the *fin de siècle*, both in secret societies and in popular culture – shaped, in their turn, by the new science of nuclear physics – Theosophy was part of that culture of 'modern alchemy' and was changed by it as well.

To grasp just how innovative Besant and Leadbeater's turn towards scientific experimentation was, one must look briefly at forms of engagement with science from the early years of the Theosophical Society. The society had been founded in 1875, in New York spiritualist and occult circles. It was led by the colourful H. P. Blavatsky (1831–91), a spirit medium – con artist and fraud to some, spiritual leader and speaker of ancient truths to others – along with Irish American lawyer and occultist William Quan Judge (1851–96) and Colonel Henry Steel Olcott (1832–1907), who had worked in the Navy Department during the American Civil War and had been a member of the commission investigating Abraham Lincoln's assassination.

A scientist synthesises gold, from Hugo Gernsback's popular monthly *Science and Invention*, November 1924.

The Theosophical Society espoused a synthesis of Western Hermeticism and Eastern religions (primarily Buddhism and Hinduism) and claimed access to wisdom that was thousands of years older than any world religion, wisdom known to the Masters, or mahatmas, of the Great White Brotherhood (occult adepts whose ranks had included the Buddha, Lao Tzu, Confucius, Moses, Plato, Jesus, Roger Bacon, Jacob Boehme and Francis Bacon, among many others more recent and still walking the earth).[8] These hidden Masters guarded spiritual secrets and ancient sciences, revealing them as needed to help shape the spiritual evolution of humanity, and they had been in touch with Blavatsky to help her launch her new movement with a series of remarkable writings. As Maria Carlson explains, Theosophy 'offered to resolve the contradiction between science and religion, knowledge and faith'.[9]

This synthesis of mysticism and spirituality with modern science met a receptive audience in later nineteenth-century America, Britain and India in the last decades of the British empire, and the Theosophical Society grew quickly in numbers and appealed to many distinguished converts, including the naturalist Alfred Russel Wallace, Thomas Edison, the chemist Sir William Crookes and many literary figures, among them Ella Wheeler Wilcox and, for a time, W. B. Yeats.[10]

Blavatsky's effort to displace modern materialism as a monistic explanation for all reality, including life and mind, was couched in gestures toward the cognitive authority of science in the nineteenth century. She didn't reject materialism by asking Theosophists to turn their backs on the modern world and on modern science. Instead, she sought to make a claim for Theosophy as an even more far-reaching science. A multiplanar universe worked by its own natural laws, which could only be explained by a science focusing well beyond the material plane. Her foundational text, the two-volume *Isis Unveiled: A Master-Key to the Mysteries of Ancient and Modern Science and Theology* (1877), offered a scientific worldview that, in some ways, might not sound objectionable to a contemporary scientist: 'We believe in no Magic which transcends the scope and capacity of the human mind, nor in "miracle", whether divine or diabolical, if such imply a transgression of the laws of nature instituted from all eternity.'[11]

The World Set Free, by H. G. Wells, 1914. Dedicated by the author 'To Frederick Soddy's *Interpretation of Radium*'.

Yet what Blavatsky meant by 'the laws of nature' was not what a scientist in 1877 would have understood by that phrase. Claiming that *Isis Unveiled* was the work of the Masters, Blavatsky portrayed herself as only the medium of its transcribing: the laws of nature would be revealed, but not by experimentation and scientific method. While she often cited contemporary science and intended Theosophy to be accepted as a rational discipline aimed toward spiritual wisdom, her claims for it very quickly veered into the oracular, drawing upon the certitude of revealed secret knowledge rather than scientific forms of argument, citation or experimentation. In short, Blavatsky didn't do experiments or engage much with those who did.

In *Isis Unveiled* and in her magnum opus, *The Secret Doctrine* (1888), the modern science that most occupied Blavatsky was the theory of evolution, rather than discoveries in chemistry and physics. When she did engage with chemistry and physics, she focused primarily on two nineteenth-century concepts: the luminiferous ether, from physics, and the protyle, a speculative and relatively marginal concept in chemistry. With light construed as a wave, most physicists could not imagine it travelling through the vacuum of space without a medium. How could there be a wave without a medium? Some physicists supplied that medium with a hypothetical, super-attenuated invisible particle they called 'ether'. Albert A. Michelson and Edward W. Morley ran an experiment in 1887 designed to detect the presence of luminiferous ether, and though it had negative results, many scientists went to their graves believing that ether must exist.

An undetectable particle permeating all space and matter allowed Blavatsky – and some Victorian scientists – to speculate wildly about it. She was often vague about the modern science she characterised and typically quoted only writings by scientists sympathetic to spiritualism. These works included Balfour Stewart and P.G. Tait's *Unseen Universe* (1875), which found in Victorian ether physics and thermodynamics explanations of contact with spirits, as well as the most speculative work of Sir William Crookes. But the very notion of an undetectable particle espoused by orthodox physics gave Blavatsky room to posit other types of matter or substance. In *The Secret Doctrine*, she proclaims that the scientists' hypothetical ether is only a lesser explanation for something subsumed into greater physical and spiritual laws.[12]

Turning to chemistry, Blavatsky's engagement with the notion of a fundamental building block of all matter followed a similar logic. In the early nineteenth century, William Prout had theorised that hydrogen was that fundamental building block, which he called the 'protyle' (since measurements of atomic weights at the time suggested that all elements might be multiples of hydrogen). Crookes picked up this notion in the later 1880s, when Blavatsky was writing *The Secret Doctrine*. To occult writers on alchemy, such as the Golden Dawn co-founder Wynn Westcott, the protyle was clearly the alchemist's *prima materia*, and transmutations would soon follow.[13] In Blavatsky's 'occult metaphysics', the protyle was only a late development, after 'the Primordial Substance passe[d] out of its precosmic latency into differentiated objectivity, [and became] the (to man, so far,) invisible Protyle of

Madame Blavatsky and Colonel Henry Steel Olcott, in a photo at the then newly founded Aryan Theosophical Society of New York, 1888.

Science'.[14] Once again, she claimed that the contemporary sciences only addressed a very small part of the larger natural laws at play in a multiplanar world of emanation from the divine.

While Blavatsky explored contemporary physics and chemistry in a limited way, she was often more engaged with the work of the medieval alchemists, and her defence of alchemy was rooted in the reputations of the medieval and early modern scientists and alchemists who claimed to have witnessed successful transmutations of base metals into gold.[15] Ultimately, she primarily understood alchemy as a spiritual practice. In 'Alchemy in the Nineteenth Century', which ran in *La Revue Théosophique* in Paris from October to December 1889, she explained ancient alchemy in terms of a spiritual and material faculty developed in ancient humans. Over time, much of this spiritual power of antediluvian science – a kind of 'magnetic Magic' – was lost, and even the alchemists of the Middle Ages began to focus primarily on its material aspect. 'Thus,' Blavatsky laments, 'came to birth modern Chemistry': 'The Virgin-Substance, or Adamic Earth, the Holy Spirit of the old Alchemists of the Rosy Cross, has now become with the Kabbalists, those flunkeys of modern science, Na_2Co_3, *Soda* and C_2H_6O or *Alcohol*. Ah! Star of the morning, daughter of the dawn, how fallen from thine high estate – poor Alchemy!'[16]

A spiritual interpretation of alchemy was common in Theosophical circles. A version of it was conveyed to Hermetic occultists in Britain by Mary Anne Atwood's *A Suggestive Inquiry into the Hermetic Mystery* (1850) and in America by the work of another founding member of the Theosophical Society, Alexander Wilder, whose *New Platonism and Alchemy* (1869) made a case for a Neoplatonic origin of alchemy in the philosophical and religious melting pot of Ptolemaic Alexandria: 'Alchemy, therefore, we believe to have been a spiritual philosophy, and not a physical science. The wonderful transmutation of baser metals into gold was a figurative expression of the transformation of man from his natural evils and infirmities into a regenerate condition, a partaker of the divine nature.'[17] Chemistry emerged when later generations construed the mystical alchemy not allegorically but literally.

This initial Theosophical engagement with contemporary science in order to subsume it under a spiritualised ancient science was a form of what sociologists have called 'sanitisation', in which practitioners of what mainstream scientists see as 'pseudoscience' strategically imitate the institutions, forms of citation, references and even methods of mainstream science to win public validation.[18] Egil Asprem has identified three historical stages in Theosophy's engagement with science: a first period, from roughly 1875 (the date of the founding of the Theosophical Society) to 1891, the year of Blavatsky's death; a second period, roughly from 1891 to the 1930s; and a later period, after the Second World War.

Blavatsky's work obviously falls into that first period. Perceptively arguing that the changes in the 'Theosophy-science discourse' can only be understood in relationship to 'broader changes in the cultures of the sciences *themselves*', Asprem notes a major problem for Theosophy's scientific discourse, tied as it was to specific scientific paradigms and research programmes: those paradigms and programmes eventually fall out of favour. 'This fundamental problem of esoteric investments in

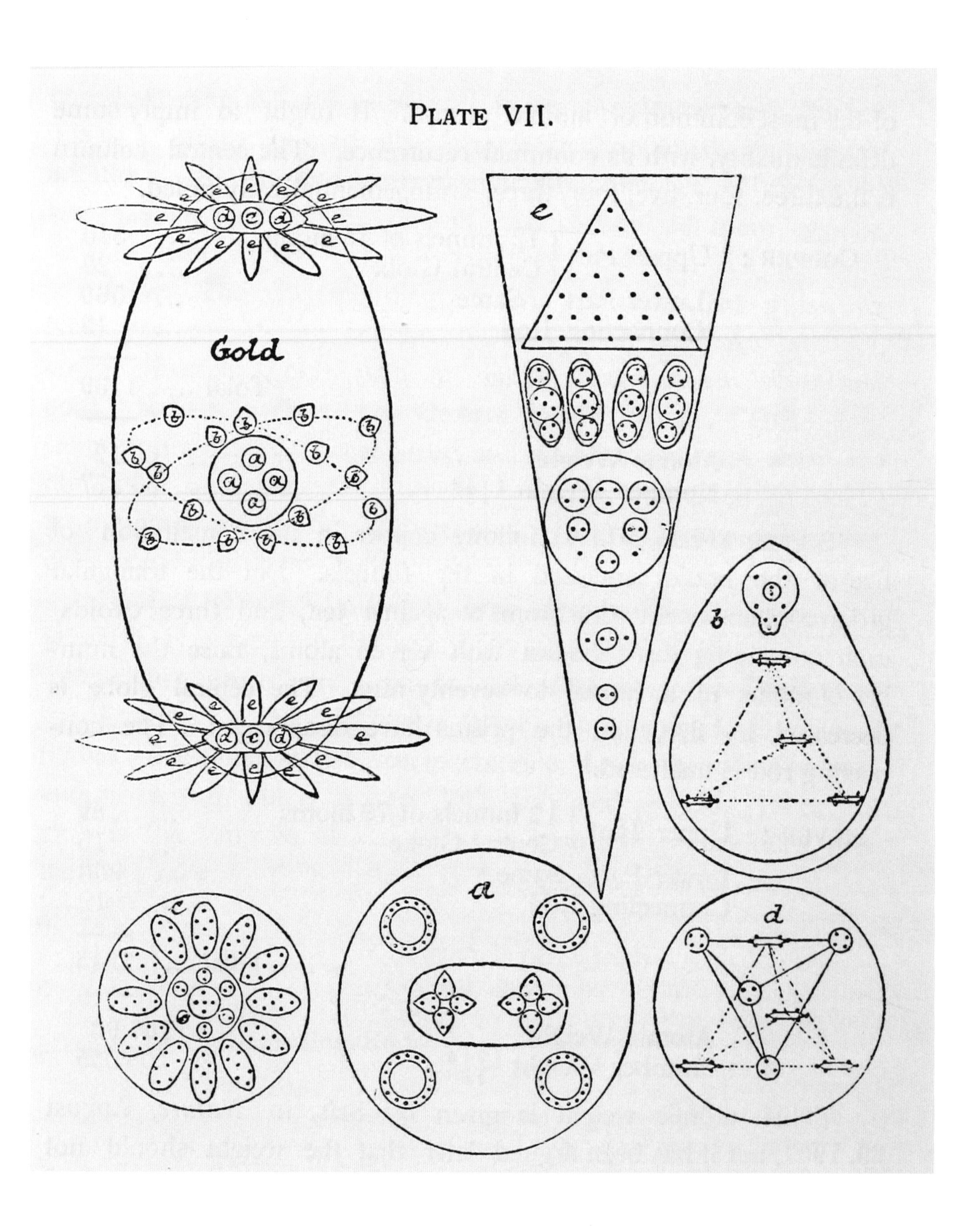

PLATE VII.

science,' he explains, 'has led to a constant recurrence of new waves of discursive updates in Theosophical science, necessitating conceptual and theoretical renewals while always passing on fossils of previous scientific cultures.'[19] By the late 1890s, physics and chemistry were both about to go through the major paradigm shifts at the heart of 'modern alchemy'.

For the first few decades of the twentieth century, an evocation of the mysterious 'alchemical' powers of radioactive substances and a fundamental shift in the conceptualisation of the atom brought cutting-edge nuclear physics into very close proximity to the occult understandings of the Theosophical Society and other forms of Western esotericism, sparking engagement between spiritual and scientific, material ways of understanding the universe. Writers in Theosophical magazines followed the scientific journals carefully and frequently made connections to alchemy, as when 'S. R.', writing for the *Theosophical Review*, followed Rutherford's publications between 1902 and 1906 on radium transformations in the most

Opposite:
Gold, plate VII from 'Occult Chemistry' by Annie Besant and Charles W. Leadbeater, 1908.

Plate 34, 'At a Funeral', from Annie Besant and C. W. Leadbeater's *Thought-Forms*, 1905.

significant physics outlet of the day, *The Philosophical Magazine*, and explicitly linked his research with alchemy.[20] Blavatsky had tried to describe an alchemy that looked back well beyond historical alchemy to a science of supposedly ancient origin, one that understood the multiplanar world in a way that nineteenth-century science could not. Theosophy's engagement with the modern alchemy of nuclear physics was no longer about chemical bonds or the composition of soda or alcohol but rather about the subatomic world – what could seem like another plane of existence. While Besant and Leadbeater's 'Occult Chemistry' article had preceded the discovery of radium and the transmutation of radioactive elements, the clairvoyance and spiritual alchemy that they had invoked across their occult chemistry experiments were undergirded by a vitalist particle cosmology – one in which the Theosophical life force, or 'Fohat', flows freely through subatomic geometries.

When radioactivity was discovered around 1896, many articles in Theosophical periodicals quickly turned to new scientific explanations, which featured a great deal of excitement about radioactivity as a bridge between life or spirit and matter. Indeed, periodical publication was an important communicative strategy for Theosophy's scientific claims. In nineteenth-century Britain, as Gowan Dawson, Richard Noakes and Jonathan Topham have explained, 'from the perspective of readers, science was omnipresent, and general periodicals probably played a greater role than books in shaping the public understanding of new scientific discoveries, theories, and practices'.[21] In her column in the *Theosophical Review*, 'On the Watch-Tower', Besant commented on many writings about radioactivity, noting, for example, an article on the origin of life by J. Butler Burke, a young physicist working in the Cavendish Laboratory at Cambridge, in a 1905 edition of the *Daily Chronicle*. There, Burke noted that radium may be 'that state of matter that separates, or perhaps unites, the organic and inorganic worlds'.[22] Burke, whose 'controversial work... comprised some of the first experimental work on the origin of life, wove radium into the history of life on the primordial earth'. He had placed radium salts in a sterilised beef broth, producing spots that resembled bacteria, which he called radiobes. Burke speculated that he had created evidence of an ancestor to life forms on earth.[23]

Theosophist Fio Hara argued that radioactivity overthrew 'all current and reputable theories of the constitution of Matter and its inherent quality'. He added, 'To radium we owe much, for it paved the way for patient investigations which absolutely revolutionized the world's ideas as regards the working of subtle forces, hitherto postulated but by a few brave pioneers in Nature's workshop and unacceptable to the rest.' Praising the research of the physicists J. J. Thomson and Joseph Larmor, which showed that radioactive emissions did not behave like normal matter (they imagined them to be electrons), Hara argued that this finding served as evidence for Theosophical claims about subtle occult particles and energies. He saw recent publications by scientists as confirming Besant and Leadbeater's claims about the structure of atoms.[24]

Radioactivity and new research on atomic structure over the next few decades gave Theosophists further confirmation that Theosophy was based on an ancient science, that it was correct and that it had much to offer contemporary science. Claims by both scientists and occultists that radiation might serve as some kind of elixir of life often contributed to new consumer fads, such as radium water – offered as a life-extender and revitaliser – and even radium cosmetics. (Luckily, much of that make-up contained no actual radium, though some of it did contain radioactive ingredients!) Within a few short years, committed Theosophists could publish accounts in Theosophical journals simply announcing developments in nuclear energy, deeming the science on this subject to be of importance to readers, even without making any direct connections to Theosophical ideas.[25]

While Blavatsky had adopted (at least in a limited way) the practice of citing other scientific works and inserting her claims into mainstream scientific contexts, Besant and Leadbeater took this even further. Especially after the turn of the century, as 'modern alchemy' became a common trope in discussions of nuclear physics, Theosophical occult chemistry played a significant role in a phenomenon that social scientists call 'boundary work'. This involves efforts by scientists to control what counts as 'science' by differentiating it from pseudoscience or other non-scientific practices, but it also involves a constant multidirectional effort to claim cognitive legitimacy. As anthropologist David J. Hess puts it, 'scientific boundaries are recursive, nested, and multiple; there are layers of scientificity that become clearer as one unfolds levels of skepticism and "pseudo-scientificity" both within and across discursive boundaries, not just in the direction of orthodox science toward religion and "pseudoscience"'.[26]

Besant and Leadbeater's occult chemistry suffered a significant legitimacy problem among many scientists. While it carefully followed conventions of scientific journal papers, including visual illustrations, tables and careful descriptions of methods, and provided the kinds of visual data that are often compelling in the construction of scientific facts, the 'instruments' used to obtain this data were purely the minds of the researchers. Other Theosophists would try to link powers of the mind to material explanations involving the new sciences – opining, for instance, that yogic powers might result from the mind's ability to control forces like those involved in radioactivity.[27] Besant herself had gestured toward the limitations of scientific instruments in comparison to the powers of the mind as the starting point for the 'Occult Chemistry' article in 1895: 'Of late years there has been much discussion among scientific men as to the genesis of the chemical elements, and as to the existence and constitution of the ether. The apparatus which forms the only instrument of research of the scientists cannot even reach the confines of the ether, and they apparently never dream of the possibility of examining their chemical atom.'[28] But, despite Besant's claim at the end of her 'paper' that 'the observations recorded have been repeated several times and are not the work of a single investigator',[29] for most scientists, no clear method of reproducibility in other labs could confirm or falsify their results. In this early period, as is the case today, nuclear physicists could only explore an invisible subatomic world indirectly, using instruments ranging from counting/logic devices, such as Geiger counters or scintillation screens, to visual image-producing devices such as cloud chambers, in which ionisation trails left by protons and other subatomic reactions could be photographed.[30] Though Leadbeater explained clairvoyance as involving something like a microscopic tube of etheric matter extending from the observer's third eye, this form of instrument was not rhetorically persuasive to chemists or physicists.

Yet the idea of Theosophical Science creating instruments based on what nuclear physicists were seeing as modern alchemy caught on in the imaginative culture of Theosophy. From its earliest decades, Theosophy had understood itself as offering followers a 'science of self-knowledge'.[31] After Besant and Leadbeater's shift toward engagement with the physics of a subatomic world Theosophists began to give a decidedly material and scientific basis to the interpretation of alchemy as spiritual alchemy (as advocated by writers such as Atwood and Wilder).

The broad cultural uptake of this synthesis of spirituality, psychology and nuclear physics can be seen in both pulp fiction magazines and books. The Theosophist and adventure writer Talbot Mundy's first major Theosophical adventure novel, *OM: The Secret of Ahbor Valley*, was serialised in *Adventure* in 1924, reaching several hundred thousand readers before being published as a book the same year. A major component of the novel was the mysterious Crystal Jade of Ahbor. To those who look at it, the giant stone reveals their true selves, both higher nature and lower, thus providing the self-knowledge needed for a spiritual self-transmutation. Characters in the novel conjecture that advanced ancient scientists, as far back as the Paleolithic period, created the stone as an instrument by harnessing the power of radiation or somehow trapping electrons in it. As one character argues, 'magic is merely science that hasn't been recognized yet by the schools'.[32]

OM: The Secret of Ahbor Valley, by Talbot Mundy, 1924. Cover artwork by Leonard Lester.

Historians and anthropologists offer two other ways we might understand the fertile interplay between science and occultism or spirituality around the trope of modern alchemy. One is the concept of 'knowledge in transit'. As Jonathan Topham summarises James A. Secord's concept, 'Questions of "how knowledge travels, to whom it is available, and how agreement is achieved" are fundamental to the making of knowledge, and in this sense the process of knowledge making *involves* communication, rather than merely being followed by it.'[33] In other words, language and writing don't simply communicate knowledge; they are involved in creating it. Moreover, the concept of 'domaining', from Marilyn Strathern's anthropological writings – adapted by Susan Squier for science studies – emphasises that ideas which 'reproduce themselves in our communications *never reproduce themselves exactly*. They are always found in environments or contexts that have their own properties or characteristics. These environments or contexts provide a range of domains...[and] insofar as each is a domain, each imposes its own logic of "natural" association.'[34] The ways in which we communicate scientific and other ideas, and the different areas of knowledge in which we discuss them, often allow new meanings to emerge in the discourse that wends across domains.

'Boundary work', 'knowledge in transit' and 'domaining' all help us see the remarkable roles that an understanding of nuclear physics as a modern alchemy could play in explaining the reality and consequences of an invisible world of mutable matter and energy. New developments in Theosophical science demonstrated by

OM
THE SECRET OF
AHBOR VALLEY
by TALBOT MUNDY
Author of
KING of the Khyber Rifles

Besant and Leadbeater's occult chemistry played a role in boundary work between science and occultism, raising questions about who got to claim knowledge as scientific, and on what terms. As Sumangala Bhattacharya has argued, 'Occult Chemistry' 'offers a type of "situated knowledge" that presents a compelling resistance to the rationalized and progressive historiography of atomic science and quantum theory, and thus to the hegemonic cultural authority of science'. The experiments challenged the universalising authority of scientific institutions and scientists of the period. Moreover, Bhattacharya concludes that in drawing Hindu epistemologies into the field of atomic science, Besant and Leadbeater's work 'appropriates a cultural moment in the British Raj and the Indian nationalist movement to articulate the yearning for an "outside" and dissident stake in knowledge production'.[35] These experiments can also be seen as part of the history of that hegemonic science, not just of occultism. As Philip Ball observed, Besant and Leadbeater 'were connected with several leading scientists of their age, some of whom read the book (with varying degrees of skepticism). Like it or not, *Occult Chemistry* is a part – a bizarre, even unsettling part – of chemical history.'[36]

At its core, the trope of 'modern alchemy' spoke to the implications of mutability at the bedrock of an invisible material reality. If atoms could transmute into other elements, could they be made to do so artificially by humans – through the use of a particle accelerator, or perhaps by the power of the mind? What would become of the basis of monetary exchange if no element could serve as a fixed token of value? What would become of society if the gold standard collapsed? What would become of religion if miracles could be shown to operate under scientific principles? Could control over a subatomic world contribute to a spiritual transmutation? What is the *meaning* of a dynamically mutable material world? That all of these disparate questions, and others, could be raised by the new science of nuclear physics when explored through the lens of alchemy shows just how important 'modern alchemy' could be in addressing fundamental questions of change in a dynamic period, with transformations in science, technology, religion, society and culture happening at a dizzying pace.

The alchemist's dream – to discover the secret of transmutation – had been achieved in physics labs in the early twentieth century. The multifaceted search for the meaning of this new science brought many different domains of knowledge and culture into intense and fertile communication, and in so doing made the alchemical production of gold one of the least consequential quests. Indeed, bismuth was transmuted into gold by a particle accelerator at the University of California at Berkeley in 1980, but at the cost of $10,000 to produce around a billionth of a cent's worth of gold – hardly a path towards alchemical riches or even a major scientific achievement. More importantly, modern alchemy saw major breakthroughs in the scientific understanding of a subatomic world that are still central to nuclear physics today and led to increasingly sophisticated instruments (including particle accelerators) to probe the nucleus and other subatomic particles. Likewise, Theosophy's efforts to propound a material, experimental science capable of bridging the apparent divide between mechanistic science and spirituality persisted across the twentieth century and beyond. It even continued to catch the attention of scientists, including Fellow of the Royal Society E. Lester Smith, the discoverer of vitamin B12, and Stephen Phillips, who graduated with a degree in theoretical physics from Cambridge University and earned a PhD from the University of California. Both published books in the early 1980s about the use of clairvoyance as a means of examining the nature of matter. Phillips even argued that Besant and Leadbeater had proven that clairvoyance exists, since they had viewed quarks decades before physicists had even theorised them.[37] Smith's and Phillips's books were published by the Theosophical Publishing House.

vocatus atque non ~ C. G. Jung ~ vocatus deus aderit

A PARTISAN OF THE SOUL

Carl Jung, the Death of God and Alchemy

KURT ALMQVIST

Western culture has been fundamentally shaped by secularisation and the idea of progress, the notion that, through the course of history, individual ideas and actions can continually improve human life, independently of divine grace. Charles Darwin's theories about natural selection, from 1859, inspired Herbert Spencer and other intellectuals, who popularised them and promoted the concept of progress. Ideas relating to evolution appeared in many guises, influencing everything from politics and philosophy to literature and art – and, not least, psychology – altogether marking the end of the Christian worldview's hegemony. Human consciousness had a still-unfinished history of development. Humans had become evolutionary beings, with apparently infinite potential to raise our consciousness.

Jung's ex libris with the Latin motto *Vocatus atque non vocatus deus aderit* ('Called or not called, God is present'), designed by the Swiss artist Claude Jeanneret and representing the Divided Self.

Enlightenment ideals had indeed long been prevalent among learned men and at enlightened courts, but it was not until the societal transformations of the nineteenth century, which were in many ways violent and brutal, that it became apparent that humans could influence and control development. Thanks to technological advances and capitalism, broad groupings in the West could now really put religious explanations to the test.

Criticism of the Bible

During the nineteenth century, the contradictions in the Bible, and its static view of the world, gradually became apparent; it lost its status as a historical source when it became the subject of critical study. The theologian David Friedrich Strauss (1808–74) used historical methodology to examine the Bible in *Das Leben Jesu* (1835–36) and found it to be an ideological product. The philosopher Ludwig Feuerbach (1804–72) showed that the Christian concept of God was a projection of humanity's own desires and beliefs in *Das Wesen des Christentums* (1841), and, in 1863, the philosopher Ernest Renan (1823–92) published his controversial book *La vie de Jésus*, arguing that Jesus was not the son of God but the presentation of an ideal human.

The attacks would continue unabated throughout the century, as expressed by the phrase often attributed to Karl Marx, that religion is an 'opium of the people' and a means for the ruling classes to exert control via an ideological superstructure.[1] *The Golden Bough* by Scottish anthropologist James George Frazer (1854–1941), published in two parts in 1890, helped to further relativise Christianity.

Most damning of all was the condemnation of Christianity by Friedrich Nietzsche (1844–1900) in *Der Antichrist* in 1888: '"Das Wort schon Christentum" ist ein Missverständnis – im Grunde gab es nur einen Christen, und der starb am Kreuz.' Nietzsche was influenced by Darwin in his work *Also Sprach Zarathustra* (1889), where he has Zarathustra preach that man is something to be overcome: we have evolved from worm to human. God is dead. Man is destined to be transformed into the Superman – a higher level than the present form of evolution. The proclamations of God's death were simply the culmination of a long period of religious criticism in the West, or, perhaps more accurately, in christendom. Still, the lost status of Christianity and the Bible not only created a fertile soil for atheists but also opened up a path for competing religions and beliefs, such as psychoanalysis.

After 1900 and Sigmund Freud's introduction of psychoanalysis, religion came to be widely regarded as a compulsive neurosis, as in *Zwangshandlungen und Religionsübungen* (1907). In *The Future of an Illusion* (1927), Freud, who examined religion in several books, writes:

> ... they are illusions, fulfilments of the oldest, strongest and most urgent wishes of mankind. The secret of their strength lies in the strength of those wishes. As we already know, the terrifying impression of helplessness in childhood aroused the need for protection – for protection through love – which was provided by the father; and the recognition that this helplessness lasts throughout life made it necessary to cling to the existence of a father, but this time a more powerful one.[2]

The romanticism of the Enlightenment

Ever since the Enlightenment, undercurrents in European culture have recognised the power of the unconscious and its impact on human life. In *The Discovery of the Unconscious: The History and Evolution of Dynamic Psychiatry*, Henri F. Ellenberger describes the background to movements that often make scientific claims and are rooted in shamanism, magic, mesmerism and other precursors of psychoanalysis and Jung's analytical psychology. Spiritism, spiritualism and turn-of-the-century movements such as Theosophy and anthroposophy believed they were in contact with an external and objective, hitherto unknown world, which could be explored in a manner equivalent to that in which natural science explored the material world. According to Ellenberger, dynamic psychiatry attempted to understand occult phenomena as intrapsychic forces, based on the perspective of reason or enlightenment. He illustrates the differences by comparing C. G. Jung and Rudolf Steiner: 'in most cases when Jung sees projected contents of the unconscious, Steiner is inclined to see independent spiritual beings'.

Attempting to establish the terms of the relationship between a spiritual and idealistic world and the physical world, during a time of crisis for post-Darwinian Christian doctrine, was part of the spirit of the age. There was great optimism that science could explore the domains of the mind and soul, as the Church's authority was slowly but surely being undermined by scientific findings, while many people were unwilling to submit to what could now be called an atheistic, materialistic, scientistic and positivist belief in science.

Jung tried to use analytical psychology to bridge the contradictions between science and religion, those caused by the acute conflict between faith and knowledge. He believed that the human soul possesses a religious function that can be described scientifically, thus also expressing the typical tendency to anchor metaphysical ideas in contemporaneous science.

Carl Gustaf Jung's paternal grandfather, of the same name, was the Grand Master of the Grand Lodge Alpina of Switzerland, and his books were probably found in Jung's childhood home. Within Freemasonry, Rosicrucian and alchemical ideas had long provided study materials and foundations for notions of spiritual transformation. Jung drew on the alchemic tradition of symbolic interpretation that had been cultivated within Freemasonry, choosing to study the alchemists whose work was characterised by the notion of *ora et labora* – 'pray and work'.

Carl Gustaf Jung, senior, *c.* 1850–64, around the time he was grand master of the Grand Lodge Alpina of Switzerland.

Jung's writings on the alchemical process describe what he, from 1916, called individuation: a natural path towards personal development, from an unconscious state to human maturity in the form of what, in Jungian terminology, is known as the Self. Individuation, Jung's integration of the Self, can be characterised as an ongoing natural process of maturation, in which previously unconscious material, including biographical memories and collective evolutionary memories, archetypal material, is integrated in consciousness.

Myth as personal destiny

Jung stated that for an individual's life to be perceived as meaningful, they must live what could be called their myth as personal destiny. Later, this idea was popularised by literature scholar Joseph Campbell in *The Hero with a Thousand Faces* (1949), which had a huge impact on the dramaturgy of Hollywood films. Campbell knew Jung and found inspiration in his 1911–12 publication *Wandlungen und Symbole der Libido: Beiträge zur Entwicklungsgeschichte des Denkens*, which examines the heroic journey into the underworld, the battle in darkness, victory over the powers of the underworld and the return to the light.

For Jung, the alchemical process thus forms a kind of objective correlate to his theories about individuation, a process of inner transformation and purification that leads to a higher level of consciousness. Jung's work amplifies and finds correspondences between the analysands' dream series and the stages of the alchemical process; perhaps best known is the amplification of the quantum physicist Wolfgang Pauli's dreams in 1936–37.[3]

The archetype theory that Jung used to understand spiritual alchemy also entails uniting contemporaneous evolutionary ideas with psychology. He also believed that archetypes originated in typical reaction patterns, which humans had acquired through natural selection as they evolved. This means he regarded archetypes as instinctual dispositions that can be activated by symbols – and he found these symbols in alchemy.

For Jung, the alchemical process of some alchemists is an example of their unconscious projection of what we can call the hero's journey into their work on the transmutation of metal from lead to gold. Here, however, one should ask how unaware these alchemists really were. We can be certain that they did not embrace a modern, psychoanalytical and secular interpretative framework, but the alchemists Jung utilised for his amplifications did embrace a Hermetic and Gnostic worldview,

Bull sacrifice to Mithras, illustrated in C. G. Jung's *Wandlungen und Symbole der Libido* (*Transformations and Symbols of the Libido*), 1911–12.

which included ideas about a spiritual journey and equivalent correspondences to this journey's stages on the path to higher consciousness. However, Jung does not launch a superhuman ideal like that of Nietzsche, but refers to the realisation of the Self as a means for achieving one's full potential, which includes operating in the world and the time into which one is born.

Valentinian Gnosticism, with its notion that humanity is a co-creator in the cosmos, is most reminiscent of Jung's ontological assumption of the *Wirklichkeit der Seele*, the reality of the soul, as the ultimate foundation for human existence and which, through a change of consciousness, can affect existence in a broad sense.

The Gnostic-Hermetic tradition

With his psychological teachings, Jung sought to recapture what contemporary materialism had conquered and plundered from Christianity. To this end, he first turned to the ancient Gnostics who, in turn, led him to Hermeticism. After 1911, there is almost no work by Jung that is not in some way influenced by the Gnostic-Hermetic tradition, or which does not have this tradition as its object.

In *Memories, Dreams, Reflections*, Jung writes that with the alchemical work *Mysterium Coniunctionis* (1955–56), he followed

'Third Picture of John' from the *Hexastichon of Sebastian Brant*, 1502. One of the woodcuts reproduced in Jung's *Mysterium Coniunctionis*, 1955–56.

> ...my original intention of representing the whole range of alchemy as a kind of psychology of alchemy, or as an alchemical basis for depth psychology. In *Mysterium Coniunctionis* my psychology was at last given its place in reality and established upon its historical foundations. Thus my task was finished, my work done, and now it can stand.[4]

Jung had been aware of the significance of alchemy from several sources from an early stage; it was also thanks to his client Sabina Spielrein who, as a child, had fantasised about the transformation of the soul in alchemical terms. He later read the psychoanalyst Herbert Silberer's book *The Hidden Symbolism of Alchemy* (1914), which argues that the images and ideas of Hermeticism clearly illustrate the psychoanalytic theory of the libido's transformation processes. In 1928, Jung became aware of a Chinese alchemical manuscript through a friend, the sinologist Richard Wilhelm (1873–1930). He wrote a preface for it when it was published the following year as *Das Geheimnis der goldenen Blüte*, edited and translated by Wilhelm (later translated into

Overleaf:
'The Eighth Key, from the *Twelve Keys of Basil Valentine* (published by Johann Thölde, 1599), and reproduced in Herbert Silberer's *The Hidden Symbolism of Alchemy*, 1914.

English as *The Secret of the Golden Flower*).[5] In his preface, Jung writes that before he was aware of alchemy, the only resemblances he had found with his own ideas were those of the Gnostics. In *Memories, Dreams, Reflections*, he states that the Gnostic principle lived on in secret, later appearing in medieval alchemy.

What Jung had found lacking in the Gnostics he found in alchemy: a direct correspondence to the images he encountered in his own dreams and visions and in the experiences of his clients. Whether the Gnostics created their own images during antiquity is unknown, and if so, they had long since been destroyed by the Church.

The alchemical process and analytical work

The Secret of the Golden Flower dates from the sixth century CE. It contains instructions for yoga exercises and an alchemical treatise; the text deals with the circulation of light and the creation of the 'diamond body'. The treatise states that the mental attitude of the alchemist is important for the outcome of the process, and it was in this process of separation and integration in Chinese alchemy that Jung saw parallels with his own analytical work.[6]

Jung was not alone in recognising that the aims and activities of alchemy were not primarily concerned with the production of gold. The historian of religion Mircea Eliade had written about alchemy in a similar manner but without linking the process to individuation. Alchemy, Eliade shows, exists all over the world; it is not only part of the Hermetic tradition. Much of Jung's interpretation of alchemy's focus is supported by historian Titus Burckhardt, who argues that the alchemists

Das Geheimnis der Goldenen Blüte: ein chinesisches Lebensbuch (The Secret of the Golden Flower), by Lü Dongbin. Translated into German by Richard Wilhelm with a commentary by Jung, 1929.

Richard Wilhelm's *The Secret of the Golden Flower* reproduced woodcuts from the well-known Chinese Ming text *Xingming guizhi*, 1622.

Ein lamaistisches Vajry-Mandala
zur Betrachtung zwecks Vorbereitung auf die Meditation

DAS GEHEIMNIS
DER GOLDENEN
BLÜTE

EIN CHINESISCHES LEBENSBUCH

Übersetzt und erläutert von
RICHARD WILHELM
mit einem europäischen Kommentar von
C.G.JUNG

DORNVERLAG GRETE ULLMANN MÜNCHEN

端拱冥心圖

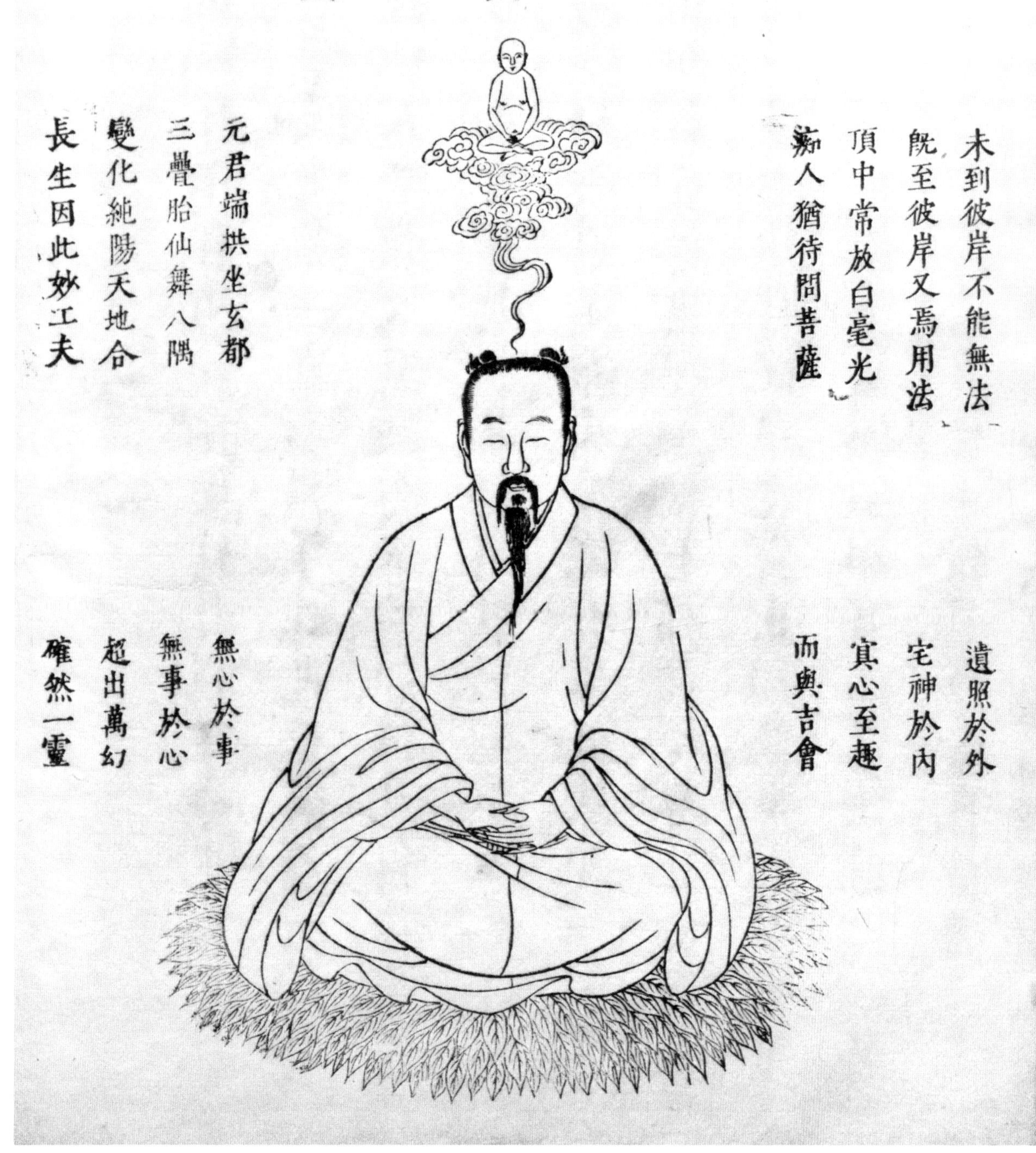

未到彼岸不能無法
既至彼岸又焉用法
頂中常放白毫光
痴人猶待問菩薩
元君端拱坐玄都
三疊胎仙舞八隅
變化純陽天地合
長生因此妙工夫
遺照於外
宅神於内
冥心至趣
而與吉會
無心於事
無事於心
超出萬幻
確然一靈

were primarily focused on inner work and only secondarily on producing gold. However, Burckhardt rejects Jung's hypothesis that the alchemists were unaware of what they were doing.[7]

Sketch of a mandala made by Jung in September 1917.

Western alchemy probably originated in the ancient cultural centre of Alexandria. Two strands can be distinguished: a lower level of artisanal metalworking and a higher level that uses metallurgical processes for the symbolic illustration of internal processes. This alchemy is highly syncretic and reflects the ancient worldview and its assumption of correspondences, how the microcosm reflects the macrocosm, the great celestial world. Similarly, alchemy contains the four humours and the four elements.

In this holistic worldview, mental attitude had a decisive significance for the occurrence of events and processes in the external; the right attitude could contribute to accelerating God's – nature's – work. The outer work, *laboratorium*, was thus linked to an internal process, *oratorium*: prayer. It is perhaps misleading to speak of an inner and outer world when trying to comprehend this worldview, as it appears that those who embraced it believed that everything was linked through a meaningful web of corresponding significances. This was a question of a heightened and deliberate adaptation to the course of nature, which is not to say that they did not also believe they could indirectly gain something for themselves. That alchemists were often aware of what they were doing is evident in the sixteenth-century treatise *Rosarium Philosophorum*, which Jung uses in his text *The Psychology of the Transference*. In the treatise, the expression *aurum nostrum non est aurum vulgi* – 'our gold is not ordinary gold' – indicates the author's desire to convey that he is not part of the lower alchemy.[8]

The alchemical process is described in various ways by different authors, with almost no two being alike. Jung chose to reproduce an ideal type that can be described in three, four and sometimes five stages, and, as such, the process can be described by the terms *solve et coagula*, i.e. dissolve and coagulate, which Jung believed correspond to the meaning of the terms 'analysis' and 'synthesis'. In the initial phase of the process, it is assumed that there is a state in which antagonistic forces do battle. The alchemical work, like analysis, then consists of a process in which these forces will ultimately be brought together in a unified, higher synthesis.

As in analysis, chaos – dissolution – must first be achieved through the processing of *prima materia*, or the analysands' life problem. The material must be separated into its various components by *solutio* and *putrefactio* to enable a new synthesis, *conjunctio*. However, the path is hazardous. Usually, the process fails repeatedly before the goal is achieved: the Self, the conjunction or marriage of sun and moon, king and queen. These stages are also denoted by a planet, a typical state and a colour, and have an almost infinite number of possible correspondences.

The final goal of the process is an entirely new quality, a qualitatively higher state that can have different names in different systems and is associated with an indestructible energy such as *lapis* (the philosopher's stone), *infans solaris* (the sun child), *filius philosophorum* (the philosopher's child), *aqua permanens* (the water of life) and other names which, according to Jung, correspond to his concept of the Self.

Carl Gustaf Jung reading in his study at his home in Küsnacht, Switzerland, 1960.

The marriage of true souls

In contrast to Freud, who interpreted incest fantasies as an expression of the Oedipus complex, Jung regards the incest motif from a religious perspective, where sexual intercourse as a symbol, as it is expressed in alchemy, plays a major role. In *The Psychology of the Transference*, Jung described how the integration of what he calls the Anima and Animus archetypes takes place in therapy, shown through references to alchemical images.[9]

The union of opposites, especially that between male and female, is critical to Jung's interpretation of alchemy. In *Memories, Dreams, Reflections*, when alluding to his concept of Anima, he describes how Salome, the most dangerous of women, finds herself in the company of an older man, Philemon. Jung says he discovered that the relationship between an older man and a younger woman was an archetypal motif, as in the case of the Gnostic Simon Magus and Helen, Klingsor and Kundry, Lao Tzu and the dancer.[10]

The unconscious produces the soul-image, said Jung, calling it Anima in men and Animus in women. This archetypal image, Anima/Animus, is a messenger from the other side. It appears in dreams and through projections onto members of the opposite sex. When Anima appears in a man's life, she does so in a way that means everything is explained and understood; Anima is the one to whom one cannot say no, for Anima's behaviour cannot be separated from the state of being in love. This figure is a complement to the conscious attitude; the person on whom you project is the bearer of qualities you have not developed and need to integrate. By recovering the projections of Anima and integrating them into your own personality, to then allow her power to be transformed into inspiration, a person individuates through creative expression.

We have long known that alchemy was not just a protochemistry, as positivist historians have tried to portray it.[11] The spiritual side of alchemy lived on in Freemasonry for centuries but was hidden from the common man. Since the eighteenth century, the king of Sweden has been the High Protector of the Swedish Order of Freemasons, and its members have been initiated into the secrets of alchemy. From the public literature on Freemasonry, we know that the Masons see alchemy as a figurative and symbolic representation of the life journey, strikingly similar to what Jung chose to call individuation's pathway to becoming human through solving the problem of alienation.[12]

Jung's analytical psychology, a form of modern theurgy that he tied to alchemy, was an attempt to provide contemporary people with a way of regaining admission to a lost inner religious landscape, the Gnostic experience of God, through an internal path of initiation, of which first the Church and then contemporary materialism and atheism had taken possession. In a letter in 1954 Jung writes:

> The cooperation of conscious reasoning with the data of the unconscious is called the 'transcendent function'. This function progressively unites the opposites. Psychotherapy makes use of it to heal neurotic dissociations, but this function had already served as the basic of Hermetic philosophy for seventeen centuries. Besides this, it is a natural and spontaneous phenomenon, part of the process of individuation. Psychology has no proof that this process does not unfold itself at the instigation of God's will...Begging you to excuse the somewhat heretical character of my thoughts ...[13]

Jung was a partisan of the soul.

Translated from Swedish by Clare Barnes.

Jung reading, 1950s. His library contained a vast collection of alchemical works.

Overleaf:
Woodcuts from *De alchimia opuscula complura veterum philosophorum* (1550), used by Jung to illustrate his *Psychology of Transference* (1969).

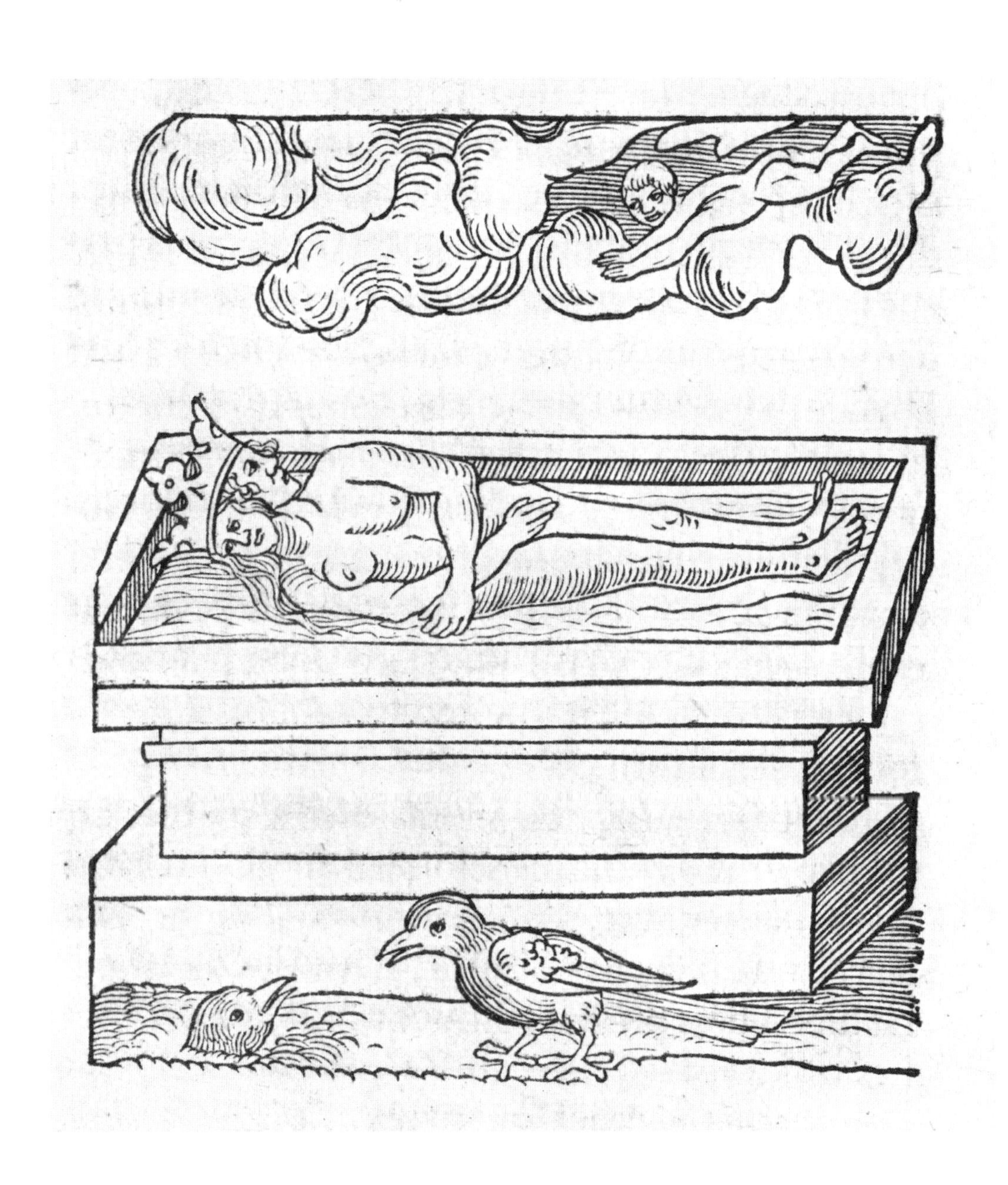

A LABOUR WITHOUT PAY

Alchemy in Literature

MATTIAS FYHR

One of the first classical examples of alchemy in literature is Dante Alighieri's (1265–1321) alchemists among forgers in Hell in his *Divine Comedy* (*c.* 1321), capturing two views of alchemists at the time: as scam artists or as people trying to emulate God, who in the words of one of Dante's alchemists, 'a skilful ape of nature was'.[1] A more complex literal example is 'The Chanouns Yemannes Tale' ('The Canon's Yeoman's Tale') in Geoffrey Chaucer's (*c.* 1343–1400) *Canterbury Tales* (*c.* 1400). Although often viewed as a criticism of alchemy, in 1924 S. Foster Damon showed examples of real historical alchemists at the time quoting from it, presenting Chaucer's alchemy as true. According to Damon, the end of the tale shows Chaucer had knowledge of real alchemy.

It might here be argued that as a literary work the ending of Chaucer's tale also follows from the previous parts, for instance in rhetorical devices such as the enumeration of alchemical materials and tools at the start of the tale. Also, it could be noted that alchemists are depicted simultaneously as forgers and as sincere: 'Yet is it fals, but ay we han good hope' ('Still it is false, but still we hope'). The tale ends with the Yeoman making a complete turn, from ridiculing alchemy to treating it seriously, with references to alchemical authorities like Hermes Trismegistus and concrete alchemical language such as: 'By the dragoun, Mercurie and noon other / He understood; and brimstoon by his brother, / That out of *sol* and *luna* were y-drawe.' True to alchemical practice at the time, Plato is mentioned, and finally it is declared that God does not want this alchemical secret revealed except 'wher it lyketh to his deitee' ('where it is pleasing to his deity') and that 'to werken any thing in contrarie / Of his wil, certes, never shal he thryve', stating the importance of submission to God.[2]

Milton's 'womb of uncreated night'

In the seventeenth century, alchemical imagery forms a natural part of the literary works of many authors, such as in Ben Jonson's (1572–1637) comedy *The Alchemist* (*c.* 1610), or when William Shakespeare (1564–1616) describes the sun of the morning kissing 'with golden face the meadows green, / Gilding pale streams with heavenly

Frontispiece of August Strindberg's alchemical *Antibarbarus*, showing the author in his laboratory, 1906.

Overleaf:
The alchemists sit among the falsifiers and forgers in Hell in Dante's *Divine Comedy* (*c.* 1321), as depicted by Sandro Botticelli in his coloured drawing for Canto XVIII, 1480s.

alchemy' in Sonnet XXXIII (1609), or in poems by the metaphysical poet John Donne (1572–1631).[3] At this time the alchemist and esotericist John Dee (1527–1608/9) was court astrologer to Elizabeth I, and physicist Isaac Newton (1642–1727) experimented in alchemy and left more than a million words in manuscripts on the subject, including a commentary on the *Emerald Tablet* of Hermes Trismegistus.[4] Alchemy was a system through which to view the world and its creation, based on Plato, Aristotle and Christianity, and used by John Milton (1608–74) in the Christian epic *Paradise Lost* (1667, 1674). Here Milton explicitly refers to alchemy, first when Satan lands on '[t]he golden sun in splendour likest Heaven', which is described in terms of metals and precious stones, and also a stone comparable to 'That stone, or like to that which here below / Philosophers in vain so long have sought, / In vain, though by their powerful art they bind / Volatile Hermes'. At 'one vertuous touch' of the sun, 'fields and regions... / Breathe forth *Elixir* pure, and Rivers run / Potable gold'. The 'arch-chemic sun', though remote, still affects earth when it 'Produces with terrestrial humour mixed, / Here in the dark so many precious things / Of colour glorious and effect so rare'.[5]

Engraving by Michael Burghers depicting Satan on the 'golden sun in splendour likest Heaven', 1688. From Book III of John Milton's *Paradise Lost.*

Milton believed matter was a form of the divine and shows this by comparing the Archangel Raphael's digestive system to that of humans, in that both 'concoct, digest, assimilate' food, thereby 'corporeal to incorporeal turn', which is again compared to alchemy, when 'by fire / Of sooty coal the empiric alchemist / Can turn, or holds it possible to turn, / Metals of drossiest ore to perfect gold'. In light of these explicit alchemical motifs and themes (and more), one interesting question concerns how prime matter, from which the world is made, is depicted here, in the form of the entities called 'eldest Night / And Chaos, ancestors of Nature', which 'hold / Eternal anarchy'. A fascinating idea in Milton is that visiting the domain of these entities means losing one's self, which is why the fallen angel Belial prefers his pain and existence in Hell to obliteration in night/Night, 'for who would lose, / Though full of pain, this intellectual being, / Those thoughts that wander through eternity, / To perish rather, swallowed up and lost / In the wide womb of uncreated night, / Devoid of sense and motion?' It also plays an important part in the role of Satan, since his offering to traverse this dimension proves his courage to the fallen angels and makes them choose him as their leader, which in turn is used by Milton to describe even these fallen angels as being above some men: 'neither do the Spirits damned / Lose all their virtue; lest bad men should boast'.[6] Satan then travels 'the void profound / Of unessential night', that 'with utter loss of being / Threatens him'.[7] Night and Chaos are described as 'The womb of nature, and perhaps her grave', that 'must ever fight, / Unless the almighty maker them ordain / His dark materials to create more worlds'. Their nature has been debated by scholars, since, as Malabika Sarkar put it 2012, 'How can good prime matter possibly be hostile?'[8]

Sarkar traces Milton's source of Night and Chaos to (the alchemist) Robert Fludd (1574–1633), who in *History of the Macrocosm and Microcosm* (1617) writes that before the creation there existed '[s]ome first state of unformed matter (*material prima*), without dimension or quantity, neither small nor large, without properties or inclination, neither moving nor still', which 'Paracelsus calls... the Great Mystery (*Mysterium Magnum*) which he says is uncreated; others claim it as God's first creation'. Sarkar, who doesn't mention alchemy in connection to this, points out that Milton differs from Fludd in that the latter describes Night and Chaos as 'entirely separable states existing at different points of time'. Milton, meanwhile, 'has brought them together as partners' which, according to Sarkar, 'deepen the horror of Chaos by imparting an added dimension of total privation to the hostility of Chaos'. She argues that possibly Milton 'regards Night as the womb of chaos, as chaos came into being from the original darkness' and concludes that if this is indeed the case, 'it creates an unnatural relationship between Night and Chaos with Night as both parent and partner of Chaos'. Finally, she notes that in 'the ultimate analysis, Milton's chaos is neither male nor female but androgynous'.[9]

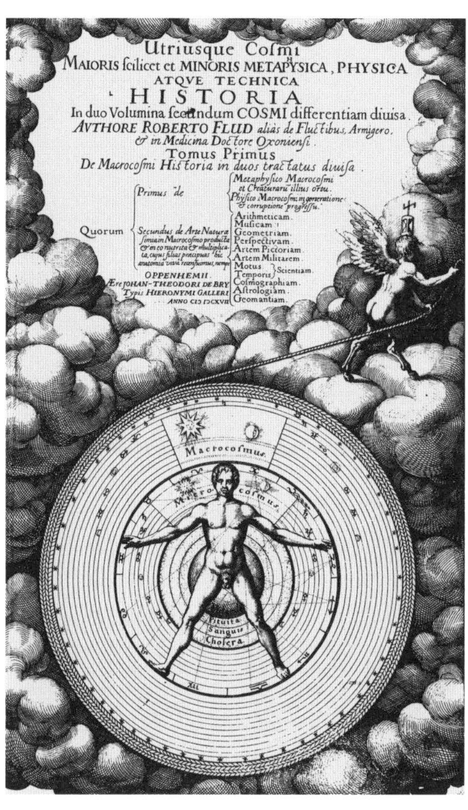

Title page of Robert Fludd's *Utriusque cosmi maioris scilicet et minoris metaphysica* (*History of the Macrocosm and Microcosm*), 1617–21.

If, however, we go back to earlier scholars, from before the present 'theoretical' turn, it was pointed out that one possible source for Milton was Neoplatonism. In 1947 the medievalist and poet Walter Clyde Curry based his view on what he called the generally accepted view, stated by C. G. Osgood, that Night is based on Orphic cosmonony as the first and oldest being. Curry reads Night and Chaos from the Neoplatonic Proclus's view that an ineffable God first manifested itself as one male and one female entity, which among later Neoplatonists came to be viewed as, for example, Night in Homer. Night and Chaos are then emanations of the One, manifesting throughout the world, for instance as the sun and the moon, and the question as to whether Night begets Chaos vanishes if both are manifestations of God. This reading would also fit Sarkar's observation of Eve's association with darkness in Paradise and her relation to Chaos as a mirror of Night, which Sarkar calls 'a tragic duality in Eve, which can be traced back to the basic materials out of which both she and the rest of the universe were created'. Curry points out that these 'male and female principles are found everywhere operative even to the last manifestation of divine power'. Neither Curry nor Sarkar discusses alchemy in connection to this, but I would venture the hypothesis that Night and Chaos, based on the alchemy in *Paradise Lost*, and Curry's Neoplatonical analysis, and these entities also manifestating as the moon (*Luna* in alchemy) and the sun (*Sol*), and Eve and Adam, might be inspired by the alchemical queen and king.[10]

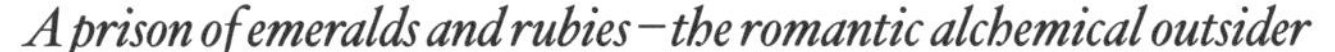

A prison of emeralds and rubies – the romantic alchemical outsider

During the eighteenth century, the Enlightenment pushed alchemy out of fashion, even though alchemical interest is still found in pre-romantic and romantic authors such as Friedrich Christoph Oetinger (1700–82) and the mystic Johann Heinrich Jung-Stilling (1740–1817). Both depict alchemy negatively compared to Christianity, the latter in his novels *Theobald, oder die Schwärmer* (1785) and *Das Heimweh* (1794–97).[11]

Alchemy did, however, make an important mark on romantic fantastic literature. William Godwin, after having published the Gothic novel *Things as They Are; or, The Adventures of Caleb Williams* (1794) and lost his wife, the famous feminist author Mary Wollstonecraft, as she gave birth to their later even more famous daughter Mary Shelley, in 1799 published the novel *St Leon: A Tale of the Sixteenth Century*. Tiring of war and luxury, Reginald, Count de St Leon, whose family have fled France and become poor, decides 'only to dedicate myself to the simplicity of nature and the genuine sentiments of the heart', as well as enjoying such beauties as 'are spread out before me by the Author of the universe'. He meets a nameless alchemist who persuades St Leon to accept his secret knowledge and vow not to tell anyone until the passing of 100 years. From this moment St Leon becomes estranged from human bonds and society, starting with his son leaving him and society casting doubt upon his sudden wealth. His secret also alienates his wife of 17 years. She guesses her husband's secret: 'The stranger who died your guest was in possession of the philosopher's stone, and he has bequeathed to you his discovery.' But she finds 'An adept and an alchemist' to be 'a low character' and not what she married: 'When I married you, I supposed myself united to a nobleman, a knight, and a soldier.' She stays with him but dies, and we follow the mistakes of St Leon attempting to use his powers on a grand scale for the good of people. His success comes at the end, when worked in complete secrecy.[12]

Alchemy also figures in Mary Shelley's *Frankenstein; Or, The Modern Prometheus*. Shelley acknowledges her father in her dedication of the novel to him and lets her protagonist Victor Frankenstein start his life reading alchemical authorities, Cornelius Agrippa, Paracelsus and Albertus Magnus. Later, Victor blames his reading of Agrippa for his ruined life, declaring that Agrippa's principles, which fired his imagination, 'were chimerical' in comparison to those of modern science, which 'were real and practical'.[13] Victor Frankenstein is not explicitly described as an alchemist, but what he creates is in line with what Paracelsus calls a homunculus and defines as a man 'generated without naturall Father, or Mother, i.e. not of a Woman in a naturall way', by 'a skillfull Alchymist'.[14] Just as in Godwin's novel, this leads to Victor losing his son and wife and to the death of an innocent servant. In describing his afflictions in Godwin's novel, the immortal alchemist St Leon might as well be describing those of the immortal monster in *Frankenstein*:

> How unhappy the wretch, the monster rather let me say, who is without an equal; who looks through the world, and in the world cannot find a brother; who is endowed with attributes which no living being participates with him; and who is therefore cut off for ever from all cordiality and confidence, can never unbend himself, but lives the solitary, joyless tenant of a prison, the materials of which are emeralds and rubies![15]

Another alchemically inspired story of this time is 'Der Sandmann' ('The Sandman', 1816) by E. T. A. Hoffmann. Here the boy Nathaniel witnesses his father's alchemical experiments, which later cause an explosion that kills his father, after which he is trapped in an obsession. Nathaniel's fiancée Clara tries to make him understand that 'The awful nightly occupation with your father, was no more than this, that both secretly made alchemical experiments', and 'your father's mind being filled with a fallacious desire after higher wisdom was alienated from his family'. Hoffmann plays with alchemy on many levels, likening Nathaniel's psychic obsession to an alchemical process with his mind as vessel, then asking us readers if we have experienced a similar passion and felt that there 'was in you a fermentation and a boiling', and much more.

One more example of the alchemist as standard literary motif in the nineteenth century is when Bram Stoker in *Dracula* (1897) has Van Helsing explain that the vampire Count Dracula 'was in life a most wonderful man. Soldier, statesman, and alchemist – which latter was the highest development of the science-knowledge of his time'.[16]

'Victor Frankenstein and his creature.' Frontispiece by Theodor von Holst to the 1831 edition of Mary Shelley's novel *Frankenstein; Or, The Modern Prometheus*.

Esoteric alchemy

During the '*Inferno* period' of his life in the 1890s, August Strindberg consciously adopted an occult worldview, and in a letter in 1896 he described a calling to become 'the Zola of the Occult'. Just as Émile Zola created the literary genre Naturalism, showing the influence of heritage and environment on the individual, Strindberg wanted to find a link between the material and immaterial, what he (and others) called Supernaturalism. Alchemy was a logical path in this and in 1894 he published what he considered a scientific work, *Antibarbarus*, on his alchemical experiments.[17] While this shows one material part of Strindberg's esoterical work, this time also saw alchemy interpreted as a purely spiritual quest and being combined with esoteric teachings on spiritual development by the Theosophical Society and the Hermetic Order of the Golden Dawn.[18] This purely esoteric view of alchemy in turn inspired the famous novel *Der Golem* (*The Golem*), published as a serial in 1913–14 and a book in 1915 by the esotericist and mystic Gustav Meyrink (1868–1932). He was born Meyer in Vienna and raised in Munich, Hamburg and Prague, possibly of Jewish descent, with a mother who was an actress and a nobleman father. The novel made him instantly famous and influences people to this day (such as in a novel by one of the crime writers behind the pen name Lars Kepler).[19] A golem is a humanoid, homunculus-like creature from Jewish legend, but it never appears in the novel, except in dream-like sightings. The narrator lies on his bed, leaves his body, enters the main character Athanasius Pernath in Prague and follows Pernath's often strange and experiences. For instance, the narrator learns that he must 'feel written letters, not just read them in books with his eyes'. At the end of the novel the narrator returns to his body and decides to go find Pernath, tracing him to 'Alchemist Alley' ('Alchimistengasse', which is a real place in Prague) and a double door, a hermaphrodite (an alchemical *rebis*), the left door male and the right female. The doors open and let him glimpse Pernath beside his love Mirjam, then close:

'and I see only the shimmering hermaphrodite'. Incidentally, Meyrink's novel was later used by psychoanalyst Carl Jung in his work on alchemy as manifestations of the subconscious in *Psychology and Alchemy* (1944).[20]

Alchemy and folktale

In the twentieth century alchemy was a motif and theme used in many ways in literature, but a knowledge of the tradition sometimes helps interpretation. 'Karln. En julsägen' ('The Man: A Christmas Tale') by the Nobel Prize winner and Academy member Selma Lagerlöf (1858–1940), written in the 1890s and submitted to a magazine but rejected and never returned to her, only to be found and published posthumously in 1949, tells of a house haunted by a male ghost. Artur, the ill, 16-year-old son of the house, who might not survive winter, returns from school for Christmas and the ghost appears more frequently. Artur says he'll solve the riddle of the ghost. For his siblings he invents a story about Anselm, a prior searching for the philosopher's stone, who brings his monks into the walled garden of the monastery during Midnight Mass, swinging a censer and singing hymns, mysteriously turning the garden into summer and attracting animals. A stork arrives, carrying a bottle of 'the

Gustav Meyrink

Der Golem

Ein Roman

Kurt Wolff Verlag München

Title page of Gustav Meyrink's *Der Golem* (*The Golem*), 1915.

water of life, the philosopher's stone...A drop of it was enough to make a human immortal, a drop was enough to turn all the cobblestones of Småland into gold.' This is said to be 'Nature's Christmas gift to the pious prior who had wanted Nature to partake in the joy of the Christmas night'. When Anselm is later about to drink a cup of wine with a drop of elixir the convent is attacked; an enemy soldier drinks the wine and smashes the bottle on the floor. The soldier later finds himself unable to die, associates this with the wine and returns to ask the prior but finds him long dead. The soldier then starts reading the books in the subterranean library of the ruined monastery, and Artur ends his story: 'He needed a hundred years to learn Latin, a hundred years to spell through Kabbalah, and had to live a hundred years more to learn the art of dying.'

Swedish author and alchemist August Strindberg. Self-portrait in Gersau, Switzerland, 1886.

To his siblings, Artur denies that the story is about the ghost and calls it an allegory of war. But he associates his story with the ghost, and when it one night beckons him to follow Artur wonders if he can bring peace to the ghost and whether it wants to show him a summer landscape. Moonlight falls on the ghost's face and the boy, terrified by its twisted expression, flees and starts to die. In spite of this, Artur defends the ghost, saying he would have died anyway, that he had misread the ghost's face and will follow it next time. He dies and the ghost never returns to the house. With the last sentence the narrator asks: 'How would one want to explain such a story?' This question has puzzled scholars.[21] From a literary alchemical perspective, however, it's about spiritual gold and the afterlife. The boy uses his story to help himself die, as in his story the soldier is said to learn the art of dying.

Alchemy is about death and resurrection, and in alchemical texts the phoenix (which is also found in Milton) is a symbol of this, of Christ and the death and rebirth from the *nigredo* stage. In Artur's story a stork delivers the philosopher's stone, and possibly Lagerlöf had in mind that the stork in Nordic folklore is often seen as the bird who brings children to mothers, even into the womb of mothers. At the end it is stated that after Artur died the nearby monastery ruins are searched but nothing is found, but in creating his story to deal with death Artur does what the narrator of Lagerlöf's tale does in the prologue when asking if 'the eternally surging flood of folktales bring any snippets of the gold of truth' and if 'all this talk about spirits and visions bear witness of the eternal life of the souls'. The narrator answers this with the mindset of an alchemist from, for instance, Chaucer: 'These are questions without answers. It is a panning with a sieve, it is a labour without pay, and still I am compelled to listen and to listen again.'[22]

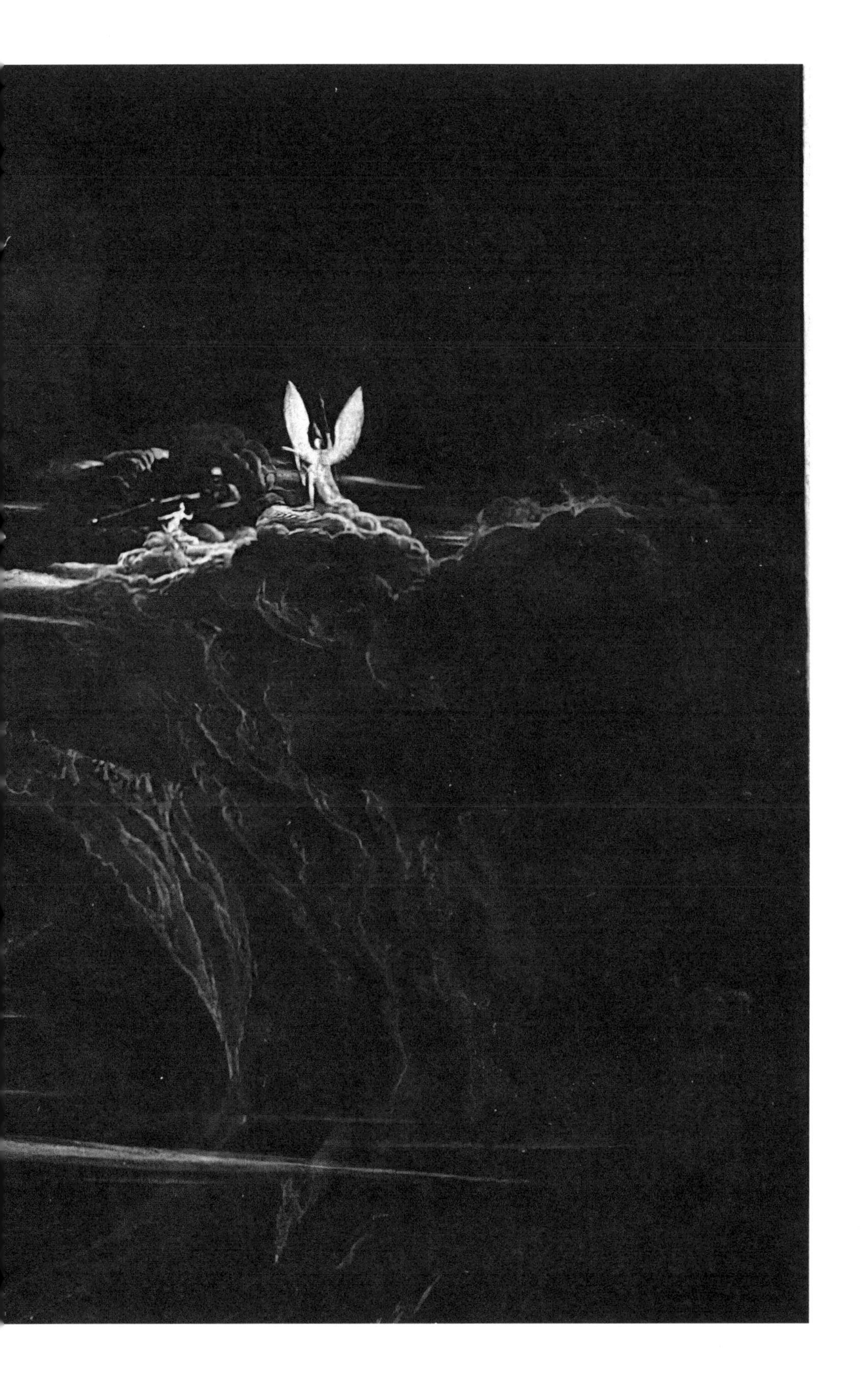

'The Bridge over Chaos'. Mezzotint by John Martin for *Paradise Lost*, Book X, 1824–26.

'A WINE THAT WAS DRUNK BY THE MOON AND THE SUN'

Alchemy in Surrealism

PER FAXNELD

Ciencia inútil o El alquimista (*Useless Science or The Alchemist*) Oil on masonite by Remedios Varo, 1955.

It is a mainstay in art criticism and scholarship to describe surrealists and related artists as alchemists.[1] At times, this is merely a vague analogue, but, as we will see, one actually reflected in alchemy's presence as a central figure of thought in key surrealist texts. Moreover, it has a long pre-history.

The Flemish painter David Teniers the Younger (1610–90) famously painted several works portraying alchemists toiling away in their laboratories. Unlike earlier artists, who tended to portray alchemists negatively, Teniers was more respectful. Furthermore, he painted a self-portrait (1680) depicting himself as an alchemist, thus establishing the notion of a parallel between the artist's work with colours and the alchemical process. Several of his colleagues were inspired and produced similarly reverential alchemical paintings.[2]

The metaphorical likening of artistic creativity to alchemy proved enduring. For example, art critic (and much else) Walter Pater's (1839–94) celebrated 1869 essay on Leonardo (whose work, it should be emphasised, was in reality not particularly influenced by esotericism) repeatedly described him as an alchemist and claimed that the pupils he chose were men possessing 'enough genius to be capable of *initiation* into his secret'.[3]

Let us now turn to the initiatory secrets of the surrealists, or, perhaps, the lack thereof in many cases. One of numerous early twentieth century avant-garde movements, surrealism took shape around 1920 in Paris. As was common practice in such undertakings, the surrealists produced bombastic manifestos, with the second, in 1929, seeing their main theorist André Breton (1896–1966) calling for 'the profound, the veritable occultation of surrealism'.[4] But what did this mean? Not that surrealists in general were, or would become, practising occultists in any conventional sense. Nor that any part of occult teachings was really accepted indiscriminately by most of them. It did, however, signify the great value Breton and his cohort attached to esoteric traditions as a reservoir of techniques and symbols that could, alongside, for example, 'primitive' art, be put to creative use in the surrealist project of ontological revolt, political societal subversion and artistic renewal.

Self-Portrait as an Alchemist, 1680. Oil on panel by David Teniers the Younger.

A key impetus came from Sigmund Freud's ideas about the primordial origins of art in magic.[5] The surrealist framing of esoteric material, therefore, consistently privileges an ultimately 'secular' interpretation (or at least one where transcendence is located in art and the unconscious), with the 1929 manifesto highlighting the 'remarkable analogy, insofar as their goals are concerned, between the Surrealist efforts and those of the alchemists' in terms of a liberation of the imagination that breaks 'the mind's domestication'.[6] Breton's understanding of the alchemical philosopher's stone is thus 'that which was to enable man's imagination to take a stunning revenge on all things'.[7] As surrealism was a multifaceted, anti-authoritarian movement (and quickly globalised to boot), not all participants adhered to this 'mainline' approach to alchemy.

During the war and in its aftermath, surrealists intensified their employment of esoteric material, as attested by key texts like *Arcane 17* (1944), in which Breton draws heavily on the tarot and occultist Eliphas Lévi (1810–75) as well as referring to the alchemists Nicolas Flamel (1330–1418) and Paracelsus (1493–1541, lauded by Breton as one of the 'adventurers of the mind').[8] The grand exhibition *Le Surréalisme en 1947* in Paris was filled with magical and occult objects on altars and included references to the tarot and Swedenborg, and the visitor's route climaxed with passing through an initiatory labyrinth. However, Breton maintained surrealists should assume an attitude of 'enlightened doubt' in relation to the realm of magic and esotericism and emphasised that the labyrinth did not offer an actual initiation but a 'guideline'. Esotericism, then, was a means to an end, not a worldview to accept in its entirety, and must, he insisted, remain subsumed under the greater goals of surrealism itself. This attitude also permeates Breton's lavishly illustrated *L'Art magique* ('Magic Art', 1957), where the rhetoric chiefly situates magic in art itself.[9] At the same time, the book is bursting with imagery from historical esoteric traditions – including, prominently, alchemy.[10]

Alchemical references appear in many surrealist works. One notable example is *The Philosopher's Stone* (1940) by Victor Brauner (1903–66).[11] The title, of course, refers to the miraculous substance that alchemists strive to create, which can cure all illnesses, bestow eternal life, transform base metals into gold, and many other things. The painting depicts the arms and torso of a woman dissolving into shimmering spheres of light as she straddles a branch of a tree (or, possibly, has a portion of the branch penetrate her) crowned by the glowing philosopher's stone. Above the stone, an indeterminate animal with glowing eyes stands or hovers – part bird, part dog or cat. In keeping with the alchemical theme, the *solve* of the human form – and the dissolution of animal taxonomy as well as animal-plant dichotomies – lead to the *coagula* of an amalgamated higher order of being.

Kurt Seligmann's (1900–62) canvas *The Alchemy of Painting* (1955), meanwhile, highlights the parallels between art and alchemy as liberatory practices, following the argument in his book *The Mirror of Magic: A History of Magic in the Western World* (1948). The painting shows a deconstructed figure – combining plant-like, animal and mechanical-looking parts – set against a geometrically fragmented background reflecting the malleability of matter central to alchemy as well as art.[12]

One of surrealism's great innovators was Max Ernst (1891–1976), who in a 1937 essay famously described collage, one of his central working methods, as 'something like the alchemy of the visual image'.[13] Often read as a purely metaphorical statement, M. E. Warlick has demonstrated conclusively that it in fact signalled how deeply affected by alchemy Ernst was in both his art and thinking, though he 'rarely created replicas of alchemical emblems'.[14]

We can, however, find laboratory equipment (a so-called pelican vessel) combined with birds likely signalling an alchemical symbolism in *The Laugh of the Cock* (one of the collages from *Une semaine de bonté*, A Week of Kindness, 1934), while Ernst's enigmatic *The Robing of the Bride* (1940) has been interpreted by Warlick as signifying the 'chemical wedding' of sulphur and mercury and 'the reddening phase of rubedo, the sexual conjunction of the King and Queen'.[15] The same (al)chemical wedding is also present in the titles of Ernst's frottage *Les Noces chimiques* ('The Chemical Wedding', 1923) and his painting *Chemical Nuptials* (1948). The latter, Warlick points out, includes a red alembic vessel of the type employed by alchemists in the background.[16]

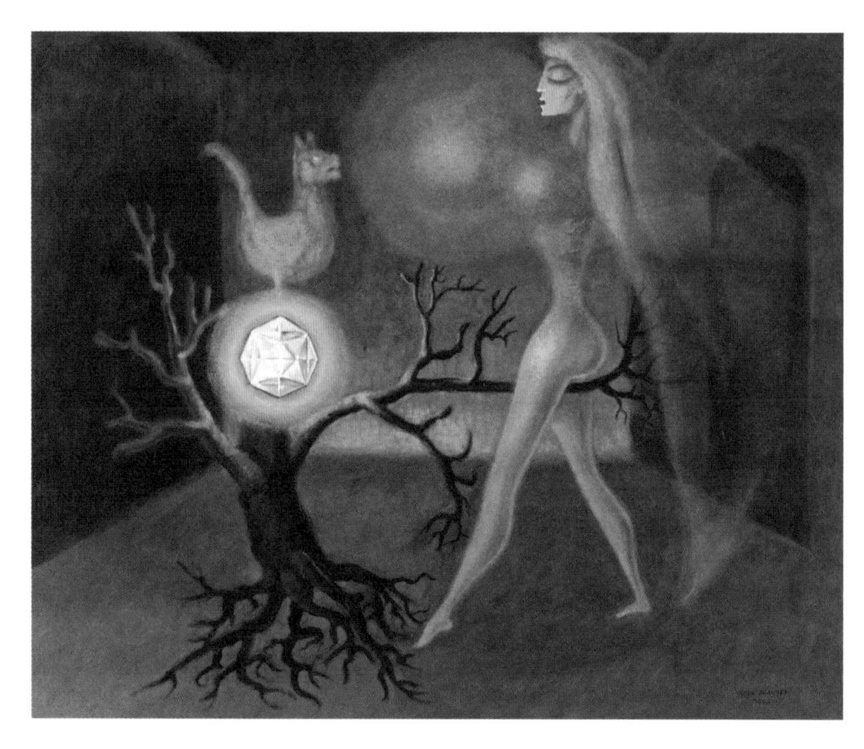

The Philosopher's Stone. Oil on canvas by Victor Brauner, 1940.

In the 1930s, Ernst had a dramatic relationship with Leonora Carrington (1917–2011), who is today considered one of the most important creative figures to have emerged from the surrealist movement. It has been said that 'Carrington, more than any other Surrealist, went on to explore a wide range of esoteric themes'.[17] She was also one of few surrealists to actually practice esotericism in a more hands-on sense, and her engagement with alchemy was deep and sustained.

Carrington's first encounter with surrealism seems to have been through Herbert Read's book *Surrealism* (1936), when she was nineteen and at art school. Interestingly, the book explicitly connects surrealist art with alchemy. Around the same time, Carrington started to read books on the history of alchemy.[18] She and Ernst subsequently explored this theme together, as evidenced by the alchemical symbolism expressed in the animal hybrids with which they decorated their shared house in France.[19]

When the war broke out, Ernst was incarcerated because he was German, and Carrington became psychotic, leading to her internment in a mental hospital. This harrowing experience resulted in the autobiographical account *Down Below* (1944), where she reinterpreted her suffering through an alchemical lens. For example, she writes: 'I was transforming my blood into comprehensive energy

– masculine and feminine, microcosmic and macrocosmic – and into a wine that was drunk by the moon and the sun.'[20] She was aided in this reframing by the French physician, occultist and Freemason Pierre Mabille (1908–52), who had worked closely with Breton and taught him geomancy and astrology. A key figure in surrealism's engagement with topics like esotericism and initiation during the war years, Mabille published the book *Mirror of the Marvelous* in 1940. In line with Breton's assertion in the first surrealist manifesto that 'only the marvelous is beautiful', Mabille's book provides text excerpts ranging from gothic novels and modern poetry to a Masonic ritual of initiation and alchemical works.[21] 'As with folklore, we are in the presence of a timeless tradition', Mabille comments on the text 'Manufacturing the Stone' by Basile Valentin.[22]

Reading *Mirror of the Marvelous* and engaging in intense conversations with its author enabled Carrington to understand her breakdown as an alchemical, initiatory experience – part of said 'timeless tradition'. As Kristoffer Noheden explains, Mabille helped Carrington recognise 'planetary symbols and alchemical imagery which correlated with the signs she used to transform mundane things with during her psychosis, a connection that now lent her delusions a heightened esoteric

Le Rire du coq 2, L'Ile de Pâques (*The Laugh of the Cock, Easter Island*). Engraving by Max Ernst after a collage from *Une Semaine de bonté* ('A Week of Kindness'), 1934.

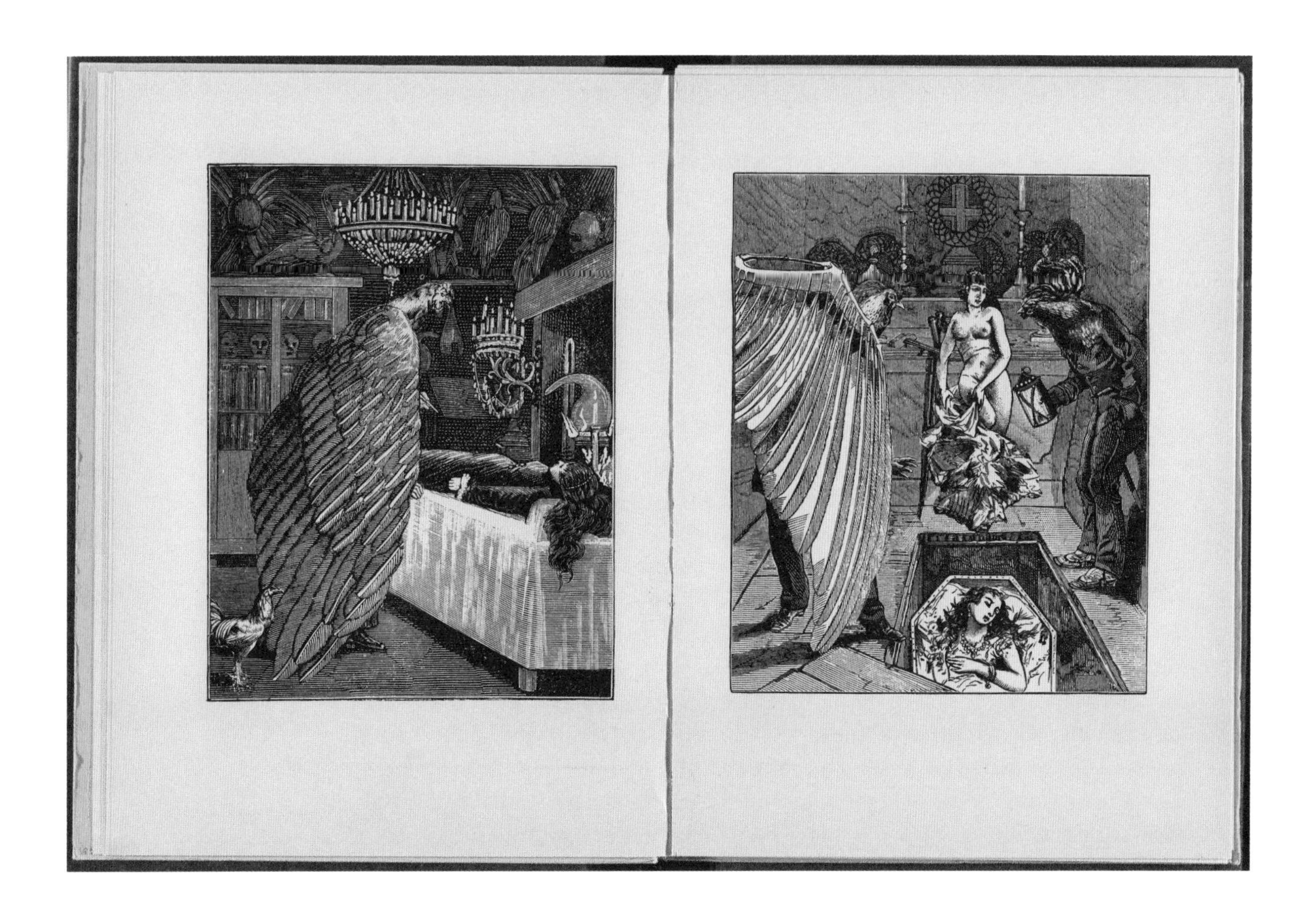

significance'.[23] Mabille fully affirmed this, and in a 1946 essay he commented that Carrington's book resembled an alchemical manuscript.[24]

In 1942, Carrington moved to Mexico, where she would remain for most of her life, and continued immersing herself deeply but eclectically in several esoteric traditions. Simultaneously, she led a family life with her Mexican husband and two children, ingeniously managing to turn that into an alchemical working as well. As Susan Aberth explains: 'The transit of food from the kitchen to the table to consumption was, in particular, likened to alchemical processes of distillation and transformation, which in turn led to associations to art production.'[25]

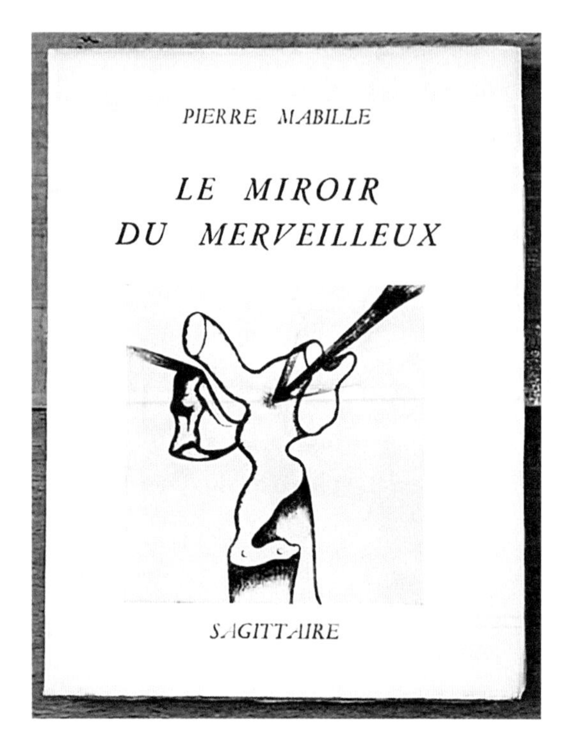

Le Miroir du Merveilleux (Mirror of the Marvelous), by Pierre Mabille, 1940. Cover illustration by Yves Tanguy.

In Carrington's paintings from the long Mexican period, the references to alchemy are generally quite vague. We can, for example, find figures like the androgyne – certainly familiar from alchemical symbolism but just as present in a great many other systems of thought. There are also many cauldrons, described by Aberth as symbols 'of fertility and abundance, mystic transformations both alchemical and magical'.[26] Even when there appears to be a quote in Latin from an actual alchemical text in a painting (*AB EO QUOD*, 1956), however, this seems to not be the case when scrutinised more closely.[27] The presence of, say, an egg and a white rose in the same work[28] certainly seems to point towards well-established alchemical symbols that Carrington would have been aware of, but it always remains difficult to attach any conclusive meaning to the imagery in her works.

On the other hand, Carrington could on rare occasions set out her understanding of specific symbols and their correspondences quite plainly, as in this esoteric explication of cabbage: 'The cabbage is a rose, the Blue Rose, The Alchemical Rose, The Blue Deer (Peyote)... The cabbage is still the alchemical Rose, for any being able to see or taste.'[29] It is likely that this alchemical-psychedelic (blue) rose is what we see in the centre of the kitchen table, placed in a magic circle, in the celebrated *Grandmother Moorhead's Aromatic Kitchen* (1975), which is also yet another example of Carrington's subversive enchantment of the domestic realm.[30]

Living in Mexico City, Carrington explored the local markets in search of magical herbs together with her equally esoterically inclined artist friend Remedios Varo (1908–63).[31] Several of Varo's precise, exquisite paintings thematise alchemy. In 1955, for example, she made the cheekily titled *Useless Science or The Alchemist*, where the alchemical apparatus – incorporating towers, bells, horns, wooden cogs and a device for collecting rain for the distilling process – is operated by an androgynous figure wrapped in a chequered cloak that metamorphoses into the chequered floor (accentuating the theme of transmutation). It has been suggested Varo's specific interest in alchemy was rooted in the history of her native country, as the 'sages of Spain influenced by Islam had been astrologers and alchemists', while the depiction of the sky in *Useless Science* was created using a chance technique related to automatism: 'repeatedly blowing or smearing wet paint along the sky to create mysterious nuances'.[32] *The Call* (1961) sees Varo combine alchemy (the female figure clutches an alchemical retort and her pendant is a mortar and pestle) and astrology (her fiery hair extending to the sky, becoming the tail of a comet or the ring around a planet).[33]

Another friend of Leonora Carrington's in Mexico was Alejandro Jodorowsky (b. 1929). In his autobiography, Jodorowsky claims his Zen master sent him to Carrington for spiritual instruction and presents a series of unlikely anecdotes about their interaction.[34] Be this as it may, she clearly exerted an influence on Jodorowsky's voluminous production, which includes some 40 books, 80 graphic novels and nine feature films. After attending the Paris surrealist group's meetings for two years as a young man, Jodorowsky had started to find Breton too dogmatic. He would later claim one of his major gripes was highbrow contempt of comic books, science fiction and rock 'n' roll.[35]

Moving into feature-length filmmaking in the late 1960s, Jodorowsky's breakthrough came with the cult classic *El Topo* (1970). This transgressive 'Acid Western' won popularity with hippies – including John Lennon, who became so fascinated he

Grandmother Moorhead's Aromatic Kitchen. Oil on canvas by Leonora Carrington, 1975.

arranged funding for Jodorowsky's next project, the alchemical epic *The Holy Mountain* (1973). The director had ambitions far beyond making entertainment or even art-house cinema, claiming 'I ask of film what most North Americans ask of psychedelic drugs' and explaining the ultimate purpose of *El Topo* had been '[t]o create mental change' and 'to reach the state of enlightenment', adding, 'This is LSD without LSD.'[36] As we will see, however, his next film was not entirely without actual LSD.

The Alchemist's Workshop, installation by Jan Švankmajer in his Kunstkamera, Czech Republic.

While surrealism scholar Abigail Susik has analysed *The Holy Mountain* as a satire of spiritual quests[37] it also typifies the ambiguity of surrealist methodologies wherein the satirical has a dimension of complete metaphysical earnestness. Whether *The Holy Mountain* is a work of 'pure' surrealism is debatable, but it decidedly inscribes itself in this cultural method in many ways.[38] The script was (very) loosely based on the unfinished novel *Le Mont Analogue* ('Mount Analogue', 1952) by René Daumal (1908–44), who was not a surrealist but interacted frequently with the movement.[39]

The film first introduces us to a Christ-like thief, who subsequently breaks into a tower containing an occult temple decorated with clear colours and geometrical shapes not unlike a three-dimensional rendering of one of the initiatory works by Hilma af Klint (1862–1944) or Eranos founder Olga Fröbe-Kapteyn (1881–1962). There is also a camel, a kabbalistic symbol of the devil figure Samael (a connection underscored by the presence of a woman with Hebrew writing on her body and an inverted pentagram on her hand) and a black pillar next to a white one (Masonic symbolism seen in some renderings of the tarot card The Priestess).

The tower's owner, an alchemist played by Jodorowsky himself, has the thief ritually defecate into a glass vessel and then transmutes the faeces into gold in his alchemical laboratory, telling the intruder: 'You are excrement. You can change yourself into gold.' The alchemist, the thief and seven affluent people representing the seven planets then embark on a quest for spiritual enlightenment, seeking the eponymous holy mountain where nine immortals live, 'having conquered death'. A series of tribulations, group rituals and the consumption of psychedelic substances follow. The mountain itself serves as a metaphor of initiatory ascent, just like the pyramid does for Hilma af Klint and Olga Fröbe-Kapteyn. Yet the stated goal here is to aggressively 'conquer the wisdom of the immortals' – far from the reverential approach to hidden, enlightened mountain sages in the Theosophy that af Klint and Fröbe-Kapteyn drew on. It is also explained that 'the philosophical stone of the alchemists was LSD', conflating lofty historical alchemy with the subcultural rebellion of Jodorowsky's time.

A recurring symbol in the film is the nine-pointed enneagram, derived from George Ivanovich Gurdjieff (d. 1949) and elaborated upon by his disciple Pyotr Ouspensky (1878–1947). Indicating, among other things, nine personality types, it constitutes a sort of macrocosmic map, which Gurdjieff stated contains the secret of the philosopher's stone.[40] Jodorowsky similarly asserted there was more to his film than meets the eye, declaring that 'In *Holy Mountain* you do not follow a story, you follow an experience with me and the actors...It is training.'[41] This ambition is

rooted in Jodorowsky's earlier work in theatre, where he stated: 'I want to reach a mystical theatre characterized by the search for self; a kind of alchemist theatre, where man changes, progresses and develops all his potentials.'[42]

The main actors prepared for three months through spiritual exercises led by Oscar Ichazo (1931–2020) of the Arica Institute, which blended Zen, Sufism, Yoga, Kabbalah, I Ching and Gurdjieffian techniques. Filming the scene where the group of seekers die and are reborn, Jodorowsky – aiming to make the production an actual initiatory journey for the actors – gave them psilocybin mushrooms, and he himself had been instructed by Ichazo to take LSD.[43]

The Holy Mountain ends with the alchemist breaking the fourth wall to address the viewer, divulging that he is in fact the director of the film, while the camera zooms back to reveal the lighting and sound technicians. Through this collapsing of the film's diegesis and the world of the audience, we become present and included not only in the fictional narrative but also in the alchemical-initiatory journey undertaken by the ensemble.

Surrealist factions appeared in many countries, but not all were as directly esoteric in their interests as the Parisian and Mexican contingents. For example, Sweden's Halmstad Group (founded in 1929) did not incorporate such material to any great extent. However, exactly two decades later, in 1949, the Dane William Freddie (1909–95) and others organised a surrealist exhibition in Stockholm – *Surrealistisk manifestation* – where magical and esoteric themes took pride of place.[44]

The Stockholm Surrealist Group, founded in 1986, came to dive deeply into the esoteric waters, drawing on the mysticism of Emanuel Swedenborg (1688–1772), the occult painting of Hilma af Klint, Norse rune magic and many other things. Alchemy also featured prominently.[45] A prime mover in these explorations was Carl-Michael Edenborg (b. 1967), who would in 2002 obtain a PhD in intellectual history at Stockholm University with a study of the marginalisation of alchemy and wrote the alchemistic novel *Alkemistens dotter: slutet på universum* ('The Alchemist's Daughter: The End of the Universe', 2014).

The Stockholm surrealists relished the marginal and outré, embracing Swedish alchemists like the bizarre Gustaf Bonde (1682–1764), whose 'excrement alchemy' entailed swallowing a golden pearl every day, picking it out of his chamber pot and then repeating the procedure perpetually – believing this would ultimately turn the gold into the philosopher's stone. Another favourite was the subversive (though periodically enjoying royal patronage) Swedish-Finnish alchemist August Nordenskiöld (1754–92), who attempted to crush the 'tyranny of money' by creating so much gold that it would become worthless. To a degree, the special fascination of such figures for Swedish surrealists was perhaps related to the pervasive rationalist conformism bound up with the country's secular welfare state.[46]

Local traditions and conditions also marked the politically repressed environment of post-war Czechoslovak surrealism, where engagement with the starkness of material reality and everyday objects became central. Joining the Prague surrealists in 1971, the filmmaker and assemblage artist Jan Švankmajer (b. 1934) would reintroduce more magical themes, mostly adhering to the traditional (Bretonian)

Still from Alejandro Jodorowsky's film *La Montagne sacrée* (*The Holy Mountain*), 1973.

ambiguously detached approach to them. Esotericism informs a large portion of his internationally celebrated body of short and feature films, and in the *Alchemy* cycle of objects he involves himself directly with this tradition.[47] Combining animal skeleton parts, cutlery and many other disparate objects, the assemblages also at times feature shapes of traditional alchemical laboratory equipment – but the vessels have grown feet, as if freed from their lifeless state and embarking on an alchemical progression of their own.

Arguably, there is a broader alchemical thrust to most of Švankmajer's works, informed as they are, in the words of Kristoffer Noheden, 'by an alchemical poetics of transmutation of base matter into poetic gold'.[48] This is reflected in the uncanny transformations and animations of everyday objects in his films, and the correspondence-based alchemical monism discernible in his recurring disintegration of limits between inanimate and living as well as animal, human and plant.

In *Faust* (1994), the protagonist ends up in an alchemical laboratory, where he accidentally effects the generation of a homunculus (a term first used by Paracelsus to describe a minuscule human being created through alchemy) from clay held in an alembic. Disgusted by the figure's tricks of metamorphosis, however, Faust smashes it and tears up the slip of paper with magic symbols that animated it. The whole film is structured by the stages of the alchemical process, beginning with *nigredo* motifs of blackening and rotting, continuing through a (subversively futile) *citrinitas* and ending in a grim *rubedo*, where a blinking red cue light signals the arrival of Satan come to collect the transgressor's soul.[49]

Švankmajer, in his irreverent use of alchemical tradition, relishes the disorienting, destabilising dimensions of the initiatory process. Yet, being a habitual blasphemer and rebel against structures and strictures, the notion of a resulting harmonious synthesis and spiritual enlightenment holds limited appeal for him. Thus Švankmajer's attitude epitomises the impertinence of surrealist approaches to alchemy, demonstrating how esoteric currents can provide creative fuel and live on in transmuted forms far beyond their traditional contexts.

Artworks from the Alchemy cycle by Jan Švankmajer. *Distillation II* and *The Fifth Essence*, 1996.

'The sun needs the moon, like the rooster needs the hens.' Emblem 30 illustrated by Matthäus Merian the Elder. From *Atalanta fugiens* by Michael Maier, 1618.

Endnotes

Foreword: The History of Alchemy
Carl Philip Passmark

1. Lawrence M Principe, 'Alchemy', in Glen A Magee (ed), *The Cambridge Handbook of Western Mysticism and Esotericism*.Cambridge: Cambridge University Press, 2016, pp. 359–71.
2. Peter Dear, *Revolutionizing the Sciences*. Princeton: Princeton University Press, 2001.
3. Antoine Faivre, *Access to Western Esotericism*. Albany: State University of New York Press, 1994, pp. 10–15.
4. ibid.

Elixirs and Immortality
Fabrizio Pregadio

1. F. Pregadio, *Great Clarity: Daoism and Alchemy in Early Medieval China* (Stanford: Stanford University Press, 2006), 68–70. For Diény's statement, see the summary of his courses in *École pratique des hautes études, 4e section, Sciences historiques et philologiques* 18.4 (1994), 167.
2. For several examples, see M. Kaltenmark, *Le Lie-sien tchouan (Biographies légendaires des Immortels taoïstes de l'antiquité)* (Peking: Université de Paris, 1953); and R. Campany, *To Live as Long as Heaven and Earth: A Translation and Study of Ge Hong's Traditions of Divine Transcendents* (Berkeley: University of California Press, 2002).
3. On the hagiographies of Laozi, see L. Kohn, *God of the Dao: Lord Lao in History and Myth* (Ann Arbor: Center for Chinese Studies, University of Michigan, 1998), 7–36. On the legends about his transmission of Neidan, see F. Pregadio, 'Laozi and Internal Alchemy', in I. Amelung and J. Kurtz (eds), *Reading the Signs: Philology, History, Prognostication: Festschrift for Michael Lackner* (Munich: Iudicium, 2018), 271–301.
4. Ngo Van Xuyet, *Divination, magie et politique dans la Chine ancienne* (Paris: Presses Universitaires de France, 1976); K. DeWoskin, *Doctors, Diviners, and Magicians of Ancient China: Biographies of Fang-shih* (New York: Columbia University Press, 1983).
5. J. Needham, *Science and Civilisation in China*, vol. 5.3 (Cambridge: Cambridge University Press, 1976), 29–33.
6. Pregadio, *Great Clarity*, 35–51.
7. Ibid., 79–99.
8. Ibid., 100–20.
9. On the alchemical vessel, see R. A. Stein, *Le monde en petit: Jardins en miniature et habitations dans la pensée religieuse d'Extrême-Orient* (Paris: Flammarion, 1987), 61–82 and *passim*.
10. On Chinese cosmology, see B. Schwartz, *The World of Thought in Ancient China* (Cambridge, MA: Harvard University Press, 1985), 350–82; A. C. Graham, *Disputers of the Tao* (La Salle: Open Court, 1989), 319–56; and M. Kalinowski, *Cosmologie et divination dans la Chine ancienne* (Paris: École Française d'Extrême-Orient, 1991).
11. See F. Pregadio, *The Seal of the Unity of the Three: A Study and Translation of the Cantong qi, the Source of the Taoist Way of the Golden Elixir* (Mountain View: Golden Elixir Press, 2011).
12. N. Sivin, 'The Theoretical Background of Elixir Alchemy', in J. Needham, *Science and Civilisation in China*, vol. 5.4 (Cambridge: Cambridge University Press, 1980), 232–3, 264–6.
13. Ibid., 266–79.

14. On early Taoist meditation, see I. Robinet, *Taoist Meditation: The Mao-shan Tradition of Great Purity* (Albany: SUNY Press, 1993); on its relation to Neidan, see F. Pregadio, 'Early Daoist Meditation and the Origins of Inner Alchemy', in B. Penny (ed.), *Daoism in History: Essays in Honour of Liu Ts'un-yan* (London: Routledge, 2006), 121–58.
15. I. Robinet, 'Shangqing: Highest Clarity', in L. Kohn (ed.), *Daoism Handbook* (Leiden: Brill, 2000), 196–224.
16. On the concepts of *xing* and *ming* in Neidan, see I. Robinet, *Introduction à l'alchimie intérieure taoïste* (Paris: Les Éditions du Cerf, 1995), 165–95; F. Pregadio, 'Destiny, Vital Force, or Existence? On the Meanings of Ming in Daoist Internal Alchemy and Its Relation to Xing or Human Nature', *Daoism: Religion, History and Society*, 6 (2014), 157–218.
17. On the stages of Neidan practice, see C. Despeux, *Zhao Bichen: Traité d'alchimie et de physiologie taoïste* (Paris: Les Deux Océans, 1979), 48–82; Robinet, *Introduction à l'alchimie intérieure taoïste*, 147–64; and Wang Mu, *Foundations of Internal Alchemy: The Taoist Practice of Neidan* (Mountain View: Golden Elixir Press, 2011).
18. E. Valussi, 'Female Alchemy: An Introduction', in L. Kohn and R. R. Wang (eds), *Internal Alchemy: Self, Society, and the Quest for Immortality* (Cambridge, MA: Three Pines Press), 141–62.
19. See F. Pregadio (trans.), *Liu Yiming: Cultivating the Tao: Taoism and Internal Alchemy* (Mountain View: Golden Elixir Press, 2013), 104.
20. Pregadio, 'Destiny, Vital Force, or Existence?', 205.

Of Alchemy, Astrology and Magic in Late Roman Egypt
Andreas Winkler

1. The text has been translated several times into multiple languages. Most recently: R. Halleux, *Les alchimistes grecs I: Papyrus de Leyde, Papyrus de Stockholm, Fragments de recettes* (Paris: Les Belles Lettres, 1981). An English translation is furnished by E. R. Radcliffe and W. B. Jensen, *The Leyden and Stockholm Papyri: Greco-Egyptian Chemical Documents from the Early 4th Century AD* (Cincinnati: Oesper Collection in the History of Chemistry, University of Cincinnati, 2008). The numbering refers to the paragraphs in Halleux's edition; H = P.Holm.; L = P.Leid. I 397.
2. See the note above. Add, e.g., O. Lagercrantz, *Papyrus Graecus Holmiensis (P.Holm.): Rezepte für Silber, Steine and Purpur* (Uppsala/Leipzig: A.-B. Akademiska Bokhandeln/Otto Harrassowitz, 1914).
3. See, e.g., M. Martelli, *The Four Books of Pseudo-Democritus* (Leeds: Maney, for the Society for the History of Alchemy and Chemistry, 2013); O. Dufault, *Early Greek Alchemy, Patronage and Innovation in Late Antiquity* (Berkeley: California Classical Studies, 2019), 1; R. G. Edmonds III, *Drawing Down the Moon: Magic in the Ancient Greco-Roman World* (Princeton: Princeton University Press, 2019), 271–6 and 284–90.
4. See, e.g., Halleux, *Alchemistes grecs I*, 24–32; Radcliffe and Jensen, *The Leyden and Stockholm Papyri*, 1–9, for overviews of earlier discussions. Recent works include Martelli, *Four Books*; L. Principe, *The Secrets of Alchemy* (Chicago: University of Chicago Press, 2013); S. Grimes, *Becoming Gold: Zosimos of Panopolis and the Alchemical Arts in Roman Egypt* (Auckland: Rubedo Press, 2018).
5. Such procedures are also attested in the magical text *GEMF* 16.923–4, which gives instructions for testing whether a mineral is real or not. The magical texts mentioned in this contribution that have been re-edited in C. A. Faraone and S. Torallas Tovar (eds), *Greek and Egyptian Magical Formularies: Text and Translation* (Berkeley: California Classical Studies, 2022), are abbreviated as *GEMF*. The remaining texts are cited as *PGM* (*Papyri Graecae Magicae*), translations of which can be found in H. D. Betz (ed.), *The Greek Magical Papyri in Translation Including the Demotic Spells* (Chicago: University of Chicago Press, 1986).
6. E.g. Dufault, *Early Greek Alchemy*, 139, and Edmonds, *Drawing Down the Moon*, 312.

7. Principe, *Secrets of Alchemy*, 11.
8. E.g. M. Mertens, *Les alchimistes grecs IV.1: Zosime de Panopolis – Mémoires authentiques* (Paris: Les Belles Lettres, 1995), xi–xx; idem, 'Alchemy, Hermetism and Gnosticism at Panopolis *c.* 300 AD: The Evidence Zosimos', in A. Egberts, B. P. Muhs and J. van der Vliet (eds), *Perspectives on Panopolis: An Egyptian Town from Alexander the Great to the Arab Conquest – Acts from an International Symposium Held in Leiden on 16, 17 and 18 December 1998* (Leiden: Brill, 2002), 165–75; Dufault, *Early Greek Alchemy*, 93–142.
9. S. Grimes, 'Defining Greco-Egyptian Alchemy', *Gnosis*, 7 (2022), 78.
10. P.Vindob. Dem. 6321+6687. The text is translated and discussed by J. F. Quack, 'Färberei, Diätetik etc.: Neue spätägyptische „Schnipsel" zur ägyptischen Verfahrenstechnik', in P. Dils et al (eds), *Wissenschaft und Wissenschaftler im Alten Ägypten: Gedenkschrift für Walter Friedrich Reineke* (Berlin: De Gruyter, 2021), 187–202, at 190–4. Also, Greek texts of the first or second centuries CE attest to the practice of astrology among the Egyptian priesthood. For instance, OMM 1229 (A. Menchetti and R. Pintaudi, 'Ostraka greci e bilingui da Narmuthis', *Chronique d'Égypte*, 82 (2007), 230–2) is a list from the priestly environment at Narmouthis in the Fayum connecting planets to different minerals, making reference to using orpiment and indigo in a process of sorts; perhaps the efficiency of dyeing would have been connected to the powers of the stars. These two texts demonstrate the wide application of such techniques in Egyptian sacerdotal circles.
11. Dufault, *Early Greek Alchemy*, 84–5.
12. Ibid., 52–9. Dufault stresses that Anaxilaos can be connected to *paignia* ('play') procedures, which were designed to amaze the audience. It is therefore of interest to note that a magical papyrus (*PGM* VII 165–186) contains such recipes attributed to Pseudo-Democritus.
13. Martelli, *Four Books*, 30 and 63.
14. Pammenes, *De natura animalium*, XVI, 42.
15. Astrological practices were indeed combined with those of alchemy (see below) and knowledge of snakes and scorpions was in the domain of the Egyptian priesthood, which was connected to the art under discussion. It is also known that Egyptian priests could combine both kinds of expertise in one person. The priest, scorpion-charmer and astronomer/astrologer Harentebo, who might have lived during the early Ptolemaic period, was one of these individuals. See, e.g., P. Derchain, 'Harkhébis, le Psylle-astrologue', *Chronique d'Égypte*, 64 (1989), 74–89; J. Dieleman, 'Claiming the Stars: Egyptian Priests Facing the Sky', in S. Bickel and A. Loprieno (eds), *Basel Egyptology Prize 1: Junior Research in Egyptian History, Archaeology, and Philology* (Basel: Schwabe, 2003), 277–89; idem, 'Stars and the Egyptian Priesthood in the Graeco-Roman Period', in S. Noegel, J. Walker and B. Wheeler (eds), *Prayer, Magic, and the Stars in the Ancient and Late Antique World* (University Park: Penn State University Press, 2003), 137–53; P. Clancier and D. Agut, 'Charming Snakes (and Kings), from Egypt to Persia', *Journal of Ancient Near Eastern History*, 21 (2021), 1–18, esp. 6–7; A. Winkler, 'Stellar Scientists: The Ancient Egyptian Temple Astrologers', *Journal of Ancient Near Eastern History*, 21 (2021), 115–21 with further references.
16. See, e.g., Edmonds, *Drawing Down the Moon*, 269–76.
17. M. Berthelot, *Collection des anciens alchimistes grecs III* (Paris: Georges Steinheil, 1888), 354 and 369–70.
18. Halleux, *Alchimistes grecs I*, 32–3.
19. K. Dosoo, 'Rituals of Apparition in the Theban Magical Library', PhD dissertation, Macquarie University, Sydney, 2014, 148.
20. One can note that a similar ingredient is occasionally found in ancient Egyptian medical texts (e.g. P.Ebers Spell 729; P.Berlin 3038 v 12, vi 3, ix 6), the urine of a woman who is menstruating (described as impure), thus not pregnant, and therefore the term in question could been understood as referring to a virgin.
21. The text is also known as *PGM* XII. See Dufault, *Early Greek Alchemy*, 139 with further references, for the recipes in question.
22. Given the Egyptian context, Dosoo, *Rituals of Apparition*, 147, tentatively suggests that the term could refer to a little cobra-shaped statue (cf *LSJ*, 259).
23. Halleux, *Alchimistes grecs I*, 137, n. 4. See M. Martelli, 'Greek Alchemists at Work: "Alchemical Laboratory" in the Greco-Roman Egypt', *Nuncius*, 26 (2011), 271–311, for a discussion on the artisan vs the alchemist at work with a particular respect to their workspaces. The author argues that the two

were active in the same workshops rather than being the same people based on, among other things, the fact that ancient craftsmen were usually specialised only in one craft and different processes required access to different types of instruments, but cf. Halleux, *Alchimistes grecs I*, 33–5.

24. One should note, however, that also dyeing fabrics was often carried out in the realm of the Egyptian temple.
25. See, e.g., M. Smith, *Following Osiris: Perspectives on the Osirian Afterlife from Four Millennia* (Oxford: Oxford University Press, 2017), 493–526, for a minimalist overview of the situation of traditional religion in the Theban region during the later Roman period with a specific reference to the cult of Osiris. See also E. O. D. Love, *Code-Switching with the Gods: The Bilingual (Old Coptic-Greek) Spells of PGM IV and Their Linguistic, Religious, and Socio-Cultural Context in Late Roman Egypt* (Berlin: De Gruyter, 2016), 264–8, 271–3 and 279–82.
26. E.g. *O.Bodl.* II 1820, 1861 and 2143; *SB* I 4340; XVIII 13622 and 13625 are attestations to some form of continuation of traditional of Egyptian religion in the third and fourth centuries CE. In the mortuary sphere, there is also some evidence for the continuation of traditional beliefs into the fourth century CE (Smith, *Following Osiris*, 509).
27. E.g., *The Letter from Isis to Horus*, for which see M. Mertens, *Un traité gréco-égyptien d'alchimie: La lettre d'Isis à Horus* (Liège: University of Liège, 1985). Recent studies of the text include M. Blanco Cesteros, '(De)Constructing an Authoritative Narrative: The Case of the Letter of Isis', *Arys*, 20 (2022), 227–69; F. Lopes da Silva, 'In the Melting Pot: Cultural Mixture and the Presentation of Alchemical Knowledge in the Letter from Isis to Horus', *Ambix*, 69:1 (2022), 49–64.
28. See, e.g., the autobiography of Ikhernofret, of which a partial translation and discussion can be found in, for instance, D. Lorton, 'The Theology of Cult Statues in Ancient Egypt', in M. B. Dick (ed.), *Born in Heaven, Made on Earth: The Making of Cult Image in the Ancient Near East* (Winona Lake: Eisenbrauns, 1999), 124–6. See further A. von Lieven, 'Im Schatten des Goldhauses: Berufsgeheimnis und Handwerkerinitiation im Alten Ägypten', *Studien zur Altägyptischen Kultur*, 36 (2007), 151–2, for a text from the Dendera temple explaining the materials of which statues were supposed to be made.
29. Grimes, *Becoming Gold*, 25 and 246; ead. 'Secrets of the God Makers: Rethinking the Origins of Graeco-Egyptian Alchemy', *Syllecta Classica*, 29 (2018), 67–89, esp. 70–8, but cf. M. Martelli, 'Alchemy, Medicine and Religion: Zosimus of Panopolis and the Egyptian Priests', *Religion in the Roman Empire*, 3 (2017), 214–15; O. Dufault, 'Was Zosimus of Panopolis a Christian?', *Arys*, 20 (2022), 135–70. C. Bull, 'Hermes between Pagans and Christians: The Nag Hammadi Hermetica in Context', in H. Lundhaug and L. Jenott (eds), *The Nag Hammadi Codices and Late Antique Egypt* (Tübingen: Mohr Siebeck, 2018), 223–5, suggests among other possibilities that the Panopolitan alchemist may have been a temple craftsman or a disgruntled priest who had turned to Christianity. In any case, he must have had some knowledge of traditional Egyptian religion and temple life, as mentioned by Dufault, *Early Greek Alchemy*, 109. M. Escolano Poveda, '*Zosimos Aigyptiakos*: Identifying the Imagery of the "Visions" and Locating Zosimos of Panopolis in His Egyptian Context', *Arys*, 20 (2022), 77–134, picks up this thread in an occasionally speculative commentary on Egyptianising elements in Zosimos's writings. Escolano Poveda points out that the alchemist never claims to be a priest, but she favours a reading which situates him in a temple environment, perhaps as a craftsman with access to the holdings of the temple library, but remains agnostic as to whether he can be regarded as a convert to Christianity or not.
30. P. Derchain, 'L'Atelier des Orfèvres à Dendara et les origines de l'alchimie', *Chronique d'Égypte*, 65 (1990), 233–4.
31. R. Birk, 'Thebanische Astronomen der Ptolemäerzeit (II): Das Dossier des Imuthes (RAFFMA EG.O 1.008.2020)', in A. Verbovsek, E. Hemauer and A. Herzberg-Beiersdorf (eds), *Diskurs: Akteure – Gegenstand – Beziehungen – Beiträge des elften Berliner Arbeitskreises Junge Aegyptologie (BAJA 11) 6.5.–8.5.2021* (Wiesbaden: Harrassowitz, 2023), 16–23.
32. See H. Beinlich, *Der Mythos in seiner Landschaft – das ägyptische Buch vom Fayum, Band II: Die hieratischen Texte* (Dettelbach: J. H. Röll, 2014), 387–8.
33. Corpus des étiquettes de momies grecques, number 44.
34. K. Geens, *Panopolis, a Nome Capital in Egypt in the Roman and Byzantine Period (ca. AD 200–600)* (Leuven: Trismegistos, 2007), 354–7; Smith, *Following Osiris*, 423–47.

35. A recent overview is offered by K. Dosoo, 'A History of the Theban Magical Library', *Bulletin of the American Society of Papyrologists*, 53 (2016), 251–74.
36. J. Gee, 'A Potential Findspot for the Theban Cache from the So-Called Greek Magical Papyri', presentation at the International Congress of Demotic Studies, Heidelberg, 2022.
37. See J. Dieleman, 'The Greco-Egyptian Magical Papyri', in D. Frankfurter (ed.), *Guide to Ancient Magic* (Leiden: Brill, 2019), 307–12.
38. See Dufault, *Early Greek Alchemy*, 138–9. In general, see also W. M. Brashear, 'The Greek Magical Papyri: An Introduction and Survey', in W. Haase (ed.), *Aufstieg und Niedergang der Römischen Welt: Geschichte und Kultur Roms im Spiegel der aktuellen Forschung – Teil 2: Prinzipat. Band 18: Religion. 5. Teilband: Heidentum (die religiösen Verhältnisse in den Provinzen)* (Berlin: De Gruyter, 1995), 3419. It should be noted, however, that the manuscript was probably penned by two scribes (R. W. Daniel (ed.), *Two Greek Magical Papyri in the National Museum of Antiquities in Leiden: A Photographic Edition of J 384 and J 395 (= PGM XII and XIII)* (Opladen: Westdeutscher Verlag, 1991), x–xi). The last pages appear to be written by another hand. Given that the change of scribes occurs in the middle of a section, there is no reason to assume that the second person copied the remainder of the text from another source or had access to materials or insights that the first one did not. The production of the manuscript is the collective effort of two individuals. The first scribe may also have added a few lines at the beginning of *PGM* IV, another magical formulary of the fourth century CE belonging to the so-called Theban Magical Library (Dosoo, 'History of the Theban Magical Library', 259, n. 27).
39. Dosoo, 'History of the Theban Magical Library', 259.
40. See note 38, above.
41. The fact that the text mentions the Sothic rising has been taken as a sign of a *terminus post quem*, given that the rising of the star coincided with the beginning of the Egyptian year in 139 CE (Brashear, 'The Greek Magical Papyri', 3419, n. 173). It is clear that the text must postdate the given year, but the reference to the star is a weak argument for this particular year. Rather, the reference is imbued with Egyptian cosmological references.
42. M. Smith, 'The Eighth Book of Moses and How It Grew', in *Atti del XVII Congresso di Papirologia (Napoli, 19–26 maggio 1983)* (Naples: Centro Internazionale per lo Studio dei Papiri Ercolanesi, 1984), 683–93. See also C. Bull, *The Tradition of Hermes Trismegistus: The Egyptian Priestly Figure as a Teacher of Hellenized Wisdom* (Leiden: Brill, 2018), 143 and 339–51; Edmonds, *Drawing Down the Moon*, 353–4.
43. Mertens, *Alchimistes grecs IV*, 21.
44. See, e.g., Bull, *Tradition of Hermes Trismegistus*, 303–5.
45. See D. Gieseler Greenbaum, *The Daimon in Hellenistic Astrology: Origins and Influence* (Leiden: Brill, 2016), 200–5; Edmonds, *Drawing Down the Moon*, 266–8.
46. Edmonds, *Drawing Down the Moon*, 366–7. For the planetary week, see I. Bultrighini and S. Stern, 'The Seven-Day Week in the Roman Empire: Origins, Standardization, and Diffusion', in S. Stern (ed.), *Calendars in the Making: The Origins of Calendars from the Roman Empire to the Middle Ages* (Leiden: Brill, 2021), 10–79, at 50. Given that the first day of the week is attributed to Helios, the authors argue for this being a Christian interpolation in the text, but its motivation could at least partially stem from the fact that the named deity features prominently in the text.
47. S. Piperakis, 'Sacrificing to the Planets: Planetary Incenses and Flowers of P.Leid. I 395 (= *PGM* XIII.16–20, 24–26)', *Symbolae Osloenses*, 96:1 (2022), 260–84.
48. It is probable that the third benefic planet is Mercury when favourably situated. In most ancient astrological systems, the planet was regarded as neutral or of mixed nature. That is, it could be either benefic or malefic.
49. To use the divine sphere as guides for astrology is also attested, for instance, in a related magical manuscript written in both Egyptian Demotic and Greek from the second century CE (*GEMF* 16.93–115). The described magical procedure includes both a ritual preparation, among other things, of an astrological board, as so-called *pinax*, with gems indicating the planets, and a ritual including an invocation to a solar deity. The god was adjured to send an emissary to instruct the practitioner about a chart for *katarchic* astrology. As pointed out by I. Moyer, 'Stars and Stones: Practice, Materiality, and Ontology in Astrological Rites', in R. G. Edmonds III, C. López-Ruiz, and S. Torallas Tovar (eds), *Magic and Religion in the Ancient Mediterranean World: Studies in Honor of Christopher A. Faraone* (London:

Routledge, 2023), 129, it is not entirely clear, however, whether the divine agent is supposed to arrange the markers on the board to illustrate a favourable position or merely to interpret the markers that the astrologer has laid out on the *pinax*.

50. Dufault, *Early Greek Alchemy*, 99. Note that also, for instance, Vettius Valens (fl. second century CE) makes similar connections in the first book of his astrological *florilegium*.
51. For more detail on this matter, see A. Pérez Jiménez, 'Hephaestion and the Consecration of Statues', *Culture and Cosmos*, 11 (2007), 111–34.
52. It is possible that the timing was inspired by traditional procedures for producing puppets of the gods, such as at the Khoiak festival, with timed procedures for manufacturing the so-called corn Osiris.
53. The magical papyri contain ample reference to various metal objects as well as figurines that are to be used or produced by the practitioner (e.g. *GEMF* 15.67–68, 18.1141; *PGM* I 21–22) so as to attain the goals of the spells (see introduction to *GEMF* 25; Edmonds, *Drawing Down the Moon*, 346–50).
54. Mertens, *Alchimistes grecs IV*, 1–10.
55. Dufault, *Early Greek Alchemy*, 131–7.
56. Martelli, 'Alchemy, Medicine and Religion', 67–89; Dufault, *Early Greek Alchemy*, 122–7. Also Dosoo, *Rituals of Apparition*, 149.
57. See A. Łajtar, *Deir el-Bahari in the Hellenistic and Roman Periods: A Study of an Egyptian Temple Based on Greek Sources* (Warsaw: Institute of Archaeology of the University of Warsaw), 2005, 94–103 and 242–64; Love, *Code-Switching with the Gods*, 266–8; Smith, *Following Osiris*, 523–5.
58. D. Wildung, *Imhotep und Amenhotep: Gottwerdung im Alten Ägypten* (Munich: Deutscher Kunstverlag, 1977). Specifically for the sanctuary, see A. Łajtar, 'The Cult of Amenhotep Son of Hapu and Imhotep in Deir el-Bahari in the Hellenistic and Roman Periods', in A. Delattre and P. Heilporn (eds), *"Et maintenant ce ne sont plus que des villages…": Thèbes et sa région aux époques hellénistique, romaine et byzantine. Actes du colloque tenu à Bruxelles les 2 et 3 décembre 2005* (Brussels: Association égyptologique Reine Élisabeth, 2008), 113–23.
59. See S. Pfeifer, 'Die religiöse Praxis im thebanischen Raum zwischen hoher Kaiserzeit und Spätantike', in F. Feder and A. Lohrwasser (eds), *Ägypten und sein Umfeld in der Spätantike: Vom Regierungsantritt Diokletians 284/285 bis zur arabischen Eroberung des Vorderen Orients um 635–646. Akten der Tagung vom 7.–9.7.2011 in Münster* (Wiesbaden: Harrassowitz, 2013), 72–3; Grimes, 'Secrets of the God Makers', 79–84.
60. Halleux, *Alchimistes grecs I*, 27.
61. Edmonds, *Drawing Down the Moon*, 312, states that the collection of texts 'indicates that one of the compilers … brought together alchemical and magical materials in his work. Such recipes were also not collected as a technical manual for a professional dyer or metalworker. Even if such a professional technician might well have employed a process similar to one of these recipes, all these various recipes – for transmuting colors, hardness, and other qualities – are assembled as a collection that suits the needs of someone interested in the processes of transmutation, not of someone making a living from using a particular process'. But, as indicated above, interest and professional practice are not mutually exclusive.

Alchemical Practice in Roman Egypt
Tobias Churton

1. M. Berthelot and C. E. Ruelle, *Collection des anciens alchimistes grecs*, 3 vols (Paris: Georges Steinheil, 1888), vol. 1, 5.
2. E. R. Caley (trans. and ed.), *The Leyden and Stockholm Papyri: Greco-Egyptian Chemical Documents from the Early 4th Century AD*, including *Note on Techniques and a Materials Index* by W. B. Jensen (Cincinnati: Oesper Collections in the History of Chemistry, University of Cincinnati, 2008), 42.
3. Ibid.
4. M. Martelli, *The Four Books of Pseudo-Democritus*, Sources of Alchemy and Chemistry: Sir Robert Mond Studies in the History of Early Chemistry, Supplement 1 (Leeds: Maney, for the Society for the History of Alchemy and Chemistry, 2014), 29–30.
5. *Catalogue des manuscrits alchimiques grecs* (*CMAG*) (Brussels: Maurice Lamertin, 1924–32), vol. 2, 1–22.
6. *CMAG*, vol. 1, 1–17.
7. Ibid, 17–62.
8. Cambridge University holds the fifteenth century Syriac Mm. 629 Ms. (all Syriac).
9. The British Library holds the sixteenth-century Egerton 709 and Oriental 1593 Mss. – probably dating from sixth-century and eighth- to ninth-century translations from the Greek. They were partially edited and translated by Berthelot and orientalist Rubens Duval in *La chimie au Moyen-Age*, vol. 2 (Paris: Imprimerie nationale, 1893).
10. M. Martelli, 'Pseudo-Democritus' Alchemical Works: Tradition, Contents and Afterlife', lecture paper for Humboldt-Universität zu Berlin [undated], 8.
11. Caley, *The Leyden and Stockholm Papyri*, 19.
12. Berthelot and Ruelle, *Collection des anciens alchimistes grecs*, vol. 1, 71.
13. Ibid., vol. 3, 5.7, 336–7.
14. Strabo, *c.* 64 BCE–c. 24 CE; *Geography*, XVII, 41.
15. S. J. Linden (ed.), *The Alchemy Reader: From Hermes Trismegistus to Isaac Newton* (Cambridge: Cambridge University Press, 2003), Chapter 5: 44–5.
16. Berthelot and Ruelle, *Collection des anciens alchimistes grecs*, vol. 1, 1.16, 38.
17. Zosimos, *On Tribicos and Tube*, ibid., vol. 2, 3.50, 228–9.
18. Ibid., vol. 2, 3.24, 180.
19. Ibid., 216.
20. Ibid., 3, sect. 13, 196.
21. See M. Beretta's fascinating *The Alchemy of Glass: Counterfeit, Imitation, and Transmutation in Ancient Glassmaking* (Sagamore Beach: Science History Publications, 2009).
22. M. Berthelot, *La chimie au Moyen-Age*, trans. R. Duval, vol. 2 (CMA, Syr. II), 1893, Book 6, *Beginning of the Book on the Work of Copper, letter vav. 1*, 223.
23. Ibid., Book 6, 1, 224–5.
24. S. L. Grimes, 'Secrets of the God Makers: Re-thinking the Origins of Greco-Egyptian Alchemy', *Syllecta Classica*, 29 (2018), 67–89, at 68–69.
25. See R. van den Broek, 'Religious Practices in the Hermetic "Lodge": New Light from Nag Hammadi', in R. van den Broek and C. van Heertum (eds), *From Poimandres to Jacob Böhme: Gnosis, Hermetism and the Christian Tradition* (Amsterdam: In de Pelikaan, 2000), 80.
26. *In Flaccum VIII*; first century CE.
27. *Tosephta Sukkah* IV, 6.
28. Pliny, *Natural History*, 37, 333.
29. Berthelot and Ruelle, *Collection des anciens alchimistes grecs*, vol. 2, 3.28, sect. 5, 190.
30. Ibid., sect. 9, 192.
31. Mary is quoted again on the subject of lead and the necessary depriving of metallic bodies of that state, in Berthelot and Ruelle, *Collection des anciens alchimistes grecs*, vol. 2, 3.29: *On the Philosophers' Stone* [an expression that does not appear until the seventh century], 194.
32. Pliny, *Natural History*, 33, 123.

33. M. Martelli, 'Greek Alchemists at Work: "Alchemical Laboratory" in the Greco-Roman Egypt', *Nuncius*, 26 (2011), 296.
34. L. Maini, M. Marchini, M. Gandolfi, L. Raggetti and M. Martelli, 'Quicksilver and Quick-thinking: Insight into the Alchemy of Mercury: A New Interdisciplinary Research to Discover the Chemical Reality of Ancient Alchemical Recipes', *ChemRxiv* (2001), 1, 6.
35. A.-J. Festugière, *La révélation d'Hermès Trismégiste*, vol. 1 (2nd edn, Paris: Gabalda, 1950), 260–2.
36. C. Jung, *Psychology and Alchemy*, *Collected Works*, vol. 12 (London: Routledge and Kegan Paul, 1953), 3.
37. S. L. Grimes, 'Zosimus of Panopolis: Alchemy, Nature and Religion in Late Antiquity', PhD dissertation, Graduate School of Syracuse University, 2006, 119–20.
38. Berthelot and Ruelle, *Collection des anciens alchimistes grecs*, vol. 2, 3.51, sects 5–6, 233–4
39. *The Chronography of George Synkellos: A Byzantine Chronicle of Universal History from the Creation*, trans. W. Adler and P. Tuffin (Oxford: Oxford University Press, 2002), 18–19; see also CMA, Syr. II.8.1, 238.
40. Berthelot and Ruelle, *Collection des anciens alchimistes grecs*, vol. 2, 3.51.7, 234–5.
41. Ibid., vol. 1, 6–7.
42. Ibid., vol. 2, 3.9, 146.
43. J. M. Robinson (ed.), *The Nag Hammadi Library in English* (Leiden: Brill, 1984), 126.
44. Berthelot and Ruelle, *Collection des anciens alchimistes grecs*, vol. 2, 3.49, 224–5.
45. Robinson, 'Second Treatise of the Great Seth', trans. R. A. Bullard and J. A. Gibbons; Codex VII, para. 9. *The Nag Hammadi Library in English, Translated by Members of the Coptic Gnostic Library Project of the Institute for Antiquity and Christianity, James M. Robinson, Director*. Second Edition, Leiden, E.J. Brill, 1984. 'Second Treatise of the Great Seth', trans. R. A. Bullard and J. A. Gibbons; Codex VII, para. 9.
46. Robinson, Codex II.2, trans. T. O. Lambdin, 121.
47. Berthelot and Ruelle, *Collection des anciens alchimistes grecs*, vol. 2, 3.24, sect. 9, 192.
48. Ibid., vol. 1, 2.3, 64–5.
49. Robinson, Codex II.3, trans. W. W. Isenberg, 137.
50. Ibid., 138.
51. Ibid., 141.
52. A. Adler (ed.), *Svidae Lexicon*, 5 vols (Leipzig: Teubner, 1928–38), heading: 'chi', 280, trans. R. Scaife.

From Alexandria to Byzantium
Cristina Viano

1. For a more detailed presentation of Byzantine alchemy see my essay 'Byzantine Alchemy, or Era of Systematisation', in P. Keyser and J. Scarborough (eds), *Oxford Handbook of Science and Medicine in the Classical World* (Oxford: Oxford University Press, 2018), 943–64.
2. *In remp.* 2.234.14–25 Kroll.
3. *Theophrastus*, 71 Barth.
4. M. Berthelot and Ch.-É. Ruelle, *Collection des anciens alchimistes grecs* (*CAAG*), 3 vols (Paris: Georges Steinheil, 1888–89; repr. Osnabrück: Otto Zeller Verlag, 1967), 2.425.4.
5. *CAAG* 2.56.20–69.11.
6. *CAAG* 2.58.22.
7. *CAAG* 2.59.17.
8. *CAAG* 2.59.25.
9. *CAAG* 2.61.1.
10. Cf. *De generatione et corruptione*, 1.10.328b 1.
11. See *Meteorologica*, 4.1, 379a4–11.
12. *CAAG* 2.62.23.
13. *CAAG* 2.63.6.

14. *In Meteorologica*, 274.25–9 Stüve.
15. *CAAG* 2.69.12–104.7.
16. *CAAG* 2.79.11–85.5.
17. *CAAG* 2.69.12–77.14.
18. *CAAG* 2.69.12–73.6.
19. *CAAG* 2.73.7–74.18.
20. *CAAG* 2.74.19–77.15.
21. *CAAG* 2.77.15–104.7.
22. *CAAG* 2.79.11–85.5.
23. *CAAG* 2.91.20.
24. J. L. Ideler, *Physici et medici graeci minores*, 2 vols (Berlin: Reimer, 1841; repr. Amsterdam: Hakkert, 1963), 2.199–253.
25. *CAAG* 2.395.1–421.5.
26. *CAAG* 2.421.8–441.25.
27. *CAAG* 2.400.9–401.16.
28. *CAAG* 2.40.5.
29. *CAAG* 2.407.6.
30. *CAAG* 2.433–438.
31. *CAAG* 2.414.13–415.9.
32. *CAAG* 2.418.4.
33. *CAAG* 2.425.4.
34. *CAAG* 2.426.7.
35. *CAAG* 2.439.21.
36. *CAAG* 2.433.11–441.25.
37. *CAAG* 2.79.16–20.
38. *Letter to Theodore*, in Ideler, *Physici et medici graeci minores*, 208.29.
39. *CAAG* 2.415.10.
40. *CAAG* 2.437.13.
41. *Praxis* 1, in Ideler, *Physici et medici graeci minores*, 201.27–33.
42. *CAAG* 2.441.21.
43. *In Meteorologica*, 331.1 Stüve.
44. Diodorus 3.12.1–14.5; Strabo 16.4.5–20, and Photius, *Library*, §250.

Bibliography

Abbreviations

CAAG = Berthelot, Marcellin, and Ch.-É. Ruelle. *Collection des anciens alchimistes grecs*. 3 vols. Paris: Steinheil, 1888–89; repr. Osnabrück: Otto Zeller Verlag, 1967. Vol. 1, Introduction, by Berthelot; vol. 2, Greek Texts; and vol. 3, French Translations.

1. Texts

Bidez *et al*, 1924–32: Bidez, Joseph, Franz Cumont, J. L. Heiberg and Otto Lagercrantz, ed. *Catalogue des manuscrits alchimiques grecs* (The 'CMAG'). 8 vol. Brussels: Lamertin, 1924–32.

Colinet, 2000: Colinet, Andrée, ed. *Alchimistes grecs*, v. 10: *Anonyme de Zuretti*. Paris: Les Belles Lettres, 2000.

Colinet, 2010: *Alchimistes grecs*, v. 11: *Recettes alchimiques (Par. Gr. 2419; Holkhamicus 109) – Cosmas le Hiéromoine – Chrysopée*. Paris: Les Belles Lettres, 2010.

Goldschmidt, 1923: Goldschmidt, Günther. *Heliodori carmina quattuor ad fidem codicis Cassellani*. Giessen: Töpelmann, 1923.

Halleux, 1921: *Alchimistes grecs*, v. 9/1: *Traités des arts et métiers*. Paris: Les Belles Lettres, 2021.

Halleux, 1981: Halleux, Robert. *Alchimistes grecs*, v. 1: *Papyrus de Leyde, Papyrus de Stockholm, Recettes*. Paris: Les Belles Lettres, 1981
Ideler, 1963: Ideler, J. L. *Physici et medici graeci minores*. 2 vols. Berlin: Reimer, 1841; repr. Amsterdam: Hakkert, 1963.
Martelli, 2011: Martelli, Matteo. *Pseudo-Democrito, Scritti alchemici con il commentario di Sinesio*. Milano: Archè; Paris: Société d'étude de l'histoire de l'alchimie, 2011 = *Textes et travaux de Chrysopœia* 12.
Martelli, 2013: *The Four Books of Pseudo-Democritus*. SHAC: 'Sources of Alchemy and Chemistry,' 1. Wakefield: Maney Publishing, 2013.
Mertens, 1995: Mertens, Michèle. *Alchimistes grecs*, v. 4: *Zosime de Panopolis, Mémoires authentiques*. Paris: Les Belles Lettres, 1995.
Papathanassiou, 2017: Papathanassiou, Maria. *Stephanos von Alexandreia und sein alchemistisches Werk. Die kritische Edition des griechischen Textes eingeschlossen*. Athens, 2017.
Taylor 1937: Taylor, F. Sherwood. The Alchemical Works of Stephanus of Alexandria. *Ambix* 1 (1937), 116–39, and 2 (1938), 39–49.

2. Scholarly Literature

Bacchi-Martelli, 2009: Bacchi, E., and Matteo Martelli. 'Il principe Halid bin Yazid e le origini dell'alchimia araba. Daniele Cevenini and Svevo D'Onofrio (eds),' In *Conflitti e dissensi nell'Islâm = `Uyûn al-Akhbâr: Studi sul mondo islamico* 3, 85–120. Bologna: Il Ponte, 2009.
Bain, 1999: Bain, David. '*Melanitis gê*. An Unnoticed Greek Name for Egypt: New Evidence for the Origins and Etymology of Alchemy.' In David R. Jordan, Hugo Montgomery and Einar Thomassen (eds), *The World of Ancient Magic*, 221–222. Bergen: Norwegian Institute at Athens, 1999.
Berthelot, 1885: Berthelot, Marcellin. *Les origines de l'alchimie*. Paris: Steinheil, 1885.
Berthelot, 1889: Berthelot, Marcellin. *Introduction à l'étude de la chimie des anciens et du moyen âge*. Paris: Steinheil, 1889 = *CAAG* vol. 1.
Brun *et al*, 2013: Brun, Jean-Pierre, Jean-Paul Deroin, Thomas Faucher, Bérangère Redon and Florian Téreygeol. 'Les mines d'or ptolémaïques:Résultats des prospections dans le district minier de Samut (désert Oriental).' *Bulletin de l'Institut français d'archéologie orientale*, 113 (2013), 111–42.
Carlotta-Martelli, 2023:Carlotta, Vincenzo, and Matteo Martelli. 'Metals as Living Bodies: Founts of Mercury, Amalgams, and Chrysocolla.' *Ambix*, 70:1 (2023), 7–30.
Festugière, 1944, 1950, 2014: Festugière, A.-J. *La révélation d'Hermès Trismégiste*, v. 1: *L'astrologie et les sciences* occultes. Paris: Les Belles Lettres, 1944; 3rd edn Paris: Lecoffre, 1950; new edn Paris: Les Belles Lettres, 2014.
Forster, 2016: Forster, Regula. 'Arabic Alchemy: Texts and Contexts.' *Al-Qanṭara*, 37:2 (2016), 269–78.
Halleux, 1979: Halleux, Robert. *Les textes alchimiques*. Turnhout: Brepols, 1979.
Halleux, 1985: Halleux, Robert. 'Méthode d'essai et d'affinage des alliages aurifères dans l'Antiquité et au Moyen-âge.' In Cécile Morrisson et al (eds), *L'or monnayé I: Purification et altérations de Rome à Byzance = Cahiers Ernest Babelon vol.* 2, 39–77. Paris: CNRS, 1985.
Hershbell, 1987: Hershbell, J. P. 'Democritus and the Beginnings of Greek Alchemy.' *Ambix* 34 (1987), 5–20.
Jung, 1937: Jung, Carl Gustav. 'Einige Bemerkungen zu den Visionen des Zosimos.' *Eranos Jahrbuch* 5 (1937), 15–54.
Magdalino-Mavroudi, 2006. Magdalino, Paul, and Maria Mavroudi, eds. *The Occult Sciences in Byzantium*. Geneva: La Pomme d'or, 2006.
Martelli, 2009: Martelli, Matteo. '"Divine Water" in the Alchemical Writings of Pseudo-Democritus.' *Ambix* 56 (2009), 5–22.
Martelli, 2019: Martelli, Matteo. *L'alchimista antico. Dalll'Egitto greco-romano a Bisanzio*. Milan: Editrice Bibliografica, 2019.
Martelli, 2023: Martelli. 'Late Byzantine Recipe Books. Metallurgy, Pharmacology and Cuisine.' In Petros Bouras-Vallianatos and Dionysios Stathakopoulos (eds), *Drugs in the Medieval Mediterranean: Transmission and Circulation of Pharmaceutical Knowledge*. Oxford: Oxford University Press 2023, 336–95.
Mertens, 2006: Mertens, Michèle.'Graeco-Egyptian Alchemy in Byzantium.' In Magdalino and Mavroudi, The Occult Sciences in Byzantium, 205–30.
Newman, 2006: Newman, William R., and Lawrence M. Principe. 'Some Problems with the

Historiography of Alchemy.' In William R. Newman and Anthony Grafton (eds), *Secrets of Nature: Astrology and Alchemy in Early Modern Europe*, 385–432. Cambridge, MA, and London: MIT Press, 2006.

Papathanassiou, 2006: Papathanassiou, Maria K. 'Stephanos of Alexandria: A Famous Byzantine Scholar, Alchemist and Astrologer.' In Magdalino and Mavroudi, The Occult Sciences in Byzantium, 163–204.

Principe, 2013: Principe, Lawrence M. *The Secret of Alchemy*. Chicago: University of Chicago Press, 2013.

Redon-Faucher, 2020: Redon, Bérangère, and Thomas Faucher, eds. *Samut Nord: L'exploitation de l'or du désert Oriental à l'époque ptolémaïque*. Cairo: Institut français d'archéologie orientale, 2020.

Roberts, 2023: Roberts, Alexandre M. 'Thinking about Chemistry in Byzantium and the Islamic World.' *Journal of the History of Ideas,* 84:4 (2023), 595–619.

Saffrey, 1995: Saffrey, H.-D. 'Historique et description du Marcianus 299.' In Didier Kahn and Sylvain Matton (eds), *Alchimie: Art, histoire et mythes*, 1–10. Paris: Archè, 1995.

Viano, 1995: Viano, Cristina. 'Olympiodore, l'alchimistes et les Présocratiques: Une doxographie de l'unité (*De arte sacra*, §18-27).' In Didier Kahn and Sylvain Matton (eds), *Alchimie: Art, histoire et mythes*. Paris: Archè, 1995, 95–150.

Viano, 2005: Viano, Cristina, ed. *L'alchimie et ses racines philosophiques: La tradition grecque et la tradition arabe*. Paris: Vrin, 2005.

Viano, 2006: Viano, Cristina. *La matière des choses: Le livre IV des Météorologiques d'Aristote, et son interprétation par Olympiodore, avec le texte grec révisé et une traduction inédite de son Commentaire au Livre IV*. Paris: Vrin, 2006.

Viano, 2015: Viano, Cristina. 'Une substance, deux natures: Les alchimistes grecs et le principe de la transmutation.' In Fabienne Jourdan and Anca Vasiliu (eds), *Dualismes: Doctrines religieuses et traditions philosophiques*, 309–25. *Paris: Editura Polirom and Chôra*, 2015.

Viano, 2018a: Viano, Cristina. 'Byzantine Alchemy, or Era of Systematisation.' In Paul T. Keyser and John Scarborough (eds), *The Oxford Handbook of Science and Medicine in the Classical World,* 943–64. Oxford: Oxford University Press, 2018.

Viano, 2018b: Viano, Cristina. 'Olympiodore l'alchimiste et la *taricheia:* La transformation du minerai d'or Technê, nature, histoire et archéologie.' In E. Nicolaïdis (ed.), Greek Alchemy from Late Antiquity to Early Modernity,Brepols, Turnhout, 2018, 59–69.

Viano, 2021: Viano, Cristina. 'Olympiodorus and Greco-Alexandrian Alchemy.' InAlbert Joosse (ed.), *Olympiodorus of Alexandria: Exegete, Teacher, Platonic Philosopher*, 14–30. Leiden and Boston: Brill, 2021.

Viano, 2022: Viano, Cristina. 'Noir alchimique : Questions d'étymologie et de transmutation.' In M. Martelli and M. M.

Sassi (eds), *Ancient Science and Technology of Colour: Pigments, Dyes, Drugs and Their Experience in Antiquity*, 115–55, *TECHNAI* 13 (2022).

Wilson, 1984: Wilson, C. Anne. 'Philosophers, *Iôsis* and Water of Life.' *Proceedings of the Leeds Philosophical and Literary Society (Lit. and Hist. Sect.)* 19 (1984), 101–219.

Hermes Trismegistus's *Emerald Tablet*

Peter J. Forshaw

1. My translation is inspired by that in D. Kahn (ed.), *Hermès Trismégiste, La Table d'Émeraude et sa tradition alchimique* (Paris: Les Belles Lettres, 2012), 3–8. For an alternative, see A. Faivre, *The Eternal Hermes: From Greek God to Alchemical Magus*, trans. J. Godwin (Grand Rapids, MI: Phanes Press, 1995), 93–4.
2. As early as the third century CE in *On the Mysteries*, VIII, 5 and X, 7, the Syrian Neoplatonist Iamblichus had written of the prophet Bitys as the one who found hieroglyphic texts of Hermes in Egyptian temples and translated them into Greek. See Iamblichus, *De mysteriis*, trans. with introduction and notes E. C. Clarke, J. M. Dillon and J. P. Hershbell (Atlanta: Society of Biblical Literature, 2003), 317–19, 353.
3. M. K. Z. Asl, 'Sirr al-Khalīqa and Its Influence in the Arabic and Persianate World: 'Awn b. al-Mundhir's Commentary and Its Unknown Persian Translation', *Al-Qantara*, 37:2 (July–December 2016), 435–73, at 436. For more detail, see an old but still extremely insightful work by J. Ruska, *Tabula Smaragdina: ein Beitrag zur Geschichte der Hermetischen Literatur* (Heidelberg: Carl Winter, 1926). On Apollonius see M. Dzielska, *Apollonius of Tyana in Legend and History*, trans. P. Pienkowski (Rome: 'L'ERMA' di Bretschneider, 1986).
4. Faivre, *The Eternal Hermes*, 95, cites Ehrd de Naxagoras, who provides an interpretation of Hermes's *Emerald Tablet* in *Supplementum Aurei Velleris*, 11ff, in *Aureum Vellus oder Güldenes Vliess* (Frankfurt, 1733). See also Ruska, *Tabula Smaragdina*, 116.
5. M. Maier, *Symbola aureae mensae* (Frankfurt, 1617), 24, citing A. Magnus, *Liber de secretis Chymicis*.
6. P. Lucentini and V. P. Compagni, 'Hermetic Literature II: Latin Middle Ages', in W. J. Hanegraaff et al. (eds), *Dictionary of Gnosis & Western Esotericism* (Leiden: Brill, 2006), 499–529, at 515. On the significance of the book, see also S. N. Haq, *Names, Natures and Things: The Alchemist Jabir ibn Hayyan and his* Kitab al-Ahjar *(Book of Stones)* (Dordrecht: Kluwer Academic, 1994), 29. For further information, see [Apollonius], *De secretis naturae*, ed. F. Hudry, 'Le *De secretis nature* du ps.-Apollonius de Tyane, traduction latine par Hugues de Santalla du Kitàb sirr al-halīqa', *Chrysopoeia*, 6 (1997–99), 1–154.
7. In the very first chapter of *De Chymico Miraculo* ('On the Chymical Miracle', 1583), 3, Bernard Trevisan discusses the 'first inventors of this art', beginning with Hermes or Mercurius, because he was the wisest of all of his time, and Trismegistus, Thrice-Great ['Termagnus'] or Thrice-Wise ['Tersapiens'], because he cleverly instructed and taught all worldly wisdom, or threefold nature, vegetable, mineral and animal, one stone.
8. M. Ficino, *Mercurii Trismegisti Liber de Potestate et Sapientia Dei* (Venice, 1481); W. J. Hanegraaff (ed.) and R. M. Bouthoorn (ed. and trans.), *Lodovico Lazzarelli (1447–1500): The Hermetic Writings and Related Documents* (Tempe: Arizona Center for Medieval and Renaissance Studies, 2005).
9. For a translation of the philosophical Hermetica, see B. P. Copenhaver, *Hermetica: The Greek 'Corpus Hermeticum' and the Latin 'Asclepius' in a New English Translation, with Notes and Introduction* (Cambridge: Cambridge University Press, 1992; repr. 1995; repr. 2000). For background, see G. Fowden, *The Egyptian Hermes: A Historical Approach to the Late Pagan Mind* (Princeton: Princeton University Press, 1986); K. Van Bladel, *The Arabic Hermes: From Pagan Sage to Prophet of Science* (Oxford: Oxford University Press, 2009); F. Ebeling, *The Secret History of Hermes Trismegistus: Hermeticism from Ancient to Modern Times*, trans. D. Lorton (Ithaca and London: Cornell University Press, 2007).
10. Ruska, *Tabula Smaragdina*, 136. See also M. Plessner, 'Neue Materialien zur Geschichte der Tabula Smaragdina', *Der Islam*, 16 (1927), 77–113; idem, 'Hermes Trismegistus and Arab Science', *Studia Islamica*, 2 (1954), 45–59.
11. E. J. Holmyard, 'The Emerald Table', *Nature*, 112 (6 October, 1923), 525–6. Ruska, *Tabula Smaragdina*, 119.
12. *Secretum secretorum Aristotelis ad Alexandrum Magnum* (Venice, 1555), Sig. Iijr 'Pater Hermogenes triplex'.
13. On the *Liber Hermetis de alchimia*, or *Liber dabessi* or *Liber Rebisī*, see R. Steele and D. Waley Singer, 'The Emerald Table', *Proceedings of the Royal Society of Medicine, Section of the History of Medicine*, 21 (1927), 485–501; A. Colinet, 'Le livre d'Hermès intitulé *Liber dabessi* ou *Liber rebis*', *Studi Medievali*, 36 (1995), 1011–52. For various Latin translations of the *Emerald Tablet*, see J.-J. Manget (ed.), *Bibliotheca chemica*

curiosa, vol. 1 (Geneva, 1702), 381–2. See also Kahn, *La Table d'Émeraude et sa tradition alchimique*, Preface, ix–xxviii; S. Gentile and C. Gilly (eds), *Marsilio Ficino and the Return of Hermes Trismegistus* (Florence: Centro Di, 1999; 2nd edn, 2001), 200–2, 'The Vulgate Edition of the Emerald Table'.

14. Ruska, *Tabula Smaragdina*, 220. See M. A. Zuber, 'Between Alchemy and Pietism: Wilhelm Christoph Kriegsmann's Philological Quest for Ancient Wisdom', *Correspondences*, 2:1 (2014), 67–104.
15. Khunrath's Latin version is probably taken from Chrysogonus Polydorus's edition of *De Alchemia* (Nuremberg, 1541), 363, 'Tabula Smaragdina Hermetis Trismegisti *Περι Χημειας*. Incerto interprete', where we read 'Verba secretorum Hermetis, quae scripta erant in tabula Smaragdi, inter manus eius inventa, in obscuro antro, in quo humatus corpus eius repertum est' (Words of the Secrets of Hermes, which were written on an Emerald tablet, found between his hands, in a dark cave, in which his body was discovered buried). The same text appears in the later *Ars chemica* (Strasbourg, 1566), 32–3. His German translation closely resembles one appearing the following year in Gerard Dorn's edition of B. Trevisanus, *Von der Hermetischen Philosophia* (Strasbourg, 1601), sig. CVIr, which also includes an almost identical Latin version, save for the variant *thelesmi* for *telesmi*, sig. CVv.
16. For a contemporary English translation of the *Emerald Tablet*, see that provided in R. Bacon, *The Mirror of Alchemy* (London, 1597), 16–17, 'The Smaragdine Table of Hermes, Trismegistus of Alchimy'.
17. H. Khunrath, *Amphitheatrum Sapientiae Aeternae* (Hanau, 1609), *Pyramid* engraving Latin and German: 'Verba secretorvm hermetis: Vervm, sine mendacio certvm & verissimvm, qvod est inferivs, est sicvt qvod est svperivs; & qvod est svperivs, est sicvt qvod est inferivs: ad perpetranda miracvla rei vnivs'; 'Warhafftig, sonder Luegen gewiss und auff das aller warhafftigste, diss so UNTEN ist, ist gleich dem OBERN; Vnd dis so OBEN ist, ist gleich dem VNTERN: damit man kan erlangen und uerrichten Miracula oder wunder-zeichen EINES EINIGEN DINGES.' Khunrath considered this statement so important that he included it on the title page of the 1609 edition of his *Amphitheatre*.
18. Ibid.: 'et sicvt omnes res fvervnt ab vno, meditatione vnivs; sic omnes res natae fvervnt ab hac vna re, adaptatione'; 'Vnd gleich wie ALLE DINGE uon EINEM DINGE ALLEINE geschaffen, durch den willen und Geboth EINES EINIGEN, der es bedacht; also entspriessen und kommen ALLE DINGE uon demselben EINEM DINGE, durch Schickung und Vereinigung zusammenfuegung.'
19. Ibid.: 'pater eivs est sol, mater eivs lvna; portavit illvd ventvs in ventre svo: nvtrix eivs terra est'; 'Die SONNE ist sein VATER, und der MOND ist seine MVTTER; der WINDT hat jn getragen in seinem Bauch: Seine ERNEHRERIN oder Amme ist die ERDE.'
20. Ibid.: 'pater omnis telesmi totivs mvndi est hic'; 'Dieser ist der UATER ALLER UOLLKOMMENHEIT dieser gantzen Weldt.'
21. Ibid.: 'vis eivs integra est'; 'SEINE MACHT Jst VOLLKOMMEN.'
22. Ibid.: 'si versa fverit in terram, separabis terram ab igne, svbtile à spisso, svaviter, cvm magno ingenio'; 'Wann Es uerwandelt wird in ERDE, So sol tú das Erdreich vom FEVVER scheiden, und das Subtile uom dicken oder groben, gantz Lieblich mit grosser bescheidenheit und verstande.'
23. Ibid.: 'ascendit à terra in coelvm, itervmqve descendit in terram, & recipit vim svperiorvm & inferiorvm'; 'Es steiget uon der ERDEN in HIMMEL, und vom HIMMEL Wider zur ERDEN, Vnd gewinnet also die Krafft der Oberen und Vnteren.'
24. Ibid.: 'sic habebis gloriam totivs mvndi!'; 'ALSO WIRSTV HABEN ALLE HERRLIGKEIT DER GANTZEN WELT.'
25. Ibid.: 'ideo fvgiat à te omnis obscvritas'; 'Derhalben weiche uon dir aller Vnuerstand und Vnuermögenheit.'
26. Ibid.: 'hic est totivs fortitvdinis fortitvdo fortis; qvia vincet omnem rem svbtilem, omnemqve solidam penetrabit'; 'Diss ist uon aller STERCKE die STERCKESTE STERCKE; Dann es kan uber winden alle Subtiligkeit, und durchdringen alle Veste.'
27. Ibid.: 'sic mvndvs creatvs est'; 'ALSO IST DIE WELT GESCHAFFEN!'
28. Ibid.: 'hinc ervnt adaptationes mirabiles, quarvm modvs hic est'; 'Dahero geschehen seltzame Uereinigungen, und werden MANCHERLEY WUNDER gewürcket; Welcher Weg, die selben züwürcken, dieser ist.'
29. Ibid.: 'itaqve vocatvs svm Hermes Trismegistvs, habens tres partes philosophiae totivs mvndi'; 'Vnd bin darumb genand HERMES TRISMEGISTVS, habende dreÿ theill der WEISHEIT der gantzen Welt.'

30. Ibid.: 'completvm est qvod dixi de operatione solis'; 'Es ist erfüllet alles was Jch gesagt habe uon dem WERCKE der SONNEN.'
31. J. Ferguson, *Bibliotheca Chemica: A Catalogue of the Alchemical, Chemical and Pharmaceutical Books in the Collection of the Late James Young of Kelly and Durris*, 2 vols (Glasgow: James MacLehose and Sons, 1906), vol. 1, 419–22. Ferguson also raises the possibility that the author is Martinus Ortholanus in vol. 2, 158.
32. 'A briefe Commentarie of Hortulanus the Philosopher, upon the Smaragdine Table of Hermes', in *The Mirror of Alchimy, Composed by the thrice-famous and learned Fryer, Roger Bachon* (London, 1597), 17–27, at 19. The first published Latin edition appeared in Chrysogonus Polydorus's above-mentioned *De Alchemia* (Nuremberg, 1541). On the Hortulanus commentary, see Ruska, *Tabula Smaragdina*, 180.
33. P. Berlekamp, 'Painting as Persuasion: A Visual Defense of Alchemy in an Islamic Manuscript of the Mongol Period', *Muqarnas: An Annual on the Visual Cultures of the Islamic World*, vol. 20 (Leiden: Brill, 2003), 35–59; P. Starr, 'Towards a Context for Ibn Umayl, Known to Chaucer as the Alchemist "Senior"', *Journal of Arts and Sciences Sayi*, 11 (May 2009), 61–77.
34. For more on the Aurora Consurgens and the Emerald Tablet, as well as other representations of Hermes, see M. Heiduk, '„Was oben ist, ist gleich dem unten." Die Inschrift der Smaragdtafel und das Imaginarium von den Anfängen der Alchemie', in L. Velte and L. Lieb (eds), Literatur und Epigraphik. Phänomene der Inschriftlichkeit in Mittelalter und Früher Neuzeit (Berlin: Erich Schmidt Verlag, 2022), 283–308.
35. 'A briefe Commentarie of Hortulanus the Philosopher, upon the Smaragdine Table of Hermes', 19.
36. Ibid., 20.
37. Ibid., 20–1.
38. Ibid., 21.
39. Ibid., 22.
40. Ibid., 23–4.
41. Ibid., 25, Chapter XI, *That this worke imitateth the Creation of the worlde*.
42. Ibid., 27.
43. Ibid.
44. Paracelsus, *De tinctura physicorum, contra sophistas natos post dilivium*, in *Opera, Bücher und Schrifften*, vol. 1 (Strasbourg, 1603 and 1616), 921. This is considered to be a pseudo-Paracelsian work, probably composed in the 1560s. For more on this subject, see D. Kahn and H. Hirai (eds), *Pseudo-Paracelsus: Forgery and Early Modern Alchemy, Medicine and Natural Philosophy* (Leiden: Brill, 2022).
45. G. Dorn, *Trismegisti Aenigmatici Sermonis et Veridici, per Mercurium Hermetem Trismegistum relicti, clara & brevis admodum expositio*, in G. Dorn, *De Naturae Luce Physica* (1583), *De Naturae Luce Physica, ex Genesi Desumpta* (Frankfurt, 1583), 77–134.
46. Dorn, *Trismegisti Aenigmatici Sermonis et Veridici*, 77.
47. Ibid., 98.
48. Ibid., 129.
49. On relations between alchemy and astrology, see P. J. Forshaw, '"Chemistry, That Starry Science": Early Modern Conjunctions of Astrology and Alchemy', in N. Campion and E. Greene (eds), *Sky and Symbol* (Lampeter: Sophia Centre Press, 2013), 143–84.
50. See M. P. Crosland, *Historical Studies in the Language of Chemistry* (London: Heinemann, 1962; repr. New York: Dover Publications, 2004), 6; J. Dee, *Monas Hieroglyphica* (Antwerp, 1564), 14v, 17r; C. H. Josten, 'A Translation of John Dee's "Monas Hieroglyphica" with an Introduction and Annotations', *Ambix*, 12 (June and October 1964), 84–221, at 165, 175; N. H. Clulee, '*Astronomia inferior*: Legacies of Johannes Trithemius and John Dee', in W. R. Newman and A. Grafton (eds), *Secrets of Nature: Astrology and Alchemy in Early Modern Europe* (Cambridge, MA: MIT Press, 2001), 173–233.
51. Dorn, *Trismegisti Aenigmatici Sermonis et Veridici*, 90–1.
52. Ibid., 89.
53. Ibid., 101. On Augustine and Hermes, see Ebeling, *The Secret History of Hermes Trismegistus*, 42.
54. See Forshaw, 'Medieval Latin Alchemy' in this volume.
55. Dorn, *Trismegisti Aenigmatici Sermonis et Veridici*, 91–2, 94.
56. Ibid., 94.

57. 'A briefe Commentarie of Hortulanus the Philosopher, upon the Smaragdine Table of Hermes', 25. As Ruska, *Tabula Smaragdina*, 210, points out, Dorn seems to be polemicising against Hortulanus's chrysopoetic interpretation in his comment that Hermes says '*a te* fugiet omnis obscuritas, non dicit *a metallis*' (all darkness will flee *from you*, he does not say *from metals*).
58. 'Jo. Trithe. Ab. Spanheymensis Germano de Ganay viro doctissimo Salutem', in J. Trithemius, *De Septem Secvndeis, id est, intelligentijs, siue Spiritibus Orbes post Deum mouentibus, reconditissimæ scientiæ & eruditionis Libellus* (Cologne, 1567), 65–76, at 66, 'Reijciatur binarius & ternarius, ad unitatem convertibilis erit.' Dorn expands on this in *De spagirico artificio Iohannis Trithemii sententia* ('Johann Trithemius's Opinion about the Spagiric Art'), in *Theatrum Chemicum*, vol. 1 (Strasbourg, 1659), 388, which was previously published as *Lapis Metaphysicus, sive philosophicus* ('The Metaphysical, or Philosophical Stone', n.p., 1570). He expressed these ideas visually through geometric figures in *Monarchia Triadis, in Unitate*, in his edition of Paracelsus's *Aurora Thesaurusque Philosophorum* (Basel, 1577), 65–17, a short work written in defence of the Paracelsian principles, where he provides representations of the Unary, Binary and Ternary.
59. Dorn, *Trismegisti Aenigmatici Sermonis et Veridici*, 79.
60. Ibid., 128.
61. Ibid., 85.
62. Ibid., 129 [mispaginated as 127].
63. A. Libavius, *Rervm Chymicarvm Epistolica Forma ad Philosophos et Medicos qvosdam in Germania, Liber primus* (Frankfurt, 1595), 144–61, Epist. XVII.
64. P. J. Forshaw, 'Alchemical Exegesis: Fractious Distillations of the Essence of Hermes', in L. Principe (ed.), *Chymists and Chymistry: Studies in the History of Alchemy and Early Modern Chemistry* (Sagamore Beach: Science History Publications, 2007), 25–38, at 29.
65. See Ruska, *Tabula Smaragdina*, 2, 117, on variants in transmission of the *Emerald Tablet*. He mentions the substitution of *perpetranda* (performing) with *penetranda* (penetrating) and *praeparanda* (preparing); of *adaptatione* (adaptation) by *adoptione* (adoption), and *meditatione* (meditation) by *mediatione* (mediation). On Newton's 1680 translation and interpretation, see B. J. Teeter Dobbs, 'Newton's *Commentary* on the *Emerald Tablet* of Hermes Trismegistus: Its Scientific and Theological Significance', in I. Merkel and A. G. Debus (eds), *Hermeticism and the Renaissance: Intellectual History and the Occult in Early Modern Europe* (Washington, DC: Folger Books, 1988), 182–191; idem, *Janus Faces of Genius: The Role of Alchemy in Newton's Thought* (Cambridge: Cambridge University Press, 1991), 66–73, 271–7. The 'mediatio' version can be found, for example, in J. G. Meerheim, *Discursus Curioser Sachen, insonderheit Hermetischer, Philosophischer* (Leipzig, 1708), 45, and J. de Nuysement, *Tractatus de vero sale secreto Philosophorum* (Frankfurt, 1716), 20; while 'meditatio' appears in earlier editions of De Nuisement, (1651), 22; (1671/72), 25; as well as *Alchemiae Gebri Arabis philosophi solertissimi, Libri* (Bern, 1545), 294, and *Ars Chemica* (1566), 32.
66. 'A briefe Commentarie of Hortulanus the Philosopher, upon the Smaragdine Table of Hermes', 20. For the Latin, see Hortulanus philosophus, *Super Tabulam Smaragdinam Hermetis Commentarius*, in *De Alchemia* (1541), 364–73, at 366.
67. Dorn, *Trismegisti Aenigmatici Sermonis et Veridici*, 95–6.
68. Ibid., 96–7.
69. Ibid., 100. Cf. Wisdom 11:21, 'but thou hast ordered all things in measure, and number, and weight' (Douay).
70. Dorn, *Trismegisti Aenigmatici Sermonis et Veridici*, 109–10.
71. My copy: Albertus Magnus, *Liber Mineralium* (1499), f. vi verso on making 'aqua rosata'.
72. Libavius, *Rervm Chymicarvm Epistolica Forma*, 152–3.
73. Ibid., 158 margin: 'Paracelsitae furiosi medici'.
74. *Aurora Consurgens* in *Auriferae artis* (1572), 241.
75. Dorn, *Trismegisti Aenigmatici Sermonis et Veridici*, 112–13.
76. Ibid., 124–5.
77. Ibid., 125.
78. D. Sennert, *Epitome vniuersam Dan. Sennerti doctrinam summa fide complectens* (Cologne, 1654), 283.

Alchemy in the Medieval Islamic World
Salam Rassi

1. S. Brentjes, *Teaching and Learning the Sciences in Islamicate Societies (800–1700)* (Turnhout: Brepols, 2018), 107–10; J. K. Stearns, *Revealed Sciences: The Natural Sciences in Islam in Seventeenth-Century Morocco* (Cambridge: Cambridge University Press, 2021), 213–30.
2. For vivid examples, see P. N. Joosse, 'Unmasking the Craft: 'Abd al-Laṭīf al-Baghdādī's View on Alchemy and Alchemists', in A. Akasoy and W. Raven (eds), *Islamic Thought in the Middle Ages: Studies in Text, Transmission and Translation in Honour of Hans Daiber* (Leiden: Brill, 2008), 301–17; Jamāl al-Dīn 'Abd al-Raḥīm al-Jawbarī, *The Book of Charlatans*, ed. M. Dengler, trans. H. Davies (New York: New York University Press, 2020), Chapter 4.
3. See L. M. Principe, 'Alchemy Restored', *Isis*, 102:2 (2011), 305–12.
4. On these seismic shifts, see W. R. Newman, 'What Have We Learned from the Recent Historiography of Alchemy?', *Isis*, 102:2 (2011), 313–21.
5. For example, M. Ullmann, *Die Natur- und Geheimwissenschaften im Islam* (Leiden: Brill, 1972); F. Sezgin, *Geschichte des arabischen Schrifttums*, vol. 4 (Leiden: Brill, 1967).
6. M. Martelli and M. Rumor, 'Near Eastern Origins of Graeco-Egyptian Alchemy', in K. Geus and M. Geller (eds), *Esoteric Knowledge in Antiquity* (Berlin: Max-Planck-Institut für Wissenschaftsgeschichte, 2014), 37–62, cited in R. Forster, 'Alchemy and the Chemical Crafts', in S. Brentjes (ed.), *Routledge Handbook on Science in the Islamicate World: Practices from the 2nd/8th to the 13th/19th Centuries* (Abingdon: Routledge, 2022), 154–65.
7. S. Grimes, *Becoming Gold: Zosimus of Panopolis and the Alchemical Arts in Roman Egypt* (Auckland: Rubedo Press, 2018).
8. R. G. Edmonds III, *Drawing Down the Moon: Magic in the Ancient Greco-Roman World* (Princeton: Princeton University Press, 2019), Chapter 9.
9. See following section.
10. Many of these works have come down from a single Greek manuscript, Marcianus Graecus 299, dated to the tenth or eleventh century; M. Berthelot and C.-É. Ruelle, *Collection des anciens alchimistes grecs*, 3 vols (Paris: Georges Steinheil, 1887); A. M. Roberts, 'Framing a Middle Byzantine Alchemical Codex', *Dumbarton Oaks Papers*, 73 (2019), 69–102.
11. See D. Gutas, *Greek Thought, Arabic Culture: The Graeco-Arabic Translation Movement in Baghdad and Early 'Abbāsid Society (2nd–4th/8th–10th Centuries)* (London: Routledge, 1998).
12. G. Strohmaier, '.Hunayn b. Is.hāq', in *Encyclopaedia of Islam* 3, fasc. 3 (Leiden: Brill, 2017), 73–83.
13. B. Hallum, 'Zosimus Arabus: The Reception of Zosimus of Panopolis in the Arabic/Islamic World', PhD dissertation, Warburg Institute, London, 2008, 114–31.
14. Ibid., 132–92.
15. Ibid., Chapter 5.
16. Ibid., 32.
17. Muḥammad ibn Isḥāq ibn al-Nadīm, *Kitāb al-Fihrist*, ed. A. Fu'ād Sayyid, vol. 2 (London: Mu'assasat al-Furqān li-l-Turāth al-Islāmī, Markaz Dirāsāt Makhṭūṭāt al-Islāmiyya, 2014), 445.17–446.4.
18. L. Saif, 'A Preliminary Study of the Pseudo-Aristotelian Hermetica: Texts, Context, and Doctrines', *al-'Uṣūr al-Wuṣṭā* 29 (2021), 20–80.
19. Ibn al-Nadīm, *Kitāb al-Fihrist*, vol. 2, 446.5–10, 447.2–448.3, 449.5–450.10.
20. R. Forster, 'Khālid b. Yazīd', *Encyclopaedia of Islam*, 3rd edn, fasc. 2 (Leiden: Brill, 2021), 75–8.
21. Ibn al-Nadīm, *Kitāb al-Fihrist*, vol. 2, 448.7–8.
22. Jābir ibn .Hayyān, *Kitāb al-Rāhib*, in P. Kraus, *Mukhtār rasā'il Jābir ibn Ḥayyān* (Cairo: Maktabat al-Khānjī, 1935), 528–32.
23. For the account in the *Chronicle of Zuqnīn*, see J.-B. Chabot (ed.), *Incerti auctoris Chronicon Pseudo-Dionysianum vulgo dictum*, II (Leuven: Peeters, 1933), 210.22–211.9; Dionysius's account is preserved in J.-B. Chabot (ed. and trans.), *Chronique de Michel Le Syrien, Patriarche Jacobite d'Antioche (1166–1199)* (Paris: Ernest Leroux, 1899), vol. 2, 523–4 (trans.), vol. 4, 473–5 (trans.). Both accounts are discussed and analysed by S. Rassi, 'Alchemy in an Age of Disclosure: The Case of an Arabic Pseudo-Aristotelian

Treatise and Its Syriac Christian "Translator"', *Asiatische Studien*, 75:2 (2021), 545–609, at 551; N. Harris, 'Better Religion through Chemistry: The World and Thought of 'Izz al-Dīī Aydemīr al-Jildakī, a 14th-Century Alchemist', PhD dissertation, University of Pennsylvania, forthcoming.

24. M. V. Mavroudi, 'Occult Science and Society in Byzantium: Considerations for Future Research', in P. Magdalino and M. Mavroudi (eds), *The Occult Sciences in Byzantium* (Geneva: La Pomme d'or, 2006), 39–95, at 73, n. 86, cited by A. M. Roberts, 'Hierotechnicians by Name and Their Middle Byzantine Fame', *Journal of Late Antique, Islamic and Byzantine Studies*, 1:1–2 (2022), 167–99, at 172–3, n. 29.
25. Mavroudi, 'Occult Science and Society', 95.
26. Sezgin, *Geschichte*, vol. 4, 107.
27. Ibid., 107–8.
28. See S. Stroumsa, 'The Makeover of Ḥayy: Transformations of the Sage's Image from Avicenna to Ibn Ṭufayl', *Oriens*, 49: 1–2 (2021), 1–34.
29. (Ps.-) Ḥunayn ibn Isḥāq, *Qiṣṣat Salāmān wa-Absāl, tarjamat* Ḥunayn *ibn Isḥāq al-'Ibādī min al-lugha al-Yūnāniyya*, in *Tis' rasā'il fī l-ḥikma wa-l-ṭabī'iyyāt* (Cairo: Maṭba'a Hindiyya, 1326/1908), 158.
30. G. Strohmaier, 'Umāra ibn Ḥamza, Constantine V, and the Invention of the Elixir', *Graeco-Arabica*, 4 (1991), 21–4.
31. On this 'flexible and pluralistic approach' in Graeco-Egyptian alchemy, see Grimes, *Becoming Gold*, 53–67.
32. P. Kraus, *Jābir ibn Ḥayyān; Contribution a l'histoire des idées scientifiques dans l'Islam, Vol. 2: Jabir et la science grecque* (Cairo: Institut français d'archéologie orientale, 1943), 170–3. Here, Kraus cautiously maintains that Jābir's approach is 'stoïcisante et non pas stoïcienne', since he displays no direct knowledge of Stoicism in his work.
33. J. Sellars, *Stoicism* (London: Routledge, 2006), 81–106.
34. A. Rinotas, 'Stoicism and Alchemy in Late Antiquity: Zosimus and the Concept of Pneuma', *Ambix*, 64:3 (3 July, 2017), 203–19.
35. Ibid., 214.
36. Stephan of Alexandria, *Letter to Theodore*, in F. Sherwood Taylor, 'The Alchemical Works of Stephanos of Alexandria', *Ambix*, 2:1 (1938), 39–49, at 41. On metals possessing three hypostases, see an iambic poem written under the name of Theophrastus; C. A. Browne, 'The Poem of the Philosopher Theophrastos upon the Sacred Art: A Metrical Translation with Comments upon the History of Alchemy', *Scientific Monthly*, 11:3 (1920), 193–214, at 203–4.
37. Kraus, *Jābir*, 170–3.
38. Ibid., 189; Galen, *On the Constitution of the Art of Medicine; The Art of Medicine; A Method of Medicine to Glaucon*, ed. and trans. I. Johnston (Cambridge, MA: Harvard University Press, 2016), xli.
39. For a detailed overview of this system, see Kraus, *Jābir*, 187–236. For a briefer summary, see Alexander of Aphrodisias, *On Aristotle on Coming-to-Be and Perishing 2.2-5*, trans. E. Gannagé (London: Bloomsbury, 2014), 10–11; L. Principe, *The Secrets of Alchemy* (Chicago: University of Chicago Press, 2013), 41–4.
40. Maslama b. Qāsim al-Qurṭubī, *Rutbat al-ḥakīm*, ed. W. Madelung (Zurich: Living Human Heritage, 2016), 50. The author of this text was once thought to be Maslama Aḥmad al-Majrīṭī (d. 1007) until correctly identified in M. Fierro, Bāṭinism in al-Andalus: Maslama b. Qāsim al-Qurṭubī (d. 353/964), Author of the *Rutbatal-ḥakīm* and the *Ghāyat al-ḥakīm (Picatrix)*', *Studia Islamica*, 2:84 (1996), 87–112.
41. D. E. Eichholz, 'Aristotle's Theory of the Formation of Metals and Minerals', *Classical Quarterly*, 43:3–4 (1949), 141–6.
42. O. Dufault, 'Transmutation Theory in the Greek Alchemical Corpus', *Ambix*, 62:3 (2015), 215–44, at 229–30.
43. U. Weisser (ed.), *Das 'Buch über das Geheimnis der Schöpfung' von Pseudo-Apollonios von Tyana* (Berlin: De Gruyter, 1980), 524.4–5.
44. Ibid., 246.1–249.5.
45. Jābir ibn .Hayyān, *Kitāb al-Īḍāḥ*, in E. J. Holmyard, *The Arabic Works of Jâbir ibn Ḥayyân* (Paris: Geuthner, 1928), 54.6–9. For an English translation of the relevant passage, see Principe, *The Secrets of Alchemy*, 35.
46. Abū Na.sr al-Fārābī, *Risāla fī wujūb al-ṣinā'at al-kīmiyā'*, in A. Sayılı, 'Fârâbî'nin Simyanın Lüzumu Hakkındaki Risâlesi', *Türk Tarih Kurumu Belleten*, 15:57 (1951), 65–79, at 77–9.

47. P. Lettinck, *Aristotle's Meteorology and Its Reception in the Arab World: With an Edition and Translation of Ibn Suwār's Treatise on Meteorological Phenomena and Ibn Bājja's Commentary on the Meteorology* (Leiden: Brill, 1999), 301.
48. Al-Fārābī, *Risāla*, 78.
49. Al-Qurṭubī, *Rutbat al-ḥakīm*, 30.
50. Ibid., 24.
51. Ikhwān al-Ṣafā', *On Composition and the Arts: An Arabic Critical Edition and English Translation of Epistles 6–8*, ed. N. El-Bizri and G. de Callataÿ (Oxford: Oxford University Press, 2018), Epistle VII.
52. Avicenna, *Kitāb al-shifā': al-fann al-khāmisa fī al-āthār al-'ulwiyya*, in E. J. Holmyard, *Avicennae De Congelatione et Conglutinatione Lapidum: Being Sections of the Kitâb al-Shifâ'* (Paris: Geuthner, 1927), 41–2 (trans.), 86 (text).
53. T. Inoue, 'Avicennian Natural Philosophy and the Alchemical Theory of al Ṭughrā'ī in .*Haqā'iq al-istishhād*', MA dissertation, McGill University, 2017, 75, 92.
54. P. Carusi, 'Faḫr al-Dīn: un teologo e l'alchimia', *Memorie di scienze fisiche e naturali*, 39:2/2 (2015), 25–35.
55. *De Generatione Animalium*, 761b–763a, cited by J. Hämeen-Anttila, 'Artificial Man and Spontaneous Generation in Ibn Waḥshiyya's *al-Filāḥa an-nabaṭiyya*, *Zeitschrift der Deutschen Morgenländischen Gesellschaft*, 153:1 (2003), 37–49, at 39, n. 13; L. Saif, 'The Cows and the Bees: Arabic Sources and Parallels for Pseudo-Plato's *Liber Vaccae* (*Kitāb al-Nawāmīs*)', *Journal of the Warburg and Courtauld Institutes*, 79:1 (2016), 1–47.
56. K. M. O'Connor, 'The Alchemical Creation of Life (*takwīn*) and Other Concepts of Genesis in Medieval Islam', PhD dissertation, University of Pennsylvania, 1994, 158–60. For this reason, Jābir describes the womb as a 'world' (*'ālam*), i.e., a microcosm, which is embraced by a macrocosm (*al-'ālam al-akbar*).
57. Ibid., 158–75.
58. Hämeen-Anttila, *The Last Pagans of Iraq: Ibn Waḥshiyya and His Nabatean Agriculture* (Boston: Brill, 2006), 290–4; Saif, 'The Cows and the Bees'.
59. Dufault, 'Transmutation', 231 (citing *Synesius to Dioscorus*); Grimes, *Becoming Gold*, 50–1.
60. Ullmann, *Die Natur- und Geheimwissenschaften*, 259–60.
61. Ibid., 234.
62. M. P. Crosland, *Historical Studies in the Language of Chemistry* (London: Heinemann, 1962), 22.
63. Berthelot and Ruelle, *Collection*, vol. 2, Section 3.10, §3.
64. P. Carusi, 'Elixir', *Encyclopaedia of Islam*, 3rd edn, fasc. 4 (2013), 109–12.
65. Jābir ibn .Hayyān, *Kitāb al-Raḥma*, in Jābir ibn .Hayyān, *Rasā'il: Thalāthūna kitāban wa-risālatan*, ed. A. F. Mazīdī (Beirut: Dār al-Kutub al-'Ilmiyya, 2006), 590.
66. [Ps.-]Maslama ibn A.hmad Majrī.tī, *Ghāyatal-ḥakīm wa-aḥaqq al-natījatayn bi-l-taqdīm = Das Ziel des Weisen*, ed. and trans. H. Ritter (Berlin: B. G. Teubner, 1933), 8.6–7.
67. Jābir ibn Ḥayyān, *Kitāb al-Tajrīd*, in Holmyard, *The Arabic Works of Jâbir ibn Ḥayyân*, 129.11–128.5.
68. Al-Qur.tubī, *Rutbat al-ḥakīm*, 36–7.
69. *Risālat al-iksīr*, in A. Ateş, 'İbn Sina, Risâlat al-Iksîr', *Türkiyat Mecmuası*, 10 (1951–53), 27–54, 36.14–16.
70. Principe, *The Secrets of Alchemy*, 40.
71. Abū Bakr Mu.hammad al-Rāzī, *Kitāb al-Asrār wa sirr al-asrār*, ed. M. T. Dānishpazhūh (Tehran: Kumīsiyūn-i Millī-i Yūniskū dat Īrān, 1964), 7.
72. P. Starr, 'Towards a Context for Ibn Umayl, Known to Chaucer as the Alchemist "Senior"', *Çankaya Üniversitesi Fen-Edebiyat Fakültesi*, 11 (2009), 61–77, at 73, 74.
73. Mu.hammad Abū 'Abdallāh ibn Umayl, *al-Mā' al-waraqī wa-l-arḍ al-najamiyya*, in H. E. Stapleton, H. . Husayn and M. Turāb 'Alī, 'Three Arabic Treatises on Alchemy by Mu.hammad Bin Umayl (10th Century AD)', *Memoirs of the Asiatic Society of Bengal*, 12:1 (1933), 1–213.
74. Al-Qur.tubī, *Ghāyat al-.hakīm*, 8.7–11.
75. Al-Rāzī, *Kitāb al-Asrār*, 13.
76. See S. Moureau, 'Alchemical Equipment', in Brentjes (ed.) *Routledge Handbook on Science in the Islamicate World*, 512–22.
77. Ullmann, *Die Natur- und Geheimwissenschaften*, 266–270; Principe, *The Secrets of Alchemy*, 18–19.
78. On this phenomenon, see Rassi, 'Alchemy in an Age of Disclosure'.

79. See R. Forster, *Das Geheimnis der Geheimnisse: die arabischen und deutschen Fassungen des pseudo-aristotelischen Sirr al-asrār, Secretum secretorum* (Wiesbaden: Ludwig Reichert, 2006). For examples from the Pseudo-Aristotelian Hermetica, see Saif, 'A Preliminary Study'.
80. Grimes, *Becoming Gold*, 70–6.
81. M. Berthelot and R. Duval, *La chimie au moyen âge, Vol. 2: L'alchimie syriaque* (Paris: Imprimerie nationale, 1893), 223–4. Note that Berthelot and Duval omit mention of statues in this passage in their French translation. I take my reading from the text's unique manuscript, Ms. Cambridge Mm 6.29, 34r.
82. G. Miles, 'Stones, Wood and Woven Papyrus: Porphyry's *On Statues*', *Journal of Hellenic Studies*, 135 (2015), 78–94.
83. H. Ulrich, 'Medieval Muslim Perceptions of Pharaonic Egypt', in A. Loprieno (ed.), *Ancient Egyptian Literature: History and Forms* (Leiden: Brill, 1996), 605–27.
84. Berthelot and Duval, *La chimie au moyen âge*.
85. On *birbā*, see G. Wiet, 'Barbā', in *Encyclopaedia of Islam*, 2nd edn, vol. 1 (Brill: Leiden, 1986), 1038–9.
86. Ibn Umayl, *al-Māʾ al-waraqī*, 1–2.
87. Ibid., 2.
88. Ibid., 100.
89. Ibn al-Nadīm, *Kitāb al-Fihrist*, vol. 2, 445.10–15.
90. al-Qurṭubī, *Rutbat al-ḥakīm*, 114.
91. Jābir, *Kitāb al-Īḍāḥ*, 55.2–4.
92. Principe, *The Secrets of Alchemy*, 44.
93. A. Siggel, *Decknamen in der arabischen alchemistischen Literatur*, Veröffentlichung (Deutsche Akademie der Wissenschaften zu Berlin, Institut für Orientforschung) 5 (Berlin: Akademie-Verlag, 1951), 45.
94. Al-Fārābī, *Risāla*, 76.
95. Muḥammad Abū ʿAbdallāh ibn Umayl, *Book of the Explanation of the Symbols / Kitāb Ḥall al-rumūz*, ed. W. Madelung, trans. S. Fuad (Zurich: Living Human Heritage Publications, 2003), 2 (text), 3 (trans.).
96. Ibn Umayl, *Book of the Explanation*, 4 (text), 5 (trans.).
97. Al-Qur.tubī, *Rutbat al-ḥakīm*, 114.
98. Discussed in S. Moureau and G. de Callataÿ, 'In Code We Trust: The Concept of *Rumūz* in Andalusī Alchemical Literature and Related Texts', *Asiatische Studien*, 75:2 (2021), 429–47, at 434–5.
99. P. Lory, 'Esotérisme shi'ite et alchimie: Quelques remarques sur la doctrine de l'initiation dans le Corpus Jābirien', in M. A. Amir-Moezzi (ed.), *L'ésotérisme shi'ite, ses racines et ses prolongements* (Turnhout: Brepols, 2016), 411–22.
100. Jābir ibn Ḥayyān, *Kitāb Usṭuqus al-uss al-thānī*, in Holmyard, *The Arabic Works of Jâbir ibn Ḥayyān*, 89.5–12.
101. S. N. Haq, 'Greek Alchemy or Shīʿī Metaphysics? A Preliminary Statement Concerning Jābir ibn Ḥayyān's ẓāhir and bāṭin', *Bulletin of the Royal Institute for Inter-Faith Studies*, 4 (2002), 19–32.
102. G. Merianos, 'The Christianity of the Philosopher Christianos: Ethics and Mathematics in Alchemical Methodology', *Arys*, 20 (2022), 271–322.
103. Ibn Umayl, *Book of the Explanation*, 94 (text), 95 (trans.); idem, *al-Māʾ al-waraqī*, 100–1.
104. Al-Qurṭubī, *Rutbat al-ḥakīm*, 107–8.
105. Ibid., 24, 26, 109. On *tarjīḥ*, see W. L. Craig, *The Kalām Cosmological Argument* (London: Macmillan, 1979), 10–13.
106. Abū Ḥāmid Muḥammad ibn Muḥammad al-Ghazālī, *The Incoherence of the Philosophers / Tahāfut al-falāsifa*, ed. and trans. M. E. Marmura, 2nd edn (Provo: Brigham Young University Press, 2000), 162, §12–16.
107. G. Saliba, 'The Role of the Astrologer in Medieval Islamic Society', *Bulletin d'études orientales*, 44 (1992), 45–67, at 47, n. 15.
108. Al-Ghazālī, *The Incoherence*, 173–4, §25.
109. P. Carusi, 'Alchimia e teologia, transmutazione e miraculo nel *Sirr al-ʿālamain* attribuito a Ġazālī', *Memorie di scienze fisiche e naturali*, 37:2/2 (2013), 239–52.
110. On al-Rāzī's discourse on astral magic, see M.-S. Noble, *Philosophising the Occult: Avicennan Psychology and The Hidden Secret of Fakhr al-Dīn al-Rāzī* (Berlin: De Gruyter, 2020). Similarly to al-Ghazālī, al-Rāzī's relationship with these disciples was complex; see, for example, his refutation of astrology, mentioned in Saliba, 'The Role of the Astrologer', 47, n. 18.

111. See, for example, 'Alī ibn Yūsuf al-Qifṭī *Tārīkh al-ḥukamā'*, ed. J. Lippert (Leipzig: Dieterich'sche Verlagsbuchhandlung, 1903), 292.7.
112. See Principe, *The Secrets of Alchemy*, 102–5.
113. T. Mayer, 'Some Salient Alchemical Features of the Mysticism of Kubrā's *Fawā'iḥ al-Jamāl*', *Maghreb Review*, 45 (2020), 744–62.
114. C. Addas, *Quest for the Red Sulphur: The Life of Ibn 'Arabī*, trans. P. Kingsley (Cambridge: Islamic Texts Society, 1993), 112, n. 10, cited by Hallum, 'Zosimus', 132, n. 29.
115. N. G. Harris, 'In Search of 'Izz Al-Dīn Aydamir al-Ǧildakī, Mamlūk Alchemist', *Arabica*, 64:3/4 (2017), 531–56.
116. T. Artun, 'Hearts of Gold and Silver: The Production of Alchemical Knowledge in the Early Modern Ottoman World', PhD dissertation, Princeton University, 2013.
117. Stearns, *Revealed Sciences*.
118. E. J. Holmyard, *Alchemy* (Harmondsworth: Penguin, 1957), 101. Elsewhere, Holmyard recalls 'seeing a practising alchemist of the old school, Al-Hajj Abdul-Muhyi Arab (once Mufti of the Shah Jehan Mosque at Woking), absorbedly bent over a heated ladle which was always going to, but never did, deliver a golden stream of transmuted lead'. Idem, 'New Light on Alchemy', *Nature*, 170 (1952), 725–6, at 726.
119. P. E. Pormann and E. Savage-Smith, *Medieval Islamic Medicine* (Edinburgh: Edinburgh University Press, 2007), 68, 175.
120. R. F. Razook, 'Studies on the Works of al-Ṭughrā'ī, PhD dissertation, University of London, 1963, 67; M. Dapsens, 'The Alchemical Work of Khālid b. Yazīd b. Mu'āwiya (d. *c.* 85/704)', *Asiatische Studien*, 75:2 (2021), 327–427, at 340–4; R. Forster, 'Alchemical Stanzaic Poetry (*muwashshaḥ*) by Ibn Arfa' Ra's (fl. Twelfth Century)', *Asiatische Studien*, 75:2 (2021), 637–64.
121. R. Todd, 'Classical Poetic Motifs as Alchemical Metaphors in the *Shudhūr al-dhahab* and Its Commentaries', *Asiatische Studien*, 75:2 (2021), 665–83.

Alchemy in India
Dagmar Wujastyk

1. See A. Waley, 'References to Alchemy in Buddhist Scriptures', *Bulletin of the School of Oriental and African Studies*, 6:4 (1932), 1102–3, for a list of passages in Buddhist literature, along with summaries and translations.
2. See ibid.
3. See S. Beal, *Si-Yu-Ki: Buddhist Records of the Western World. Translated from the Chinese of Hiuen Tsiang (AD 629)*, vol. 2 (London: Trübner & Co, 1884), 210, and L. Rongxi, *The Great Tang Dynasty Record of the Western Regions. Translated by Li Rongxi*, Taishō vol. 51, no. 2087 (Moraga: BDK America, 1996), 273, on longevity preparations; and Beal, 216, and Rongxi, 276, on making gold. On the question of the identity of Nagarjuna in these sources, see D. G. White, *The Alchemical Body: Siddha Traditions in Medieval India* (Chicago and London: University of Chicago Press, 1996), 66–70.
4. The passage is discussed in N. Balbir, '*Scènes d'alchimie* dans la littérature jaina', *Journal of the European Ayurvedic Society* (Hamburg), 1 (1990), 136–137 and 143–147. Also see Balbir, 'La fascination jaina pour l'alchimie', *Journal of the European Āyurvedic Society*, 2 (1992), for other references to alchemy in Jain literature.
5. See Balbir, '*Scènes d'alchimie* dans la littérature jaina', 135.
6. The term *dhātuśāstra* is found in Book 2, Chapter 12, verse 1 in the *Arthaśāstra*. See P. Olivelle, *King, Governance, and Law in Ancient India Kau.tilya's: Arthaśāstra. A New Annotated Translation by Patrick Olivelle* (Oxford and New York: Oxford University Press, 2013), 127. For the difficulties in dating the work with any precision, see ibid., 25–8.

7. W. Doniger and S. Kakar, *Vatsyayana Mallanaga. Kamasutra. A New, Complete English Translation of the Sanskrit Text with Excerpts from the Sanskrit Jayamangala Commentary of Yashodhara Indrapada, the Hindi Jaya Commentary of Devadatta Shastri, and Explanatory Notes by the Translators* (Oxford: Oxford University Press, 2002), 15.
8. Translation by White (*The Alchemical Body*, 49) of a longer passage in the *Kādaṃbari*. The term translated as 'mercurial elixir [*rasāyana*]' could also just refer to a medicinal tonic without mercury.
9. 'Shaiva' refers to beliefs, practices and literature relating to the god Shiva. On the significance of this figure being South Indian (Dravidian), see C. Ferstl, 'Bāṇa's Literary Representation of a South Indian Śaivite', in N. Mirnig et al. (eds), *Tantric Communities in Context* (Vienna: Austrian Academy of Sciences Press, 2019).
10. Jinas ('conquerors') and Siddhas ('perfected ones') are persons that have overcome human imperfections and attained spiritual liberation according to Jain doctrine.
11. Translation of Chapter 4.3.195.29–30 in C. Chojnacki and H. Nagarajaiah (eds), *Uddyotanasūri's Kuvalayamālā: A Jain Novel From 779 AD. Translated from the French by Alexander Reynolds and Largely Revised by the Author*, vol. 2 (Bangalore: Sapna Book House, 2018), 884. The term translated as 'blows' should rather be 'smelts'.
12. Ibid., 883.
13. Edition of the *Rasahṛdayatantra* by Ācārya (Lahore: Motilal Banarsidass/The Bombay Sanskrit Press, 1927), 18–27. Translation my own.
14. Edition of the *Rasārṇava* by Rāy (Calcutta: Asiatic Society, 1985). Translation my own.
15. On this final set of procedures, see D. Wujastyk, 'Acts of Improvement: On the Use of Tonics and Elixirs in Sanskrit Medical and Alchemical Literature', *History of Science in South Asia*, 5:2 (2017), 1–36.
16. Translation of *Rasahṛdayatantra* 19.60–4. Translation my own.
17. 'Shakta' refers to beliefs etc. relating to Shakti in various forms as the supreme goddess and primordial power. The Book of the Heart of Mercury describes mercury as 'produced by Hara', i.e. Shiva, but also states that it 'conquers like Hari', ie Vishnu (*Rasahṛdayatantra* 1.2), while also making reference to the attainment of *brahmabhāva* – absorption into Brahman/the Absolute (*Rasahṛdayatantra* 1.26).
18. The Kaula tantric tradition is characterised by distinctive rituals and symbolism connected with the worship of Shakti and Shiva.
19. On the subject of mercury as the semen of Shiva, see the translation of *Rasārṇava*, Chapter 1, in D. G. White, 'The Ocean of Mercury: An Eleventh-Century Alchemical Text', in D. S. Lopez (ed.), *Religions of India in Practice* (Princeton: Princeton University Press, 1995), 281, and in this volume.
20. See White, *The Alchemical Body*, 142–5, on some of the tantric elements in the *Rasār.nava* and other alchemical works.
21. See the timeline of Sanskrit alchemical literature (http://ayuryog.org/AlchemyTimeline) for an overview of these broad developments.
22. See for example D. Wujastyk, 'Iron Tonics: Tracing the Development from Classical to Iatrochemical Formulation in Ayurveda', *HIMALAYA*, 39:1 (2019), on the introduction of alchemical techniques for making iron tonics in ayurvedic literature.
23. See Wujastyk, 'Acts of Improvement', on the subject of the recipient of the elixir.
24. Translation of *Rasār.nava* 2.2–11 in D. G. White, 'The Practice of Alchemy and Its Practitioners: Chapters One and Two of the *Rasār.nava*', in D. Wujastyk (ed.), *Indian Alchemy: Sources and Contexts* (New York: Oxford University Press, 2025).
25. Translation of *Rasār.nava* 2.20–6 in ibid.
26. See P. Sauthoff, 'Potency, Virility, Sexual Pleasure: Chapters Nine and Ten of the *Rasamañjarī*', in ibid., on the subject of aphrodisiacs, contraception and abortive measures in Sanskrit alchemical literature.
27. On the entangled disciplines of yoga, medicine and alchemy, see I. B. Kędzia, 'Alchemical Metaphors and Their Yogic Interpretation in Selected Passages of the Tamil Siddha Literature', *Religions of South Asia*, 17:2 (2023), and T. N. Ganapathy, *The Philosophy of the Tamil Siddhas* (New Delhi: Indian Council of Philosophical Research, 1993). On modern forms of Siddha medicine, see R. S. Weiss, *Recipes for Immortality: Medicine, Religion and Community in South India* (New York: Oxford University Press, 2009), and B. Sébastia, 'Preserving Identity or Promoting Safety? The Issue of Mercury in Siddha Medicine: A Brake on the Crossing of Frontiers', *Asiatische Studien – Études Asiatiques*, 69:4 (2015), 933–69.

28. Also see H. Scharfe, 'The Doctrine of the Three Humors in Traditional Indian Medicine and the Alleged Antiquity of Tamil Siddha Medicine', *Journal of the American Oriental Society*, 119:4 (1999), 609–29, on the links between Siddha medicine and Ayurveda.
29. See I. B. Kędzia, 'Mastering Deathlessness: Some Remarks on Karpam Preparations in the Medico-Alchemical Literature of the Tamil Siddhus', *History of Science in South Asia*, 5:2 (2017), 125, on this subject.
30. See ibid. on these and other materials used in Tamil Siddha practices.
31. Translation of the 'Kuru Nūl Muppu 50', 5:1–2 and 4:1–2 in ibid., 131.
32. See K. Natarajan, '"Divine Semen" and the Alchemical Conversion of Iramatevar', *Medieval History Journal*, 7:2 (2004), 255–78, and I. B. Kędzia, 'Global Trajectories of a Local Lore: Some Remarks about Medico-Alchemical Literature of the Two Tamil Siddha Cosmopolites', *Cracow Indological Studies*, 18 (2016), 93–118.
33. See Sébastia, 'Preserving Identity or Promoting Safety?', 946–8, on such practitioners.

Jewish Alchemy and Kabbalah
John M. MacMurphy

1. R. Patai, *The Jewish Alchemists: A History and Source Book* (Princeton: Princeton University Press, 1994); G. Ferrario, 'The Jews and Alchemy: Notes for a Problematic Approach', in M. L. Pérez, D. Kahn and M. R. Bueno (eds), *Chymia: Science and Nature in Medieval and Early Modern Europe* (Newcastle upon Tyne: Cambridge Scholars Publishing, 2010); G. Scholem, *Alchemy and Kabbalah*, trans. K. Ottmann (Putnam: Spring Publications, 2006); K. Burmistrov, *'For He Is Like a Refiner's Fire': Kabbalah and Alchemy* [Russian] (Moscow: Institute of Philosophy of the Russian Academy of Sciences, 2009), 24–30; T. Langermann, review of R. Patai, *The Jewish Alchemists: A History and Source Book* (Princeton: Princeton University Press, 1994), *Journal of the American Oriental Society*, 116:4 (1996); G. Mentgen, 'Jewish Alchemists in Central Europe in the Later Middle Ages: Some New Sources', *ALEPH: Historical Studies in Science and Judaism*, 9:2 (2009), 345-352; D. Jütte, 'Trading in Secrets: Jews and the Early Modern Quest for Clandestine Knowledge', *Isis*, 103:4 (2012), 668-686; A. Schwarz, *Kabbalah and Alchemy: An Essay on Common Archetypes* (Northvale: Jason Aronson, 2000); S. L. Drob, *Kabbalistic Visions: C. G. Jung and Jewish Mysticism* (Oxford: Routledge, 2023), ff. (esp. 32-45).
2. G. Ferrario, 'The Jews and Alchemy: Notes for a Problematic Approach', in M. L. Pérez, D. Kahn and M. R. Bueno (eds), *Chymia: Science and Nature in Medieval and Early Modern Europe* (Newcastle upon Tyne: Cambridge Scholars Publishing, 2010), 20-22; G. Freudenthal, 'Medieval Alchemy in Hebrew: A Noted Absence', in G. Freudenthal (ed.), *Science in Medieval Jewish Cultures* (New York: Cambridge University Press, 2011).
3. G. Bohak, *Ancient Jewish Magic: A History* (Cambridge: Cambridge University Press, 2008); Y. Harari, *Jewish Magic: Before the Rise of Kabbalah*, trans. B. Stein (Detroit: Wayne State University Press, 2017).
4. R. Patai, *The Jewish Alchemists: A History and Source Book* (Princeton: Princeton University Press, 1994).
5. Out of the numerous book reviewers, Gad Freudenthal and Tzvi Langermann, stand out as their scholarship continuously engaged with the subject of alchemy and Jewish culture, see T. Langermann, review of R. Patai, *The Jewish Alchemists: A History and Source Book* (Princeton: Princeton University Press, 1994), *Journal of the American Oriental Society*, 116:4 (1996); G. Freudenthal, review of R. Patai, *The Jewish Alchemists: A History and Source Book* (Princeton: Princeton University Press, 1994), *Isis*, 86:2 (1995). Langermann has since recanted some of the harshness of his criticism; see T. Langermann, 'An Alchemical Treatise Attributed to Joseph Solomon Delmedigo', *Aleph: Historical Studies in Science and Judaism*, 13:1 (2013), 87, n. 17.
6. Some scholars take exception to the term 'occult sciences' as it is claimed that such usage contains an inherent 'assumption of unity' among the various practices, including alchemy: see W. J. Hanegraaff,

Esotericism and the Academy: Rejected Knowledge in Western Culture (Cambridge: Cambridge University Press, 2012), 177–91; L. M. Principe, *The Secrets of Alchemy* (Chicago: University of Chicago Press, 2013), 2–3, 89. However, it can also be argued that a radical adoption of the perspective that these arts are *sui generis* systems can create a scholarly blind spot – especially considering the historical interactions, influences and hybridity of such methods.

7. On Maria the Jewess, see R. Patai, 'Maria the Jewess-Founding Mother of Alchemy', *Ambix*, 29: 3 (1982), 177-197; M. Berthelot, *La chimie au moyen âge*, vol. iii (Paris:Imprimerie nationale, 1893), 125.
8. N. Brosh, 'Introduction', in N. Brosh (ed.), *Islamic Jewelry*, Catalogue no. 281 (Jerusalem: Israel Museum, 1987).
9. Ibid.; N. Brosh, 'Some Remarks on the Islamic Jewellery Exhibition Held in Israel Museum in 1987', in N. Brosh (ed.), *Jewellery and Goldsmithing in the Islamic World: International Symposium, Jerusalem, 1987*, Catalogue no. 320 (Jerusalem: Israel Museum, 1991), 3; M. Rosen-Ayalon, 'The Islamic Jewellery from Ashkelon', in Brosh, *Jewellery and Goldsmithing in the Islamic World*, 9.
10. A. Ghabin, 'Jewellery and Goldsmithing in Medieval Islam: The Religious Point of View', in Brosh, *Jewellery and Goldsmithing in the Islamic World*; Brosh, 'Introduction'.
11. Brosh, 'Introduction'.
12. G. Bohak, 'Towards a Catalogue of the Magical, Astrological, Divinatory and Alchemical Fragments from the Cambridge Genizah Collections', in B. Outhwaite and S. Bhayro (eds.), *'From a Sacred Source': Genizah Studies in Honour of Professor Stefan C. Reif* (Leiden: Brill, 2011).
13. Ibid., 72–75.
14. Y. Yinon (P. B. Fenton), 'R. Makhluf Amsalem, an Alchemist and Kabbalist from Morocco', *Pe'amim*, 55 (1993), 93, n. 3.
15. Scholem, *Alchemy and Kabbalah*, 17, 21–22, 47; G. Scholem, *Sefer ha-Tamar: Das Buch von de Palme des Abu Aflah aus Syracuse*, Heft I: Der hebräische text, nach drei Handschriften (Jerusalem: Workmens Printing Press, 1926); G. Scholem, *Sefer ha-Tamar: Das Buch von de Palme des Abu Aflah aus Syracuse*, Heft II: Uebersetzung (Jerusalem: Workmens Printing Press, 1927); S. Pines, 'Le Sefer ha-Tamar: Et les Maggidim des Kabbalistes', in G. Nahon and C. Touati (eds.), *Hommage à Georges Vajda* (Louvain, 1980); R. Patai, *The Jewish Alchemists: A History and Source Book* (Princeton: Princeton University Press, 1994), 98.
16. R. Roshdi and M. Régis, *Encyclopedia of the History of Arabic Science*, vol. 1 (London: Routledge, 1996), 85–7.
17. Patai, *The Jewish Alchemists*, 98.
18. Scholem, *Sefer ha Tamar*, 4.
19. M. Maier, *Symbola Aureae Mensae Duodecim Nationum* [Symbols of the Golden Table of the Twelve Nations] (Frankfurt: Anton Humm, 1617).
20. Ibid., 57.
21. Oxford, Bodleian Libraries, MS. Reggio 23 (Neubauer 2234), fol. 64v; M. Idel, 'The Study Program of R. Yohanan Alemanno' [Hebrew], *Tarbiz*, 48 (1979).
22. Technically, an 'alembic' is the top part of an alchemical still configuration, see M. Rulando, *Lexicon Alchemiae sive Dictionarium Alchemisticum* [Lexicon of Alchemy or Alchemical Dictionary] (Frankfurt: Zachariah Palthenus, 1612), 28 ; M. Ruland, *A Lexicon of Alchemy*, trans. A. E. Waite (London: Westminster Press, 1893), 21. However, the term is sometimes used to reference the entire apparatus.
23. *Auriferae artis, quam chemiam vocant, antiquissimi authores, sive Turba Philosophorum* [Goldmaking arts, which the most ancient authorities or the assembly of the philosophers call Chemia] (Basel: Peter Perna, 1572).
24. Idel, 'The Study Program of R. Yohanan Alemanno', 307–12, 324.
25. Y. Alemanno, *Sefer Sha'ar ha-Cheshek* [The Book of the Gate of Desire] (Livorno, 1790), 16b; cf., Idel, 'The Study Program of R. Yohanan Alemanno', 324.
26. On this work and its various versions and editions, see A. McLean, 'Introduction', in A. McLean (ed.), *The Book of Abraham the Jew: A Very Ancient Work of Rabbi Abraham Eleazar*, trans. R. Muschter, Magnum Opus Hermetic Sourceworks no. 55 (Kilbirnie: McLean, 2015); Patai, *The Jewish Alchemists*, 219–20, 225–30 (for entries about 'kabbalah', see 228–29), 238–57.
27. On the subject of kabbalah and alchemy as it relates to the European occult traditions, see Burmistrov, *'Kabbalah and Alchemy*; A. B. Kilcher, 'Cabbala chymica. Knorrs spekulative Verbindung von Kabbala und Alchemie', *Morgen-Glantz: Zeitschrift der Christian Knorr von Rosenroth-Gesselschaft*, 13 (2003); P.

Forshaw, '*Cabala Chymica* or *Chemia Cabalistica* – Early Modern Alchemists and Cabala', *Ambix*, 60:4 (2013); P. Forshaw, 'Oratorium – Auditorium – Laboratorium: Early Modern Improvisations on Cabala, Music, and Alchemy', *Aries*, 10:2 (2010); Scholem, *Alchemy and Kabbalah*; G. Scholem, *Kabbalah* (New York: Meridian, 1978), 200–01.

28. For a translation and commentary of this work, see S. A. Farmer, *Syncretism in the West: Pico's 900 Theses (1486): The Evolution of Traditional Religious and Philosophical Systems* (Tempe: Medieval and Renaissance Texts and Studies, 1998).
29. Ibid., 498–99 (9>15).
30. The title alludes to biblical verses that have two separate, albeit related, verbs; see, for example, Isaiah 9:2 and Proverbs 4:18. As such, the term *nogah* can also be transliterated and pronounced as *nagah*. Both meanings refer to a brilliant, shining or even splendorous light.
31. A. W. Sonnenfels, *Or Nogah (אור נגה), Splendor Lucis, oder Glantz des Lichts, Pars I. Enthaltend eine kurtze Physico-Cabalistische Auslegung Des grösten Natur-Geheimnuss; Insgemein Lapis Philosophorum Genannt* Vienna: Leopold Johann Kaliwoda, 1745).
32. On this work and its author, see R. Evers, 'The Quest for the Philosophers' Stone: Alois von Sonnenfels' "אור נגה — Splendor Lucis", Vienna 1745', *Year Book – Leo Baeck Institute*, 68:1 (2023), 7-35, https://doi.org/10.1093/leobaeck/ybad001.
33. On this work, see Patai, *The Jewish Alchemists*, 322–35; Burmistrov, *Kabbalah and Alchemy*; Scholem, *Alchemy and Kabbalah*, 26, 31, 49, 52, 62–63, 68–80, 92.
34. C. K. Rosenroth, *Kabbala Denudata* [Kabbalah Unveiled] (Sulzbach: Abraham Lichtenthaler, 1677/78); C. K. Rosenroth, *Kabbalae Denudatae Tomus Secundus* [The Second Volume of the Kabbalah Unveiled] (Frankfurt: B. C. Wust, 1684). On Rosenroth's relationship with alchemy and kabbalah, see Kilcher, 'Cabbala chymica'.
35. W. Newman, *Newton the Alchemist: Science, Enigma, and the Quest for Nature's 'Secret Fire'* (Princeton: Princeton University Press, 2018), 503.
36. On Newton's specific interest in certain pages of the *Kabbala Denudata*, see J. Harrison, *The Library of Isaac Newton* (Cambridge: Cambridge University Press, 1978), H873; A. Coudert, 'Leibniz, Locke, Newton and the Kabbalah', in J. Dan (ed.), *The Christian Kabbalah: Jewish Mystical Books & Their Christian Interpreters: A Symposium* (Cambridge, MA: Harvard College Library, 1997), 179, n. 53.
37. Rosenroth, *Kabbala Denudata*, 483.
38. Scholem, *Alchemy and Kabbalah*, 58–9.
39. M. A. Zuber and R. T. Prinke, 'Mistaken Identity, Forged Prophecies, and a Paracelsus Commentary: The Elusive Jewish Alchemist Mardochaeus de Nelle', *ALEPH: Historical Studies in Science and Judaism*, 20:1–2 (2020), 263–99; R. T. Prinke and M. A. Zuber, '"Learn to Restrain Your Mouth": Alchemical Rumours and Their Historiographical Afterlives', *Early Science and Medicine*, 25:5 (2020), 413–52.
40. Scholem, *Alchemy and Kabbalah*, 72; Patai, *The Jewish Alchemists*, 578, n. 22.
41. Scholem, *Alchemy and Kabbalah*, 61; Patai, *The Jewish Alchemists*, 400–01.
42. Evers, 'The Quest for the Philosophers' Stone, 3; M. Kahana, 'An Esoteric Path to Modernity: Rabbi Jacob Emden's Alchemical Quest', *Journal of Modern Jewish Studies*, 12:2 (2013), 255–58.
43. On this account, see Y. A. Modena, *Chayey Yehudah* [Life of Yehudah], ed. A. Kahana (Kiev, 1912), 34; Scholem, *Alchemy and Kabbalah*, 61–2; Patai, *The Jewish Alchemists*, 400–01.
44. For a consolidated view of some of the common commentaries of this work, see A. Kaplan, *Sefer Yetzirah: The Book of Creation*, revised edn (San Francisco: Wieser Books, 1997). For a more historical analysis, see P. A. Hayman, *Sefer Yessira* (Tübingen: Mohr-Siebeck, 2004.); T. Weiss, *Sefer Yessirah and Its Contexts: Other Jewish Voices* (Philadelphia: University of Pennsylvania Press, 2018).
45. On Abulafia's techniques, see M. Idel, *The Mystical Experience in Abraham Abulafia* (Albany: SUNY Press, 1988), 13–55; J. M. MacMurphy, 'Abraham Abulafia and the Academy: A Reevaluation', Research MA thesis, University of Amsterdam, 2015, 55–71.
46. For some of the scholarly debates on the use of metaphors and the historiography of alchemy, see L. M. Principe and W. R. Newman, 'Some Problems with the Historiography of Alchemy', in W. R. Newman and A. Grafton (eds), *Secrets of Nature: Astrology and Alchemy in Early Modern Europe* (Cambridge, MA: MIT Press, 2001); B. Vickers, 'The "New Historiography" and the Limits of Alchemy', *Annals of Science*, 65:1 (2008); W. R. Newman, 'Brian Vickers on Alchemy and the Occult: A

Response', *Perspectives on Science*, 17:4 (2009); H. Tilton, '*Alchymia Archetypica*: Theurgy, Inner Transformation and the Historiography of Alchemy', *Quaderni di Studi Indo-Mediterranei*, 5 (2012); M. A. Zuber, *Spiritual Alchemy: From Jacob Boehme to Mary Anne Atwood* (New York: Oxford University Press, 2021).

47. For an overview of the Lurianic school, see L. Fine, *Physician of the Soul, Healer of the Cosmos: Isaac Luria and His Kabbalistic Fellowship* (Stanford: Stanford University Press, 2003).
48. For the current scholarly views on Chayim Vital and alchemy, see G. Scholem, *Alchemy and Kabbalah*, 2006), 50–2, 67, 74; Patai, *The Jewish Alchemists*, 340–64; G. Bos, 'Hayyim Vital's "Practical Kabbalah and Alchemy": A 17th Century Book of Secrets', *Journal of Jewish Thought and Philosophy*, 4:1 (1995); U. Safrai, 'Hayim Vital the Goldsmith of God' [Hebrew], *Pe'amim*, 157 (2019).
49. J. M. MacMurphy, PhD dissertation, University of Amsterdam (forthcoming).
50. C. Vital, *Sefer ha-Pe'ulot* [The Book of Operations], 2nd edn (2014); C. Vital, *Ta'alumot Chochmah* [Mysteries of Wisdom] (Jerusalem: Yarid Hasfarim, 2017).
51. H. Tilton, *The Quest for the Phoenix: Spiritual Alchemy and Rosicrucianism in the Work of Count Michael Maier (1569–1622)* (Berlin: De Gruyter, 2003), 71–7 (esp. 72–3).
52. G. B. Nazari, *Il Metamorfosi Metallico et Humano* (Brescia: Francesco Manchetti, 1564).
53. P. J. Forshaw, 'Subliming Spirits: Physical-Chemistry and Theo-Alchemy in the Works of Heinrich Khunrath (1560–1605)', in S. J. Linden (ed.), *Mystical Metal of Gold: Essays on Alchemy and Renaissance Culture* (New York: AMS Press, 2007), 262.
54. Cf. note 45.
55. On this unique form of alchemy, see MacMurphy, PhD dissertation; J. M. MacMurphy, 'Sex as a Spiritual Practice in Jewish Kabbalah', paper given at the ESOGEN Symposium: Esotericism, Gender and Sexuality, University of Amsterdam, Netherlands (16 April, 2021); J. M. MacMurphy, 'Between Chymia and Alchemy: Chayim Vital's Intimate Relationship with Western Occult Praxis', paper given at 'For there I, Hayyim, was Living': R. Hayyim Vital and His World, conference hosted by Ben Gurion University of the Negev, Ben Zvi Institute, and the World Union of Jewish Studies, Israel (14–16 September, 2020). During the writing of my dissertation project, an excellent article by Uri Safrai was published, outlining some key alchemical principles in Lurianic thought; see Safrai, 'Hayim Vital the Goldsmith of God'. For an example of the double pelican iconography, see G. B. Della Porta, *De Distillatione* [On Distillation], lib. IX (Rome: Camera Apostolica, 1608), 40.

Medieval Latin Alchemy
Peter J. Forshaw

1. Morienus, *Liber de Compositione Alchemiae, quem edidit Morienus Romanus, Calid Regi Aegyptiorum: quem Robertus Castrensis de Arabico in Latinum transtulit, in Artis Auriferae, quam Chemiam vocant*, vol. 2 (Basel, 1593), Robert of Chester's Preface, 4, 'His est namque liber divinus, & divinitate plenißimus. In eo enim duorum testamentorum (veteris scilicet & novi) continetur vera & perfecta probatio.'
2. A. Lo Bello, *The Commentary of al-Nayrizi on Book 1 of Euclid's Elements of Geometry* (Leiden: Brill, 2003), 38–9.
3. Morienus, *De transfiguratione metallorum* (Paris, 1559); Morienus, *De transfiguratione metallorum et occulta summaque antiruorum philosophorum medicina libellus* (Hanau, 1593).
4. On Stephanos of Alexandria, see F. Sherwood Taylor, 'The Alchemical Works of Stephanos of Alexandria', Part 1, *Ambix*, 1:1 (May 1937), 116–39; Part 2, *Ambix*, 2 (1938), 39–49.
5. L. M. Principe, *The Secrets of Alchemy* (Chicago: University of Chicago Press, 2013), 51f.
6. Morienus, *Liber de Compositione Alchemiae*, Preface, 5. See also C. Burnett, 'The Astrologer's Assay of the Alchemist: Early References to Alchemy in Arabic and Latin Texts,' *Ambix*, 39, Part 3 (1992), 103–9, at 104–5.

7. For discussion of the stone, see Morienus, *Liber de Compositione Alchemiae*, 37. For reference to the elixir, see 47, 50. On the power of 'Alchymia' to transmute silver into gold, see 54.
8. Ibid., 38: 1. coitus, 2. conceptio, 3. praegnatio, 4. ortus, 5. nutrimentum.
9. Ibid., 46, together with a red, yellow and white smoke (fumus), identified on p. 52 as orpiment, sulphur, quicksilver, and 49, 51, where it is identified as 'vitrum' (glass). The Green Lion is often identified as vitriol. See J. M. Rampling, *The Experimental Fire: Inventing English Alchemy, 1300–1700* (Chicago: University of Chicago Press, 2020), 93.
10. Morienus, *Liber de Compositione Alchemiae*, 37.
11. Ibid., 22.
12. P. J. Forshaw, '"Chemistry, that Starry Science": Early Modern Conjunctions of Astrology and Alchemy', in N. Campion and E. Greene (eds), *Sky and Symbol* (Lampeter: Sophia Centre Press, 2013), 143–84, at 147.
13. Principe, *Secrets of Alchemy*, 59–60.
14. Ibid., 60; W. R. Newman, 'An Overview of Roger Bacon's Alchemy', in J. Hackett (ed.), *Roger Bacon and the Sciences: Commemorative Essays* (Leiden: Brill, 1997), 317–36, at 324, 333.
15. P. Bonus, *Pretiosa margarita novella* (Venice, 1546), *[vv].
16. Principe, *Secrets of Alchemy*, 55; W. R. Newman, *The Summa Perfectionis of Pseudo-Geber: A Critical Edition, Translation & Study* (Leiden: Brill, 1991), 70.
17. Principe, *Secrets of Alchemy*, 55; Newman, *The Summa Perfectionis of Pseudo-Geber*, 86.
18. Newman, *The Summa Perfectionis of Pseudo-Geber*, 633.
19. Ibid., 648.
20. Ibid., 750.
21. Ibid., 726.
22. Ibid., 671.
23. Ibid., 725.
24. Ibid., 676.
25. L. M. Principe, 'Theory and Concepts: Conceptual Foundations of Early Modern Chymical Thought and Practice', in B. T. Moran (ed.), *A Cultural History of Chemistry in the Early Modern Age* (London: Bloomsbury Academic, 2022), 23–40.
26. Newman, *The Summa Perfectionis of Pseudo-Geber*, 679.
27. Ibid., 687. B. Obrist, 'Visualisation in Medieval Alchemy', *Hyle: International Journal for Philosophy of Chemistry*, 9:2 (2003), 131–70, comments on Geber's systematic descriptions but points out that he does not provide figures, which are drawn in the margins of the oldest manuscripts, such as Bibliothèque nationale de France ms. lat. 6514, fol. 68r–71r.
28. Newman, *The Summa Perfectionis of Pseudo-Geber*, 689; solid wood gives a strong fire, spongy wood a weak fire, dry wood a large, short-lived fire, green wood a small, long-lasting fire.
29. Ibid., 768.
30. Ibid., 784.
31. Ibid., 768.
32. Ibid., 652.
33. Ibid., 784.
34. Ibid.
35. Principe, *Secrets of Alchemy*, 47, 72. For more on Pseudo-Lullian alchemy, see M. Pereira, *The Alchemical Corpus Attributed to Raymond Lull* (London: Warburg Institute, 1989). On the *Testamentum*, see Pereira (ed.), *Il Testamentum alchemico attribuito a Raimondo Lullo* (Florence: Sismel, 1999).
36. R. Lull, *Testamentum, duobus libris universam artem Chymicam complectens, antehac nunquam excusum. Item, eiusdem Compendium animae Transmutationis artis metallorum, absolutum iam & perfectum* (Cologne, 1566), 28v.
37. Ibid., 7r.
38. Ibid., 6v.
39. Ibid., 31r.
40. Ibid., 140v, 141r.
41. Ibid., 7v.

42. F. A. Yates, 'The Art of Ramon Lull: An Approach to It through Lull's Theory of the Elements', *Journal of the Warburg and Courtauld Institutes*, 17:1/2 (1954), 115–73, at 116.
43. Lull, *Testamentum*, 169r–v.
44. Ibid., 12r–v.
45. Ibid., 16r.
46. Ibid., after 210v a large circular diagram of the four elements and processes 'Significationes Literarum Huius Testamenti'; 241r.
47. Ibid., 56v.
48. Ibid., 26r.
49. Ibid., 23r.
50. Ibid., 22v.
51. Ibid., 21v.
52. Ibid., 17r.
53. Ibid., 20r.
54. Ibid., 58r.
55. J. de Rupescissa, *De consideratione quintae essentiae omnium rerum* (Basel, 1597). On Rupescissa, see L. DeVun, *Prophecy, Alchemy, and the End of Time: John of Rupescissa in the Late Middle Ages* (New York: Columbia University Press, 2014).
56. Rupescissa, *De consideratione quintae essentiae,* 12, 104.
57. Ibid., 14.
58. Ibid., 17.
59. Ibid., 29.
60. Ibid., 103.
61. Ibid., 31.
62. Ibid.
63. Ibid., 35–7.
64. Ibid., 66.
65. Ibid., 105.
66. Ibid., 112.
67. Ibid., 114.
68. Ibid., 119.
69. Ibid., 21–4.
70. Ibid., 79.
71. Ibid., 91, 93.
72. Ibid., 100.
73. Ibid., 101.
74. Ibid., 66.
75. Ibid., 46.
76. P. Bonus, *Pretiosa Margarita Novella de Thesauro, ac Pretiosissimo Philosophorum Lapide*, ed. J. Lacinius (Venice: Aldine Press, 1546). On Bonus's identity, see M. Soranzo, *Giovanni Aurelio Augurello (1441–1524) and Renaissance Alchemy: A Critical Edition of Chrysopoeia and Other Alchemical Poems, with an Introduction, English Translation and Commentary* (Leiden: Brill, 2020), 37. For more on Bonus, see C. Crisciani, 'The Conception of Alchemy as Expressed in the *Pretiosa Margarita Novella* of Petrus Bonus of Ferrara', *Ambix*, 20:3 (1973), 165–81.
77. Bonus, *Pretiosa Margarita Novella*, 22r.
78. Ibid., 1–2.
79. Ibid., 26v.
80. Ibid., 30r.
81. Ibid., 30v.
82. Ibid., 38r.
83. Ibid., 38v.
84. Ibid.
85. Ibid., 39r

86. Ibid., 39v, 120v.
87. Ibid., 52v.
88. Ibid., 120v. See Principe, *Secrets of Alchemy*, 68.
89. Bonus, *Pretiosa Margarita Novella*, 47r.
90. Ibid., 42v.
91. Ibid., 46v.
92. Ibid., 48v.
93. Ibid., 47r.
94. Ibid., 44v–45r.
95. Ibid., 45v–46r.
96. Ibid., 47v.
97. Ibid., 58r.
98. Ibid., 57v–58r.

The Renaissance, Esotericism and Alchemy
Peter J. Forshaw

1. H. Khunrath, *Amphitheatrum Sapientiae Aeternae* (Hanau: Wilhelm Anton, 1609), II, 202.
2. J. Van Lennep, *Art & alchimie, Étude de l'iconographie hermétique et de ses influences* (Brussels: Éditions Meddens, 1966), 37–8. For detailed discussions of the *Book of the Holy Trinity*, see B. Obrist, *Les débuts de l'imagerie alchimique (XIVe–XVe siècles)* (Paris: Éditions le Sycomore, 1982), Chapter 3; M. Gabriele, *Alchimia e Iconologia* (Udine: Forum, 1997), Chapter 5. See also U. Junker, *Das „Buch der Heiligen Dreifaltigkeit" in seiner zweiten, alchemistischen Fassung (Kadolzburg 1433)*, Arbeiten der Forschungsstelle des Instituts für Geschichte der Medizin der Universität zu Köln, Band 40 (Cologne: Kölner medizinhistorische Beiträge, 1986).
3. Obrist, *Les débuts de l'imagerie alchimique*, 188, dates the oldest copy of the *Aurora*, in Zurich, to the second decade of the fifteenth century; Gabriele, *Alchimia e Iconologia*, 33, proposes the first half of the fifteenth century.
4. Obrist, *Les débuts de l'imagerie alchimique*, 126, 139. *Das Buch der heiligen Dreifaltigkeit, und Beschreibung der Heimligkeit von Veränderung der metallen offenbahret*. Anno Christi M.CCCC. See also W. Ganzenmüller, 'Das Buch der heiligen Dreifaltigkeit. Eine deutsche Alchemie aus dem Anfang des 15. Jahrhunderts', *Archiv für Kulturgeschichte*, 29 (1939), 93–146, especially 116–21 concerning the symbols and signs.
5. Obrist, *Les débuts de l'imagerie alchimique*, 127.
6. Ibid.
7. Ibid., 118; on 129 Obrist states that it is fairly clear that the Antichrist in question is the Czech religious reformer Jan Hus (1369–1415) with his 'heretical' followers (*Ketzereifer*). This is an early instance of the devil also being an alchemist, a notion that appears occasionally in later literature, such as the *Hydrolithus Sophicus, seu Aquarium Sapientum*, with its reference to 'Satan, that grim pseudo-alchymist', in *Musaeum hermeticum reformatum et amplificatum* (Frankfurt, 1749), 125, 'Illic enim illico terribilis Sathan et Pseudo-Chymicus sese repraesentat ...'; *The Sophic Hydrolith or, Water Stone of the Wise*, in *The Hermetic Museum* (1893), trans. A. E. Waite (London: James Elliot & Co, 1893), vol. 1, 107.
8. Obrist, *Les débuts de l'imagerie alchimique*, 132.
9. Ibid., 119, 141.
10. Munich, Bayerische StaatsBibliothek BSB-Hss Cgm 598 Buch der heiligen Dreifaltigkeit (second half of fifteenth century), f.24r. Cf Junker, *Das „Buch der Heiligen Dreifaltigkeit"*, 112–14.
11. Obrist, *Les débuts de l'imagerie alchimique*, 178.
12. Ibid., 150.

13. Ibid., 186.
14. Ibid.
15. P. J. Forshaw, 'Alchemical Images', in D. Jalobeanu and C. Wolfe (eds), *Encyclopedia of Early Modern Philosophy and the Sciences* (Cham: Springer, 2022), https://doi.org/10.1007/978-3-319-20791-9_472-1.
16. For a detailed analysis, see Obrist, *Les débuts de l'imagerie alchimique*, 189–208; Gabriele, *Alchimia e Iconologia*. See also I. Ronca, 'Religious Symbolism in Medieval Islamic and Chrisian Alchemy', in A. Faivre and W. J. Hanegraaff (eds), *Western Esotericism and the Science of Religion* (Leuven: Peeters, 1998), 95–116.
17. M.-L. Von Franz (ed.), *Aurora Consurgens: A Document Attributed to Thomas Aquinas on the Problem of Opposites in Alchemy*, trans. R. F. C. Hull and A. S. B. Glover (Toronto: Inner City Books, 2000), 111, 113.
18. D. Waterland, *A Critical History of the Athanasian Creed* (London: SPCK, 1850), 178: 'This is the Catholic faith, which except a man believe faithfully and firmly, he cannot be saved.'
19. John 20:29: 'Jesus saith to him: Because thou hast seen me, Thomas, thou hast believed: blessed are they that have not seen, and have believed'; John 3:18: 'He that believeth in him is not judged. But he that doth not believe, is already judged: because he believeth not in the name of the only begotten Son of God.'
20. Romans 9:33: 'As it is written: Behold I lay in Sion a stumblingstone and a rock of scandal; and whosoever believeth in him shall not be confounded.'
21. See *Turba Philosophorum*, in G. Gratarolo (ed.), *Artis Auriferae* (Basel, 1593), vol. 1, text 2; J. J. Heilman (ed.), *Theatrum Chemicum* (Strasbourg, 1660), vol. 5, text 1.
22. See P. J. Forshaw, 'Marsilio Ficino and the Chemical Art', in S. Clucas, P. Forshaw and V. Rees (eds), *Laus Platonici Philosophi: Marsilio Ficino and His Influence* (Leiden: Brill, 2011), 249–71.
23. Ibid., 259.
24. Ibid., 259–60.
25. Ibid., 260.
26. M. Soranzo, *Giovanni Aurelio Augurello (1441–1524) and Renaissance Alchemy: A Critical Edition of* Chrysopoeia *and Other Alchemical Poems, with an Introduction, English Translation and Commentary* (Leiden: Brill, 2020), 24, 26, 32.
27. Forshaw, 'Marsilio Ficino and the Chemical Art', 257–8.
28. Soranzo, *Giovanni Aurelio Augurello (1441–1524) and Renaissance Alchemy*, 5, 20–1, 50.
29. Ibid., 31. On mythoalchemy, see P. J. Forshaw, 'Michael Maier and Mythoalchemy', in T. Nummedal and D. Bilak (eds), *Furnace and Fugue: A Digital Edition of Michael Maier's Atalanta fugiens (1618) with Scholarly Commentary* (University of Virginia Press, 2020), https://furnaceandfugue.org/essays/forshaw/.
30. Soranzo, *Giovanni Aurelio Augurello (1441–1524) and Renaissance Alchemy*, 44.
31. M. V. Dougherty, 'Three Precursors to Pico della Mirandola's Roman Disputation and the Question of Human Nature in the *Oratio*', in M. V. Dougherty (ed.), *Pico della Mirandola: New Essays* (Cambridge: Cambridge University Press, 2008), 114–51, at 140; S. J. Rabin, 'Pico on Magic and Astrology', in ibid., 152–78, at 156.
32. On Pico more generally, see Dougherty, *Pico della Mirandola: New Essays*.
33. G. Lloyd Jones, 'Introduction', in J. Reuchlin, *De Arte Cabalistica: On the Art of the Kabbalah*, trans. M. and S. Goodman (New York: Abaris Books, 1983; repr. Lincoln and London: University of Nebraska Press, Bison Books, 1993), 16.
34. On Kabbalistic exegesis, see B. P. Copenhaver, 'Number, Shape, and Meaning in Pico's Christian Cabala: The Upright *Tsade*, the Closed *Mem*, and the Gaping Jaws of Azazel', in A. Grafton and N. Siraisi (eds), *Natural Particulars: Nature and the Disciplines in Renaissance Europe* (Cambridge, MA: MIT Press, 1999), 25–76. See also P. J. Forshaw, 'The Genesis of Christian Kabbalah: Early Modern Speculations on the Work of Creation', in C. Vander Stichele and S. Scholz (eds), *Hidden Truths from Eden: Esoteric Readings of Genesis 1–3*, Semeia Studies (Atlanta: Society of Biblical Literature, 2014), 121–44.
35. E. Morlok, *Rabbi Joseph Gikatilla's Hermeneutics* (Tübingen: Mohr Siebeck, 2011), 72, 225. See also Reuchlin, *On the Art of the Kabbalah*, 299.

36. G. A. Pantheo, *Ars transmutationis metallicae* (Venice, 1519); idem, *Voarchadumia contra Alchi'miam: Ars distincta ab Archimi'a, et Sophia* (Venice, 1530). See D. Kahn, *Alchimie et Paracelsisme en France (1567–1625)* (Geneva: Librairie Droz, 2007), 64. For more on Pantheo, see P. J. Forshaw, '*Cabala Chymica* or *Chemia Cabalistica* – Early Modern Alchemists and Cabala', *Ambix*, 60:4 (November 2013), 361–89, at 371–6.
37. Pantheo, *Voarchadumia* (Paris, 1550), 27v. Pantheo uses an idiosyncratic transliteration of YHVH as *Iud He Voph He*, but to avoid confusion I have stuck to the more standard *Yod He Vav He*.
38. Pantheo, *Ars et theoria transmutationis metallicae* (Venice, 1550), 28v–29r, *De Metallorum Spiritus Generatione*.
39. Pantheus, *Voarchadumia contra alchimiam* (Paris, 1550), sig. 19r.
40. N. Séd, 'L'or enfermé et la poussière d'or selon Moïse ben Shémtobh de Léon (c. 1240–1305)', *Chrysopoeia*, 3:2 (1989), 121–34, at 131.
41. Forshaw, 'Marsilio Ficino and the Chemical Art', 263; A. Weeks, *Paracelsus: Speculative Theory and the Crisis of the Early Reformation* (Albany: State University of New York Press, 1997), 6, 57, 71.
42. F. Sherwood Taylor, *The Alchemists: Founders of Modern Chemistry* (London: The Scientific Book Club, 1953), 196.
43. J. R. Partington, *A History of Chemistry*, 4 vols (London: Macmillan & Co., 1961–62), vol. 2, 135.
44. W. Pagel, *Paracelsus: An Introduction to Philosophical Medicine in the Era of the Renaissance* (Basel: S. Karger, 1958), 88.
45. G. Hedesan, 'Alchemy, Potency, Imagination: Paracelsus's Theories of Poison', in O. P. Grell, A. Cunningham and J. Arrizabalaga (eds), *It All Depends on the Dose: Poisons and Medicines in European History* (Abingdon: Routledge, 2018), 81–102.
46. A. Weeks (ed. and trans.), *Paracelsus (Theophrastus Bombastus Von Hohenheim, 1493–1541): Essential Theoretical Writings* (Leiden: Brill, 2008), 8–13.
47. D. Pickering Walker, *Spiritual & Demonic Magic from Ficino to Campanella* (London: Warburg Institute, 1958; repr. University Park: Penn State University Press, 2000), 102.
48. *De Tinctura Physicorum*, in Paracelsus, *Opera, Bücher und Schrifften*, ed. J. Hüser (Strasbourg, 1603), vol. 1, 922–5, at 923.
49. Paracelsus, *Aurora Philosophorum* (Basel, 1577), 14.
50. M. Toxites, *Onomastica II* (Strasbourg, 1574), 410–12; G. Dorn, *Dictionarium Theophrasti Paracelsi* (Frankfurt, 1583), 25–6; M. Ruland, *Lexicon Alchemiae* (Frankfurt, 1612), 108.
51. H. Khunrath, *Amphitheatrum Sapientiae Aeternae, Solius Verae: Christiano-Kabalisticvm, Divino-Magicvm, nec non Physico-Chymicvm, Tertriunum, Catholicon* (Hanau: Guilielmus Antonius, 1609).
52. H. Khunrath, *De Igne Magorum Philosophorumque secreto externo & visibili* (Strasbourg, 1608), 87. For more, see P. J. Forshaw, 'A Necessary Conjunction: Cabala, Magic, and Alchemy in the Theosophy of Heinrich Khunrath (1560–1605)', in G. D. Hedesan and T. Rudbøg (eds), *Innovation in Esotericism from the Renaissance to the Present* (Basingstoke: Palgrave Macmillan, 2021), 97–134.
53. G. Scholem, *Alchemy and Kabbalah*, trans. K. Ottmann (Putnam: Spring Publications, 2006), 88–91.
54. Khunrath, *Amphitheatrum*, II, 158.
55. All four circular engravings bear the inscription 'Henricus Khunrath Lips Theosophiae amator, et Medicinae Doctor, Dei gratia, inventor. Paullus von der Doort Antverpien scalpsit. Hamburgi. Anno a Christo nato 1595 Mense Aprili (Maio, Iulio, Septembri)'; the fourth, the most famous of the adept in his Laboratory-Oratory, also includes the words 'H.F. Vriese pinxit'.
56. M. Widmer, *Moses, God and the Dynamics of Intercessory Prayer: A Study of Exodus 32–34 and Numbers 13–14* (Tübingen: Mohr Siebeck, 2004), 286, distinguishes the 'tent of meeting' located outside the Israelite camp and used for oracular purposes from the tabernacle located in the middle of the camp and serving for cultic purposes.
57. H. Khunrath, *Von hylealischen, Das ist Pri-materialischen catholischen oder algemejnem natürljchen Chaos* (Magdeburg, 1597), 416.
58. Khunrath, *Amphitheatrum* (1609), II, 204–5.

Alchemy and Rosicrucianism
Christopher McIntosh

1. *Fama Fraternitatis* (Kassel: Wilhelm Wessel, 1614).
2. R. Edighoffer, 'Rosicrucianism', in W. J. Hanegraaff et al., *Dictionary of Gnosis and Western Esotericism*, vol. 2 (Leiden: Brill, 2005), 1009–10.
3. Ibid., 1010.
4. *Fama Fraternitatis: Manifesto of the Most Praiseworthy Order of the Rosy Cross of Europe*, trans. and annotated C. McIntosh and D. Pahnke McIntosh, intro. C. McIntosh (Germany: Vanadis Texts, 2014), 26–7.
5. J. V. Andreae, *Fama Fraternitatis, Confessio Fraternitatis, Chymische Hochzeit Christiani Rosenkreuz*, ed. and intro. R. van Dülmen (Stuttgart: Calwer Verlag, 1973), 20.
6. R. Patai, *The Jewish Alchemists* (Princeton: Princeton University Press, 1994), 207.
7. Ibid., 207–8.
8. *Fama Fraternitatis* (McIntosh edition), 31.
9. *Chymische Hochzeit Christiani Rozenkreuz* (Strasbourg: Lazarus Zetzner, 1616).
10. Edighoffer, 'Rosicrucianism', 1012.
11. J. J. C. Bode (writing anonymously), *Starke Erweise aus den eigenen Schriften des hochheiligen Ordens Gold- und Rosenkreuzer* ('Rome', i.e. Leipzig: Göschen, 1788), 25.
12. R. Caron, 'Alchemy V: 19th–20th Century', in W. J. Hanegraaff et al., *Dictionary of Gnosis and Western Esotericism*, vol. 1 (Leiden: Brill, 2005), 53.
13. Ibid.
14. R. Steiner, *Theosophy of the Rosicrucians* (London: Rudolf Steiner Publishing Company, 1953), 10.
15. Ibid., 13.
16. C. Rodríguez Galilea, *Alchemy in Anthroposophy: A Fractal Narrative towards a Present Practice of Alchemy*, master's thesis, University of Amsterdam, 2019, abstract.
17. Caron, 'Alchemy V', 55.
18. F. Wittemans, *A New and Authentic History of the Rosicrucians* (London: Rider, 1938), 156.
19. M. Stavish, 'The History of Alchemy in America', posted on website of the Hermetic Library, https://hermetic.com/stavish/alchemy/history, accessed 1 March, 2023.
20. Caron, 'Alchemy V', 55.

Alchemy, Chemistry and Medicine in the Early Modern Era
Georgiana D Hedesan

1. For short introductions to alchemy, see L. M. Principe, *The Secrets of Alchemy* (Chicago: University of Chicago Press, 2013), and the concise G. D. Hedesan, 'Alchemy', in C. Partridge (ed.), *The Occult World* (London: Routledge, 2015), 552–63.
2. W. R. Newman, *Promethean Ambitions: Alchemy and the Quest to Perfect Nature* (Chicago: University of Chicago Press, 2004), 91–7.
3. J. M. Rampling, *Experimental Fire: The Making of English Alchemy* (Chicago: University of Chicago Press, 2020), 64ff.
4. On Johannes de Rupescissa, see L. DeVun, *Prophecy, Alchemy, and the End of Time: John of Rupescissa in the Late Middle Ages* (New York: Columbia University Press, 2009).
5. Rampling, *Experimental Fire*, 45.
6. On Ripley, see J. M. Rampling, 'Establishing the Canon: George Ripley and His Alchemical Sources', *Ambix*, 55:3 (2008), 189–208. Christopher of Paris is still awaiting research.
7. On the topic, see the rather dated article by R. Multhauf, 'The Significance of Distillation in Renaissance Medical Chemistry', *Bulletin for the History of Medicine*, 30:4 (1956), 329–46.

8. T. Taape, 'Hieronymus Brunschwig and the Making of Vernacular Medical Knowledge in Early German Print', PhD dissertation, University of Cambridge, 2017, 110–18; idem, 'Distilling Reliable Remedies: Hieronymus Brunschwig's *Liber de arte distillandi* (1500) between Alchemical Learning and Craft Practice', *Ambix*, 61:3 (2014), 236–56.
9. Taape, 'Hieronymus Brunschwig', 15.
10. Ibid.
11. G. D. Hedesan, 'Plant Alchemy, Paracelsianism and Internal Signature Theory in the Writings of Guy de La Brosse (1586–1641)', *Notes and Records* (2023), https://royalsocietypublishing.org/doi/epdf/10.1098/rsnr.2023.0031.
12. The subject of potable gold will be treated in my upcoming book, *Universal Medicine in Early Modern Alchemy*.
13. T. Russell, *Diacatholicon Aureum* (London: John Flasker, 1602).
14. The story of syphilis has been told many times. For a recent treatment, see V. Nutton, *Renaissance Medicine: A Short History of European Medicine in the Sixteenth Century* (London: Routledge, 2021), 11–41.
15. Recent biographies of Paracelsus include B. Moran, *Paracelsus: An Alchemical Life* (London: Reaktion Books, 2019) and C. Webster, *Paracelsus: Medicine, Magic and Mission at the End of Time* (New Haven: Yale University Press, 2008).
16. See D. Kahn and H. Hirai (eds), *Pseudo-Paracelsus: Forgery and Early Modern Alchemy, Medicine and Natural Philosophy* (Leiden: Brill, 2021).
17. On the Italian scene, see D. Gentilcore, *Medical Charlatanism in Early Modern Italy* (Oxford: Oxford University Press, 2006); on the English, M. Pelling and F. White, *Medical Conflicts in Early Modern London: Patronage, Physicians, and Irregular Practitioners, 1550–1640* (Oxford: Oxford University Press, 2003).
18. W. Eamon, *Science and the Secrets of Nature: Books of Secrets in Medieval and Early Modern Culture* (Princeton: Princeton University Press, 1994).
19. See H. Khunrath, *Trewhertzige Warnungs-Vermahnung* (Magdeburg, 1597) and M. Maier, *Examen fucorum pseudochymicorum detectorum et in gratiam veritatis amantium succincte refutatorum* (Frankfurt, 1617).
20. J. Shackelford, *A Philosophical Path for Paracelsian Medicine: The Ideas, Intellectual Context, and Influence of Petrus Severinus (1540/2–1602)* (Copenhagen: Museum Tusculanum Press, 2004).
21. On the origins and rise of the term of 'adepts' for accomplished alchemical philosophers, see G. D. Hedesan, 'The Transformation of the Notion of "Adept": From Medieval Arabic Philosophy to Early Modern Alchemy', in G. D. Hedesan and T. Rudbøg (eds), *Innovation in Western Esotericism from the Renaissance to the Present* (London: Palgrave Macmillan, 2021), 63–96.
22. On the subject of corpuscularianism in early modern alchemy, see W. R. Newman, *Atoms and Alchemy: Chymistry and the Experimental Origins of the Scientific Revolution* (Chicago: University of Chicago Press, 2006).
23. C. Gunnoe, *Thomas Erastus and the Palatinate: A Renaissance Physician in the Second Reformation* (Leiden: Brill, 2011).
24. D. Kahn, 'La Création *ex nihilo* et la notion d'*increatum* chez Paracelse', in E. Mehl and I. Pantin (eds), *De mundi recentioribus phaenomenis: Cosmologie et science dans l'Europe des temps modernes, XVe-XVIIe siècles* (Turnhout: Brepols, 2022), 207–28.
25. D. Kahn, *Alchimie et Paracelsisme en France (1567–1623)* (Paris: Droz, 2007), 569–83.
26. A. Libavius, *Syntagmatis arcanorum et commentationem chymicorum partis tertiae* (Frankfurt: Petru Kopf, 1613–15), quoted in B. Moran, *Distilling Knowledge: Alchemy, Chemistry and the Scientific Revolution* (Cambridge, MA: Harvard University Press, 2005), 81.
27. J. V. Andreae, *Fama Fraternitatis, Confessio Fraternitatis, Chymische Hochzeit Christiani Rosenkreutz, Anno 1459*, ed. R. van Dülmen (Stuttgart: Calwer, 1973), 20, 26, 29; *Fama Fraternitatis* condemned gold-making, upheld metallic transmutation and praised Paracelsus's work.
28. Moran, *Distilling Knowledge*, 80–1.
29. On Van Helmont's life and writings, see G. D. Hedesan, *An Alchemical Quest for Universal Knowledge: The 'Christian Philosophy' of Jan Baptist Van Helmont (1579–1644)* (London: Routledge, 2016).
30. W. R. Newman, 'The Corpuscular Theory of J. B. Van Helmont and Its Medieval Sources', *Vivarium*, 21:1 (1993), 161–91.

31. On Boyle's alchemical quest and his contact with the Asterism, see L. M. Principe, *The Aspiring Adept: Robert Boyle and His Alchemical Quest* (Princeton: Princeton University Press, 1998).
32. C. McIntosh, *The Rose Cross and the Age of Reason: Eighteenth-Century Rosicrucianism in Central Europe and Its Relationship to the Enlightenment* (Leiden: Brill, 1992).
33. Principe, *The Aspiring Adept*, 197–201.
34. On this topic see M. Zuber, *Spiritual Alchemy: From Jacob Boehme to Mary Anne Atwood* (Oxford: Oxford University Press, 2021). It is notable that Zuber defines the notion of 'spiritual alchemy' in a very specific way, as 'describing the physical transfiguration of the human body through rebirth in this life, culminating in resurrection at the Last Judgement', 11.
35. On Joachim Poleman, see J. T. Young, *Faith, Medical Alchemy and Natural Philosophy: Johann Moriaen, Reformed Intelligencer, and the Hartlib Circle* (Aldershot: Ashgate, 1998). On Theodore Graw, see W. Poole, 'Theodoricus Gravius (fl. 1600–1661): Some Biographical Notes on a German Chymist and Scribe Working in Seventeenth-Century England', *Ambix*, 56:3 (2009), 239–52.
36. Zuber, *Spiritual Alchemy*, 147–52.
37. On Newton's alchemy, see W. R. Newman, *Newton the Alchemist* (Princeton: Princeton University Press, 2018).
38. L. M. Principe, *The Transmutations of Chymistry: Homberg and the Académie Royale des Sciences* (Chicago: University of Chicago Press, 2020), 368–9.
39. Hedesan, *An Alchemical Quest*, 183–9.
40. Principe, *The Transmutations of Chymistry*, 377–87.
41. C. Wentrup, 'The Alchemist, Metal-Divider and Transmuter Carl F. Wenzel and His 1776 Award from the Royal Danish Academy of Sciences through Professor *C. G.* Krantzenstein', *ChemPlusChem*, 88:5 (2023), 1–10.

Alchemy and Chemistry
Hjalmar Fors

1. W. Newman and L. Principe, 'Alchemy vs Chemistry: The Etymological Origins of a Historiographic Mistake', *Early Science and Medicine*, 3:1, (1998), 32–65: 38–41. W. Newman, 'What Have We Learned from The Recent Historiography of Alchemy', *Isis*, 102 (2011), 313–21. On usage in Swedish, H. Fors, 'Kemi & alkemi', in J. Hansson and K. Savin (eds), *Svenska begreppshistorier: Från antropocen till åsiktskorridor* (Stockholm: Fri Tanke, 2022), 247–56.
2. The long-term continuity of the laboratory space is argued by F.L. Holmes in *Eighteenth-Century Chemistry as an Investigative Enterprise*. (Berkeley: Office for the History of Science and Technology, University of California, 1989), 17, 19–20, 123–6. It is still, almost 30 years later, a tour de force account of the topic and of eighteenth-century chemistry in general. On laboratories see also: H. Fors, 'J G Wallerius and the Laboratory of Enlightenment', in H. Fors, E. Baraldi and A. Houltz (eds), *Taking Place: The Spatial Contexts of Science, Technology and Business* (Sagamore Beach: Science History Publications, 2006), 3–33; M. Crosland, 'Early Laboratories *c.* 1600–*c.* 1800 and the Location of Experimental Science', *Ambix*, 62:2 (2005), 233–53; M. Beretta, 'Laboratories and Technology', in M. Eddy and U. Klein (eds), *A Cultural History of Chemistry in the Eighteenth Century* (London: Bloomsbury Academic, 2022), 71–91.
3. Two important works in what can be called a broader cultural history approach to early modern alchemy are P. Smith, *The Business of Alchemy: Science and Culture in the Holy Roman Empire* (Princeton: Princeton University Press, 1994) and T. Nummedal, *Alchemy and Authority in the Holy Roman Empire* (Chicago: University of Chicago Press, 2007). See also H. Fors, *The Limits of Matter: Chemistry, Mining and Enlightenment* (Chicago: University of Chicago Press, 2015). A recent overview is provided by

M. Piorko, M. Hendriksen and S. Werrett, 'Alchemical Practice: Looking towards the Chemical Humanities', *Ambix*, 69 (2022), 1–18.

4. For a recent exposé, B. Bensaude-Vincent, 'Culture and science: Science: Chemistry in its Golden Age', in Eddy and Klein (eds), *A Cultural History of Chemistry in the Eighteenth Century*, 93–112. For this author's position, Fors, *The Limits of Matter*.
5. Nummedal, *Alchemy and Authority*. See also Salam Rassi's contribution to this volume, 93-111.
6. L. M. Principe, 'The End of Alchemy? The Repudiation and Persistence of Chrysopoeia at the Académie Royale des Sciences in the Eighteenth Century', *Osiris*, 29 (2014), 96–116; on Fontenelle, 102–5, 111. Idem, *The Transmutations of Chymistry: Wilhelm Homberg and the Académie Royale des Sciences* (Chicago: University of Chicago Press, 2020). For how this process played out in the Swedish context, H. Fors, 'Speaking about Other Ones: Swedish Chemists on Alchemy, *c.* 1730–70', in J. R. Bertomeu-Sánchez, D. T. Burns and B. Van Tiggelen (eds), *Neighbours and Territories: The Evolving Identity of Chemistry Proceedings of the 6th International Conference on the History of Chemistry* (Leuven, 2008), 283–9.
7. For a comprehensive critique of this 'standard view' and examples of its proponents, see Holmes, *Eighteenth-Century Chemistry*, and M. D.Eddy, S. H. Mauskopf and W. R. Newman, 'An Introduction to Chemical Knowledge in the Early Modern World', *Osiris*, 29 (2014), 1–15.
8. 'Om lapis Philosophorum kunde winnas igenom Arbete, så hade then aldrig flugit min sahl fader fel i thes yngre år; ty hans skarpsinniga efter-Tanka gick wida, och hans insikt uti naturen War stor.' Letter from E. F. Hjärne to C. C. Gjörwell, 14 December 1757. In Bref til Carl Christoffer Gjörwell Åren 1750–1757. Manuscript collections of Kungliga Biblioteket, the Royal Library of Sweden. Shelf mark Ep. G. 10:1.
9. 'Om Alchemistiska försök misslyckats för många mindre kunnige om metallernes egenskaper, så har likwist Chemien mest at tacka alchemister för de betydeligaste uptäckter. Torde henda fördenskull, de, med mindre rättighet, kallas, utan exception, griller. Best woro at skilja de förnuftiga Alchemisters Speculationer ifrån ignoranters gissningar, eller, at skilja Alchemister ifrån bedragare.' J. G. Wallerius to T. Bergman, undated but likely written in 1768. In Svensk brevväxling till Torbern Bergman. Manuscript collections of Uppsala University Library. Shelf mark G21, on 675.
10. For overviews, see Newman and Principe, 'Alchemy vs Chemistry'. L. M. Principe, *The Secrets of Alchemy*. (Chicago: University of Chicago Press, 2012). Idem. *The Transmutations*. H. Fors, L.M. Principe and H. O. Sibum, 'From Library to the Laboratory and Back Again: Experiment as a Tool for Historians of Science', *Ambix*, 63 (2016), 85–97. Criticism of this perspective has been levelled by scholars active in the field of Western esotericism, who claim that it neglects aspects of alchemy which were less connected to the scientific enterprise. See M. A. Zuber, *Spiritual Alchemy: From Jacob Boehme to Mary Anne Atwood* (Oxford: Oxford University Press, 2021), For a more comprehensive discussion, see Georgiana Hedesan, 'Alchemy in the Academy' in this volume, 6.
11. B. J. Teeter Dobbs, *The Janus Faces of Genius: The Role of Alchemy in Newton's Thought* (Cambridge: Cambridge University Press, 1991). W. R. Newman and L. M. Principe, *Alchemy Tried in the Fire: Starkey, Boyle, and the Fate of Helmontian Chymistry* (Chicago: University of Chicago Press, 2002). J. C. Powers, *Inventing Chemistry: Herman Boerhaave and the Reform of the Chemical Arts* (Chicago: University of Chicago Press, 2012). J. M. Rampling, *The Experimental Fire: Inventing English Alchemy, 1300–1700* (Chicago: University of Chicago Press, 2022).
12. See, e.g., S. Lindroth, *Svensk lärdomshistoria 2 Stormaktstiden* (Stockholm: Norstedt, 1975).
13. Dobbs's studies of Newton may be considered seminal texts which initiated the re-evaluation of alchemy's relationship to science. See especially Dobbs, *The Janus Faces of Genius*. In this context must also be mentioned Hélène Metzger's influential *Les doctrines chimiques en France du début du XVIIe à fin du XVIIIe siècle* (Paris, 1923). See Holmes, *Eighteenth-Century Chemistry*, 8–9. On the differences between Swedish idéhistoria and the Anglo-American traditions of the history of ideas and the history of science, see K. Raj, 'History of Science, Intellectual History, and the World, 1900–2020', in S. Geroulanos and G. Sapiro (eds), *The Routledge Handbook of the History and Sociology of Ideas* (New York: Routledge, 2024), 185–204, at 188.
14. Many of the following arguments were previously proposed in H. Fors, 'Elements in the Melting Pot: Merging Chemistry, Assaying and Natural History, *c.* 1730–1760', *Osiris*, 29 (2014), 230–44. Idem, *The Limits of Matter*. Idem, 'Alchemy, Chemistry and Metallurgy', in D. Jalobeanu and C .T. Wolfe (eds),

Encyclopedia of Early Modern Philosophy and the Sciences (New York: Springer, 2020), 50–5. I refer to these publications for more thorough references.

15. A more comprehensive introduction to the Jābirian framework is found in Principe, *The Secrets of Alchemy*, 33–7, 44, 46, 57–8, 109, 128–9.
16. W. R. Newman, 'Robert Boyle, before Lavoisier', *Osiris*, 29 (2014), 63–77; on Becher, 72–3. See also Newman and Principe, *Alchemy Tried in the Fire*.
17. For reasons of brevity, I omit a discussion of atomism in chemistry. For a discussion of atomism and chemistry in the seventeenth century, see Newman, 'Robert Boyle, Transmutation'; for the first half of the eighteenth century, Fors, *The Limits of Matter*.
18. On Boerhaave's influence, Powers, *Inventing Chemistry*. M. M. A. Hendriksen, 'Boerhaave's Mineral Chemistry and Its Influence on Eighteenth-Century Pharmacy in the Netherlands and England', *Ambix*, 65:4 (2018), 303–23, at 304–5, 322.
19. K. Chang, 'Communications of Chemical Knowledge: Georg Ernst Stahl and the Chemists at the French Academy of Sciences in the First Half of the Eighteenth Century ', *Osiris*, 29 (2014), 135–57, at 138–9, 146, 149–50, 154–7.
20. Holmes, *Eighteenth-Century Chemistry*, 49.
21. L. Roberts, 'Setting the Table: The Disciplinary Development of Eighteenth-Century Chemistry as Read through the Changing Structure of Its Tables', in P. Dear (ed.), *The Literary Structure of Scientific Argument: Historical Studies* (Philadelphia: University of Pennsylvania Press, 1991), 93–192. Quotation on 119.
22. A. M. Alfonso-Goldfarb and Marcia H. M. Ferraz, 'Gur, Ghur, Guhr or Bur? The Quest for a Metalliferous Prime Matter in Early Modern Times', *British Journal for the History of Science*, 46:1 (2013), 23–37. J. A. Norris, 'Auss Queckselber und Schwefel Rein: Johann Mathesius (1504–65) and Sulfur-Mercurius in the Silver Mines of Joachimstal', *Osiris*, 29 (2014), 35–48.
23. There are, however, only a few cases in which we can say with certainty that alchemical theories of metallic generation were grounded in, and fruitfully interplayed with, practical experience of large-scale smelting processes. H. Fors, '"Away! Away to Falun!": J. G. Gahn and the Application of Enlightenment Chemistry to Smelting', *Technology and Culture*, 50:3 (2009), 549–68. Idem, *The Limits of Matter*.
24. Newman, 'Robert Boyle, Transmutation'.
25. Thus we see that the three major eighteenth-century mineralogies, by Johan Gottschalk Wallerius (1747), Axel Fredrik Cronstedt (1758) and Torbern Bergman (1782), were deeply influenced by Linnean thinking. This is unsurprising, as Bergman and Wallerius were colleagues of Linnaeus at Uppsala University and Cronstedt an Uppsala alumnus and former student of Wallerius.
26. W. R. Newman, 'The Significance of "Chymical Atomism"', in E. D. Sylla and W. R. Newman (eds), *Evidence and Interpretation in Studies on Early Science and Medicine* (Boston: Brill, 2009), 248–64.
27. [A.F. Cronstedt], *Försök til Mineralogie, eller mineral-rikets upställning*. (Stockholm, 1758). Quotation from the English translation, *An Essay towards a System of Mineralogy: Translated from the Original Swedish, with Notes by Gustav von Engestrom....* (London, 1770), 241. The quotation has been slightly edited for spelling
28. Cronstedt's criticism of the search for constituent parts of metals clearly indicates that he considered metals pure, simple substances. Whether this also meant that he objected to Stahl's theory of phlogiston is difficult to say. See Fors, *The Limits of Matter*.
29. Holmes, *Eighteenth-Century Chemistry*, 107–12, 118. See also Eddy, Mauskopf and Newman, 'An Introduction to Chemical Knowledge in the Early Modern World'.
30. M. Beretta, 'Lavoisier and the History of Chemistry', *Ambix*, 71:2 (2024), 209–4. Quotation on 219. Note that Beretta and Holmes are prominent scholars of Lavoisier who are generally appreciative of his achievements. Holmes regards Lavoisier as perhaps the most probing, systematic and critical theoretician chemistry had seen up to that time and affirms that 'Lavoisier recognized that his own leadership did not extend to all parts of chemistry' (*Eighteenth-Century Chemistry*, 110). For a comprehensive argument in support of the view that Lavoisier was the father of modern chemistry and founder of an entirely new discipline where none existed before, see A.Donovan, 'Lavoisier and the Origins of Modern Chemistry', *Osiris*, 4 (1988), 214–31. For a critique of Donovan, see Holmes,

Eighteenth-Century Chemistry, 103–7. More recently, Ursula Klein and Matthew Eddy argued the centrality of Lavoisier's contribution by claiming that concepts of 'chemical combination, compound, composition and affinity were initially elaborated in a context that was independent of theories about chemical elements and atoms' and that this 'changed only in the late eighteenth century, when Antoine Laurent Lavoisier introduced what became more or less the modern concept of a chemical element'. Here is not the place to unpack this statement; let it just suffice to say that I politely disagree. M. Eddy and U. Klein, 'Introduction: The Core Concepts and Cultural Context of Eighteenth-Century Chemistry', in Eddy and Klein, *A Cultural History of Chemistry in the Eighteenth Century*, 1–21, at 11, 13–14. Quotation on 11.

31. Lavoisier, *Œuvres*, vol. 2, 104. Quoted in Beretta, 'Lavoisier and the History of Chemistry', 223.
32. For the nineteenth-century debate on the elemental status of metals, S. N. Hijmans, 'The Tantalum Metals (1801–1866): Nineteenth-Century Analytical Chemistry and the Identification of Chemical Elements', *Ambix*, 69 (2022), 399–419.
33. Eddy et al, 'An Introduction', 8.

Alchemy in the Academy
Georgiana D Hedesan

1. J. F. Gmelin, *Geschichte der Chemie*, 3 vols (Göttingen: n.p., 1797–99).
2. T. Thomson, *History of Chemistry*, 2 vols (London: Colburn and Bentley, 1830); H. Kopp, *Geschichte der Chemie*, 4 vols (Braunschweig: Friedrich Wieweg und Sohn, 1843–47).
3. H. Kopp, *Die Alchemie in älterer und neuerer Zeit: ein Beitrag zur Culturgeschichte*, 2 vols (Heidelberg: Carl Winter, 1886).
4. M. Berthelot, *Collection des anciens alchimistes grecs*, 4 vols (Paris: Georges Steinheil, 1887–88).
5. E. O. von Lippmann, *Entstehung und Ausbreitung der Alchemie* (Dordrecht: Springer, 1919, 1931, 1954).
6. J. Ferguson, *Bibliotheca Chemica*, 2 vols (Glasgow: MacLehose and Sons, 1906).
7. See for instance F. Soddy, 'Transmutation: The Vital Problem of the Future', *Scientia*, 6:11 (1912), 186–202, at 195.
8. M. S. Morrisson, *Modern Alchemy: Occultism and the Emergence of Atomic Theory* (Oxford: Oxford University Press, 2007), 56.
9. Ibid., 186–7.
10. W. H. Brock, 'Exploring Early Modern Chymistry: The First Twenty-Five Years of the Society for the Study of Alchemy & Early Modern Chemistry 1935–1960', *Ambix*, 58:3 (2011), 191–214, http://doi.org/10.1179/174582311X13129418298947.
11. G. Sarton, 'Boyle and Bayle: The Sceptical Chemist and the Sceptical Historian', *Chymia*, 3 (1950), 155–89, at 161–2.
12. H. Butterfield, *The Origins of Modern Science, 1300–1800* (New York: Macmillan, 1952), 98.
13. L. M. Principe and W. R. Newman, 'Some Problems with the Historiography of Alchemy', in W. R. Newman and A. Grafton (eds), *Secrets of Nature: Astrology and Alchemy in Early Modern Europe* (Cambridge, MA: MIT Press, 2001), 385–431, at 388.
14. Ibid., 389.
15. M. A. Atwood, *A Suggestive Inquiry into the Hermetic Mystery* (Belfast: Tait, 1918), 162.
16. C. G. Jung, 'The Idea of Redemption in Alchemy', in S. Dell (ed.), *The Integration of the Personality*, (New York: Farrar & Reinhart, 1939), 205–80. See for instance his statements that 'every insight into the nature of chemistry and its limitations was still barred to them' and that the reason for alchemy's existence was only as a precursor to chemistry', 210. The ideas of Jung on alchemy were further elaborated in *Psychology and Alchemy* (*Collected Works of C. G. Jung*, Princeton: Princeton University Press, 1968).

17. L. M. Principe, *The Aspiring Adept: Robert Boyle and His Alchemical Quest* (Princeton: Princeton University Press, 1998), 19.
18. Koyré's ideas are chiefly articulated in his books *Galileo Studies* (*Études galiléennes*, 1939, trans. 1978) and *From the Closed World to the Infinite Universe* (Baltimore: Johns Hopkins University Press, 1957).
19. A. R. Hall, *The Scientific Revolution, 1500–1800: The Formation of a Modern Scientific Attitude* (Boston: Beacon Press, 1962), 310.
20. Boas Hall wrote a number of papers and books on Boyle, most prominently *Robert Boyle and Seventeenth-Century Chemistry* (Cambridge: Cambridge University Press, 1958).
21. M. Foucault, *Les mots et les choses: Une archéologie des sciences humaines* (Paris: Gallimard, 1966), translated to English as *The Order of Things: An Archeology of the Human Sciences* (London: Pantheon Books, 1970).
22. B. Vickers, 'Introduction', in B. Vickers (ed.), *Occult and Scientific Mentalities in the Renaissance* (Cambridge: Cambridge University Press, 1984), 1–55, at 44.
23. F. Yates, *Giordano Bruno and the Hermetic Tradition* (London: Routledge and Kegan Paul, 1964), 155–6, 447–55.
24. See M. Hesse, 'Hermeticism and Historiography: An Apology for an Internal History of Science', in R. H. Stuewer (ed.), *Historical and Philosophical Perspectives of Science* (Minneapolis: University of Minnesota Press, 1970), 134–60; and E. Rosen, 'Was Copernicus a Hermetist?', in ibid., 163–71. More balanced criticism was offered by R. S. Westman, 'Magical Reform and Astronomical Reform: The Yates Thesis Reconsidered', in R. S. Westman and J. E. McGuire, *Hermeticism and the Scientific Revolution: Papers Read at a Clark Library Seminar, 9 March 1974* (Los Angeles: Clark Memorial Library, 1977), 1–91.
25. B. Vickers, 'Frances Yates and the Writing of History', *Journal of Modern History*, 51:2 (1979), 287–316 at 315.
26. Vickers, 'Introduction', 6.
27. Interestingly, Wouter J. Hanegraaff, in *Esotericism and the Academy: Rejected Knowledge in Western Culture* (Cambridge: Cambridge University Press, 2012), attributed the popularity of the notion of Hermeticism amongst Renaissance scholars to alchemy, in spite of Yates's cold shoulder: 'With hindsight, if "hermeticism" nevertheless assumed a predominant profile in the imagination of later scholars of the Renaissance, this seems to have been caused not just by the reception of the *Corpus Hermeticum* and the *Asclepius* alone, but probably even more by the influence of a quite different tradition strongly linked to the name of Hermes: that of alchemy.'
28. W. R. Newman, *Newton the Alchemist* (Princeton: Princeton University Press, 2018), 90–1.
29. R. Westfall, 'Newton and the Hermetic Tradition', in A. G. Debus (ed.), *Science, Medicine, and Society in the Renaissance*, vol. 2 (New York: Science History Publications, 1972), 183–98.
30. B. J. T. Dobbs, *The Foundations of Newton's Alchemy, or The Hunting of the Greene Lyon* (Cambridge: Cambridge University Press, 1975), 26.
31. B. J. T. Dobbs, *The Janus Faces of Genius: The Role of Alchemy in Newton's Thought* (Cambridge: Cambridge University Press, 1991), 6.
32. Ibid., 18.
33. Ibid., 77n.
34. H. M. Collins and R. G. Harrison, 'Building a TEA Laser: The Caprices of Communication', *Social Studies of Science*, 5 (1975), 441–50; B. Latour and S. Woolgar, *Laboratory Life: The Construction of Scientific Facts* (Princeton: Princeton University Press, 1979); B. Latour, *Science in Action: How to Follow Scientists and Engineers through Society* (Cambridge, MA: Harvard University Press, 1987).
35. S. Shapin and S. Schaffer, *Leviathan and the Air-Pump: Hobbes, Boyle, and the Experimental Life* (Princeton: Princeton University Press, 1985).
36. C. Webster, *The Great Instauration* (London: Duckworth, 1975).
37. A. G. Debus, *The Chemical Philosophy: Paracelsian Science and Medicine in the Sixteenth and Seventeenth Centuries*, 2 vols (New York: Science History Publications, 1977).
38. W. R. Newman and L. M. Principe, 'Alchemy vs Chemistry: The Etymological Origins of a Historiographic Mistake', *Early Science and Medicine*, 3:1 (1998), 32–65.
39. W. R. Newman, *Gehennical Fire: The Lives of George Starkey, an American Alchemist in the Scientific Revolution* (Cambridge, MA: Harvard University Press, 1994).
40. Principe and Newman, 'Some Problems with the Historiography of Alchemy'. Newman and Principe also worked together on two books, *Alchemy Tried in the Fire: Starkey, Boyle, and the Fate of Helmontian*

Chymistry (Chicago: University of Chicago Press, 2002) and *George Starkey: Alchemical Laboratory Notebooks and Correspondence* (Chicago: University of Chicago Press, 2004).

41. W. R. Newman, *The 'Summa Perfectionis' of Pseudo-Geber* (Leiden: Brill, 1991); idem, *Atoms and Alchemy: Chymistry and the Experimental Origins of the Scientific Revolution* (Chicago: University of Chicago Press, 2006); idem, 'The Significance of Chymical Atomism', *Early Science and Medicine*, 14 (2009), 248–64.
42. Newman and Principe, *Alchemy Tried in the Fire*, 81, 88, n. 144.
43. L. M. Principe, 'Chemical Translation and the Role of Impurities in Alchemy: Examples from Basil Valentine's *Triumph-Wagen*', *Ambix*, 34:1 (1987), 21–30; idem, *The Secrets of Alchemy* (Chicago: University of Chicago Press, 2012), 143–66.
44. H. Fors, L. M. Principe and H. O. Sibum, 'From the Library to the Laboratory and Back Again: Experiment as a Tool for Historians of Science', *Ambix*, 63:2 (2016), 85–97, http://doi.org/10.1080/00026980.2016.1213009; H. O. Sibum, 'Experimental History of Science', in S. Lindqvist (ed.), *Museum of Modern Science* (Sagamore Beach: Science History Publications, 2000), 77–86.
45. J. MacLachlan, 'Galileo's Experiments with Pendulums: Real and Imaginary', *Annals of Science*, 33:2 (1976), 173–85, and 'Experiments in the History of Science', *Isis*, 89:1 (1998), 90–2; T. B. Settle, 'An Experiment in the History of Science', *Science*, 133:3445 (6 January, 1961), 19–23, and 'La rete degli esperimenti Galileiani', in P. Bozzi, C. Maccagni, C. Olivieri and T. B. Settle (eds), *Galileo e la scienza sperimentale* (Padua: Dipartimento di fisica Galileo Galilei, 1995), 11–62; R. Stuewer, 'A Critical Analysis of Newton's Work on Diffraction', *Isis*, 61:2 (1970), 188–205; for pharmacology, see the experiments carried out under the guidance of Professor Wolfgang Schneider in Braunschweig, published in *Veröffentlichtungen aus dem pharmaziegeschichtlichen Seminar der Technischen Hochschule Braunschweig*.
46. H. O. Sibum, 'Reworking the Mechanical Value of Heat: Instruments of Precision and Gestures of Accuracy in Early Victorian England', *Studies in the History and Philosophy of Science*, 26:1 (1995), 73–106; idem, 'Working Experiments: A History of Gestural Knowledge', *Cambridge Review*, 116:2325 (1995), 25–37.
47. S. Dupré, A. Harris, J. Kursell, P. Lulof and M. Stols-Witlox, 'Introduction', in S. Dupré et al. (eds), *Reconstruction, Replication and Re-enactment in the Humanities and Social Sciences* (Amsterdam: Amsterdam University Press, 2020), 13.
48. Fors et al., 'From the Library', 89.
49. See also Principe, *The Secrets of Alchemy*, 143–66.
50. For instance, see Making and Knowing Project et al. (eds), *Secrets of Craft and Nature in Renaissance France: A Digital Critical Edition and English Translation of BnF Ms. Fr. 640* (New York: Making and Knowing Project, 2020), https://edition640.makingandknowing.org; T. Hagendijk, M. Vilarigues and S. Dupré, 'Materials, Furnaces and Texts: How to Write about Making Glass Colours in the Seventeenth Century', *Ambix*, 67:4 (2020), 323–45; L. Maini, M. Marchini, M. Gandolfi, L. Raggetti and M. Martelli, 'Quicksilver and Quick-Thinking: Insight into the Alchemy of Mercury. A New Interdisciplinary Research to Discover the Chemical Reality of Ancient Alchemical Recipes', *ChemRxiv* (Cambridge: Cambridge Open Engage, 2021), https://chemrxiv.org/engage/chemrxiv/article-details/610133d91f990cf878a53b5a.
51. On this topic, see M. Zuber, *Spiritual Alchemy: From Jacob Boehme to Mary Anne Atwood* (Oxford: Oxford University Press, 2021). Zuber, like Wouter J. Hanegraaff in his *Esotericism and the Academy*, has taken issue with what he considers to be Newman and Principe's 'essentialist' views of alchemy.
52. Various initiatives have been contributing to the task of transcription of alchemical texts, such as Urs Leo Gantenbein and Didier Kahn's database of texts attributed to Paracelsus, *THEO* (https://www.paracelsus-project.org) and several texts edited under the NOSCEMUS ERC project (https://www.uibk.ac.at/projects/noscemus). Moreover, an ongoing project I am part of, TOME (*The Origins of Modern Encyclopaedism*, https://tome.flu.cas.cz, 2023–25, financed by the Czech Republic's ERC-CZ scheme), is set to transcribe the majority of alchemical texts published in the Latin language between 1500 and 1700.
53. See Fabrizio Pregadio's contribution in this volume.

Theosophical Science and Modern Alchemy
Mark S. Morrisson

1. P. Washington, *Madame Blavatsky's Baboon: Theosophy and the Emergence of the Western Guru* (London: Secker & Warburg, 1993), 120.
2. A. Faivre, *Access to Western Esotericism* (Albany: SUNY Press, 1994), 10–14.
3. The *Emerald Tablet* was attributed to an ancient legendary figure, Hermes Trismegistus. But, working its way from Arabic texts to Latin translations in the Middle Ages, it was a key text for medieval and early modern alchemists, and its presentation of correspondences between the material and spiritual animated occult revival Hermeticists and Theosophists, whose understanding of it owed much to the works of the seventeenth-century alchemist Heinrich Khunrath and eighteenth-century scientist and mystic Emanuel Swedenborg.
4. Though Rutherford has been credited as achieving this transmutation, in 1919 he was only able to understand the result as a nuclear disintegration ejecting a proton (what he understood as a hydrogen nucleus). His young colleague Patrick Blackett used a cloud chamber over four years of experiments, beginning in 1921, to determine that the nitrogen atom had lost a proton but also bound the alpha particle itself, becoming an isotope (Oxygen-17). See S. Krivit, 'Rutherford's Reluctant Role in Nuclear Transmutation', *New Energy Times*, 18 May, 2019, https://news.newenergytimes.net/2019/05/18/rutherfords-reluctant-role-in-nuclear-transmutation.
5. S. Weart, *Nuclear Fear: A History of Images* (Cambridge, MA: Harvard University Press, 1988), 5–6.
6. See, for example, F. Soddy, *The Interpretation of Radium: Being the Substance of Six Free Popular Experimental Lectures Delivered at the University of Glasgow* (London: John Murray, 1908); and F. Soddy, 'Transmutation: The Vital Problem of the Future', *Scientia*, 6:11 (1912), 186–202.
7. For an in-depth account of the alchemical tropes used during the period and their interweaving of occult revival uses and scientific discourse, see M. Morrisson, *Modern Alchemy: Occultism and the Emergence of Atomic Theory* (New York: Oxford University Press, 2007).
8. Washington, *Madame Blavatsky's Baboon*, 34–6.
9. M. Carlson, *'No Religion Higher than Truth': A History of the Theosophical Movement in Russia, 1875–1922* (Princeton: Princeton University Press, 1993), 12–13.
10. Washington, *Madame Blavatsky's Baboon*, 68. See also C. J. Ryan, *H. P. Blavatsky and the Theosophical Movement: A Brief Historical Sketch* (Pasadena: Theosophical University Press, 1975); J. D. Lavoie, *The Theosophical Society: The History of a Spiritualist Movement* (Irvine: BrownWalker Press, 2012); and T. Rudbøg and E. R. Sand (eds), *Imagining the East: The Early Theosophical Society* (New York: Oxford University Press, 2019).
11. H. P. Blavatsky, *Isis Unveiled: A Master-Key to the Mysteries of Ancient and Modern Science and Theology*, vol. 1 (1877; Pasadena: Theosophical University Press, 1972), v.
12. H. P. Blavatsky, *The Secret Doctrine: The Synthesis of Science, Religion, and Philosophy*, vol. 1 (London: Theosophical Publishing Company, 1888), 296.
13. 'Sapere Aude' [W. Westcott], *The Science of Alchymy: Spiritual and Material* (London: Theosophical Publishing House, 1893), 9.
14. Blavatsky, *The Secret Doctrine*, vol. 1, 58.
15. Blavatsky, *Isis Unveiled*, 503–4.
16. H. P. Blavatsky, 'Alchemy in the Nineteenth Century', *Collected Writings*, vol. XI, October 1889, Theosophy World eBook, 529–30. https://www.theosophy.world/sites/default/files/ebooks/Collected_Writings_Volume_XI_(1889).pdf.
17. A. Wilder, *New Platonism and Alchemy: A Sketch of the Doctrines and Principal Teachers of the Eclectic or Alexandrian School; also An Outline of the Interior Doctrines of the Alchemists of the Middle Ages* (Albany: Weed, Parsons and Company, 1869), n.p.
18. R. Wallis, 'Science and Pseudo-Science', *Social Science Information*, 24:3 (1985), 585–601.
19. E. Asprem, 'Theosophical Attitudes towards Science: Past and Present', in O. Hammer and M. Rothstein (eds), *Handbook of the Theosophical Current* (Leiden: Brill, 2013), 405.
20. S. R., 'The Progress of Radium', *Theosophical Review*, 37:222 (February 1906), 559–60.

21. G. Dawson, R. Noakes and J. Topham, 'Introduction', in G. Cantor et al. (eds), *Science in the Nineteenth-Century Periodicals: Reading the Magazine of Nature* (Cambridge: Cambridge University Press, 2004), 1–2.
22. Quoted in A. Besant, 'Radio-Activity Again?', *Theosophical Review*, 36:216 (August 1905), 481.
23. L. Campos, *Radium and the Secret of Life* (Chicago: University of Chicago Press, 2015), 4.
24. F. Hara, 'The Advance of Science towards Occult Teachings', *Theosophical Review*, 37:222 (February 1906), 552–3.
25. H. Coryn, 'Intra-Atomic Energy', *Theosophical Path*, 1:6 (December 1911), 417–18.
26. T. Gieryn, *Cultural Boundaries of Science: Credibility on the Line* (Chicago: University of Chicago Press, 1999); and D. Hess, *Science in the New Age: The Paranormal, Its Defenders and Debunkers, and American Culture* (Madison: University of Wisconsin Press, 1993), 145–6.
27. For example, G. E. S., 'Electrons and Clairvoyance', *Theosophical Review*, 35:210 (February 1905), 518–19.
28. A. Besant, 'Occult Chemistry', *Lucifer*, 3 (November 1895), 211.
29. Ibid., 219.
30. P. Galison, *Image and Logic: A Material Culture of Microphysics* (Chicago: University of Chicago Press, 1997).
31. M. Collins, *Light on the Path: A Treatise Written for the Personal Use of Those Who Are Ignorant of the Eastern Wisdom, and Who Desire to Enter Within Its Influence*, 3rd edn (Boston: Cupples, Upham, and Company, 1886), 6.
32. T. Mundy, *OM: The Secret of Ahbor Valley* (Indianapolis: Bobbs-Merrill, 1924), 388 and 301.
33. J. Topham, 'Introduction (Focus: Historicizing "Popular Science")', *Isis*, 100:2 (June 2009), 311. See also J. Secord, 'Knowledge in Transit', *Isis*, 95:4 (December 2004), 654–72.
34. M. Strathern, *Reproducing the Future: Essays on Anthropology, Kinship, and the New Reproductive Technologies* (New York: Routledge, 1992), 6; S. Squier, *Babies in Bottles: Twentieth-Century Visions of Reproductive Technology* (New Brunswick: Rutgers University Press, 1994).
35. S. Bhattacharya, 'The Victorian Occult Atom: Annie Besant and Clairvoyant Atomic Research', in L. Karpenko and S. Claggett (eds), *Strange Science: Investigating the Limits of Knowledge in the Victorian Age* (Ann Arbor: University of Michigan Press, 2017), 198, 209.
36. P. Ball, 'Clairvoyant Chemistry', *Chemistry World*, 14 March, 2013, https://www.chemistryworld.com/opinion/clairvoyant-chemistry/5984.article.
37. E. L. Smith, *Occult Chemistry Re-evaluated* (Wheaton: Theosophical Publishing House, 1982); and S. Phillips, *Extra-Sensory Perception of Quarks* (Wheaton: Theosophical Publishing House, 1980).

A Partisan of the Soul
Kurt Almqvist

1. G. Eriksson, *Västerlandets idéhistoria 1800–1950* (Stockholm: Gidlunds, 1983/86), the chapter 'Angreppen på kyrkan', 214. In fact, it is Lenin who should be credited with the formulation: 'Religion is the opium of the people. Religion is a kind of spiritual intoxication, in which the slaves of capital drown their human form, their demands for a somewhat humane existence.' Quotation from D. Brolin, 'Opium för folket – ett omstritt uttryck', *Nytid*, 7 September, 2012, https://www.nytid.fi/2012/09/opium-for-folket-ett-omstritt-uttryck; see also H. Höjer, 'Marx var inte först med folkets opium', *Forskning & Framsteg*, 15 December, 2012, https://fof.se/artikel/2013/1/marx-var-inte-forst-med-folkets-opium.
2. S. Freud, *The Future of an Illusion*, trans. and ed. J. Strachey (New York: W. W. Norton & Co., 1961), https://ia802907.us.archive.org/17/items/SigmundFreud/Sigmund%20Freud%20%5B1927%5D%20The%20Future%20of%20an%20Illusion%20%28James%20Strachey%20translation%2C%201961%29.pdf, 30.

3. S. Gieser (ed.), *Dream Symbols of the Individuation Process: Notes of C. G. Jung's Seminars on Wolfgang Paulis' Dreams*, Philemon Foundation Series (Princeton: Princeton University Press, 2021).
4. C. G. Jung, *Memories, Dreams, Reflections*, ed. A. Jaffé, trans. R. and C. Winston, 26970, available at https://archive.org/details/MemoriesDreamsReflectionsCarlJung_201811.
5. See C. G. Jung, *Collected Works*, 13.
6. See M. Eliade, *The Two and the One*, 47ff, for a phenomenological analysis of religious experience in *The Secret of the Golden Flower*.
7. T. Burckhardt, *Alchemy* (Shaftesbury: Element, 1986), 8.
8. In Jung, *The Soul and Death*, *Collected Works*, 8.
9. This text is found in ibid.
10. Jung, *Memories, Dreams, Reflections*, 221.
11. For a study of alchemy as protochemistry, see M. Beretta, *The Enlightenment of Matter* (Uppsala: Science History Publications, 1993).
12. Jung only mentions Freemasons in a few places, and not in a positive light.
13. Jung, *Collected Works*, 18, sections 1554–7. Letter to Pastor Lachat (1954).

A Labour without Pay
Mattias Fyhr

1. D. Alighieri, *The Divine Comedy*, trans. H. Wadsworth Longfellow (Boston: Houghton Mifflin, 1867), 95.
2. G. Chaucer, *The Canterbury Tales* (New York: Thomas Y. Crowell & Company, 1903). S. F. Damon, 'Chaucer and Alchemy', *PMLA*, 39:4 (December 1924), 782–8. Damon's examples of treatises: 'The earliest claim of this sort that I have traced is in Thomas Norton's *Ordinal of Alchemy* (1477); the latest is on page 154 of General Ethan Allen Hitchcock's anonymous *Remarks upon Alchemy and the Alchemists* (1857). Elias Ashmole in the seventeenth century asserted that Chaucer "is ranked amongst the *Hermetick Philosophers*, and his *Master* in this *Science* was *Sir John Gower*, whose familiar and neere acquaintance began at the *Inner Temple* upon *Chaucer's* returne into England…He that Reads the latter part of the *Chanon's Yeoman's Tale*, wil easily perceive him to be a *Iudicious Philosopher*, and one that fully knew the *Mistery*.' (782–3). Damon points to the first of Chaucer's quotations being from another treatise by Villanova than is said in the tale, and to the other as being true to the original but not from the part cited, suggesting that Chaucer quoted from memory and was interested enough in these obscure texts to know their main points, but omitted the parts that might appeal to a literary reader: 'Chaucer's quotations … could not have been better chosen, for they present in a surprisingly compact form the entire alchemical formula', which is, in Damon's words: 'Take sulphur from Sol for the fire; with it roast Luna, from which will issue a water called Mercury. This water is the substance of which the Stone is made.' (785). On Chaucer's enumeration, see for instance: 'That we hadde in our matires sublyming. / And in amalgaming and calcening / Of quik-silver, y-clept Mercuric crude? / For alle our sleightes we can nat conclude. / Our orpiment and sublymed Mercurie, / Our grounden litarge eek on the porphurie' (252). Ibid, 252, 262f.
3. B. Jonson, *The Alchemist. A Comedy* (1770). W. Shakespeare, *The Sonnets of Shakespeare*, intro and notes by H. C. Beeching (Boston and London: Ginn & Company, 1904), 19. On Donne and Paracelsus as well as alchemy, see for instance J. A. Mazzeo, 'Notes on John Donne's Alchemical Imagery', *Isis*, 2:48 (1957), 103–23.
4. I. Newton, 'The Key (Keynes MS 18); The Commentary on the *Emerald Tablet* (Keynes MS 28)', in S. J. Linden (ed.), *The Alchemy Reader: From Hermes Trismegistos to Isaac Newton* (Cambridge: Cambridge University Press, 2014), 243ff. Linden writes that 'The late Professor Dobbs has stated that "most of [Newton's] great powers were poured out upon church history, theology, *the chronology of ancient*

kingdoms, prophecy, and alchemy"(*Foundations* 6) – not upon "scientific" interests. Estimates are that Newton's alchemical studies comprise more than a million words in manuscript.'

5. John Milton, *Paradise Lost* (Oxford: Oxford University Press, 2005), 95f.
6. Ibid, 150f, 70, 49, 57f. For instance in Satan's courage of traversing Night but also in that they do not quarrel: 'O shame to men! Devil with devil damned / Firm concord holds, men only disagree ... live in hatred, enmity, and strife' (59). It showed the fallen angels that 'for the general safety he despised / His own'. Here Milton also stresses the courage even of fallen angels over some men: 'for neither do the spirits damned / Lose all their virtue; lest bad men should boast / Their specious deeds on earth' (58).
7. Ibid., 57.
8. Ibid, 70. M. Sarkar, *Cosmos and Character in Paradise Lost* (New York: Palgrave Macmillan, 2012), 47.
9. Ibid., 51, 57.
10. W. C. Curry, 'Milton's Chaos and Old Night', *Journal of English and Germanic Philology*, 46:1 (January 1947), 38. Curry starts in what the Neoplatonicist Proclus calls the *intelligible triad*, later used by Neoplatonists when interpreting ancient cosmogonies about Night and Chaos. The intelligible triad is the emanations from God, or from an unknowable 'primary Unity' sometimes called 'the One' or 'the Good', 'the first cause of all causes', manifesting in six stages of which this is the first (39). It has classical literary antecedents in how Night (*Nyx*) is portrayed as more powerful and old than Zeus in Homer's *Iliad*, making for instance Damascius state that 'Homer, too, begins his genealogy of the Gods from Night' (42). Curry (46, n. 49): 'We must observe here that, for Milton and the English Platonists of the seventeenth century, such theories of emanation as that of Proclus – involving a timeless, necessary and eternal process – did not conflict with the conception of an historical creation consummated by an act of divine will. Having derived matter and chaos from God by the emanative process, Milton proceeds to represent the ordering of a part of chaos – as if it were the first composite body – into the cosmos by a special volitional act of God.' Sarkar, *Cosmos and Character in Paradise Lost*, 58; Curry, 46. On *Sol* and *Luna*, king and queen, see for instance M. Haeffner, *Dictionary of Alchemy: From Maria Prophetissa to Isaac Newton* (London: Aquarian/HarperCollins, 1994/91), 62.
11. C.-M. Edenborg, *Alkemins skam: den alkemiska traditionens utstöning ur offentligheten*, PhD dissertation, Stockholm University (Stehag/Gdansk: No Fun, 2002), 193–200.
12. W. Godwin, *St Leon: A Tale of the Sixteenth Century* (London: Henry Colburn and Richard Bentley, 1831), 101, 179, 208, 210.
13. She, however dedicates it 'To William Godwin, Author of *Political Justice*, *Caleb Williams*, &c', and she quotes Milton's *Paradise Lost*: 'Did I request thee, Maker, from my clay / To mould me man? Did I solicit thee / From darkness to promote me? – '. M. Shelley *Frankenstein; Or, The Modern Prometheus*, ed. M. Hindle (London: Penguin, 1992/85), 3, 1, 38. Without Agrippa, '[i]t is even possible that the train of my ideas would never have received the fatal impulse that led to my ruin', ibid, 38f.
14. Paracelsus, 'From *Of the Nature of Things*', in Linden, *The Alchemy Reader*, 152.
15. Godwin, *St Leon*, 211. The same goes for his description of the stranger and alchemist which in effect equals what the monster has done to Victor Frankenstein: 'He had robbed me of my son; he had destroyed my domestic peace; he had undermined the tranquillity and health of the partner of my life.' Miserable as St Leon is, he is also an antihero of the Romantic era, and the same is true of Mary Shelley's monster.
16. E. T. A. Hoffmann, *Nachtstücke: Herausgegeben von dem Verfasser der Fantasiestücke in Callots Manier* (Berlin: Gesammelte Schriften, Fünfter Band Druck und Verlag von Georg Reimer, 1872), 16, available at https://www.gutenberg.org/files/32046/32046-h/32046-h.htm#sandman. B. Stoker, *Dracula: Authoritative Text, Contexts, Reviews and Reactions, Dramatic and Film Variations, Criticism*, ed. N. Auerbach and D. J. Skal, (New York: W. W. Norton, 1997), 263.
17. E. Sprinchorn, 'The Zola of the Occult', in Strindberg Society, *Strindberg and Modern Theatre* (Stockholm: Strindbergssällskapet, 1975), 110f. E. Sprinchorn, *Strindberg as Dramatist* (New Haven: Yale University Press, 1982), 53f, 59. 'The Zola of the Occult', my translation of 'Ockultismens Zola', in A. Strindberg, *August Strindbergs brev 11. Maj 1895–november 1896* (Stockholm: Strindbergssällskapet, 1969), 307. A. Strindberg *Antibarbarus* (Lund: Bakhåll, 2003; facsimile of the 1906 edition).
18. One exponent of these ideas was Florence Farr, who was a member of both. See A. Owen, *The Place of Enchantment: British Occultism and the Culture of the Modern* (Chicago: University of Chicago Press, 2002), 124f.

19. G. Meyrink, *Golem*, trans. E. Klein (Stockholm: Bonniers, 1975). Biography of Gustav Meyrink in M. Fyhr, 'Gustav Meyrink och mystikens språk', *Minotauren. Tidskrift för skräck, fantasy och science fiction*, 7 (August 2000), 7–12. One example of its modern influence is that it is an important and explicit intertext in the novel *Jaromir* (1995) by Alexander Ahndoril (internationally known under the pen name Lars Kepler); for an analysis of this see M. Fyhr, *De mörka labyrinterna. Gotiken i litteratur, film, musik och rollspel*, PhD dissertation, Stockholm University (Lund: ellerströms, 2003/2009/2017), 225–98.
20. Meyrink, *Golem*, 128, 360. G. Meyrink, *Der Golem. Roman* (Leipzig: Kurt Wolff Verlag, 1915), 342, 344. C. G. Jung, *Psychology and Alchemy*, ed. and trans. G. Adler and R. F. C. Hull, in *The Collected Works of C. G. Jung, Complete Digital Edition*, vol. 12. Bollingen Series XX (Princeton: Princeton University Press, 1980, 2nd edn), passim.
21. S. Lagerlöf, 'Karln. En julsägen', *BLM*, 10 (1949). For the history of the story, see N. Afzelius, 'Selma Lagerlöfs "Karln". En kommentar', *BLM*, 10 (1949), 757ff. The quotation is my translation (755). One puzzled scholar was Afzelius, who can't explain why she wrote it: 'Why did she bring up a tale from the world of superstition? Did she believe in it herself?' (my translation, 760). The Lagerlöf scholar Sofia Wijkmark writes that 'The prior eventually succeeds in what is the goal of every alchemist, to conquer [sic] the philosopher's stone, or the water of life as the elixir is also called, which seems like something of a paradox in relation to his Christian belief and monasterial vows', and finds his 'relation to nature ... more in line with folklore than Christian doctrine', stating that we never receive a solution to the 'mystery' of the story. See S. Wijkmark, *Hemsökelser. Gotiken i sex berättelser av Selma Lagerlöf*, Karlstad University Studies 2009:20, PhD dissertation, Karlstad (my translation, 57f). She touches upon alchemy (73–5) but leaves its process and settles on alchemy as just as a theme of 'forbidden knowledge', thereby reading the ending as a statement saying that 'Knowledge about what is beyond life is a mystery – and must so remain'. I disagree: I think Lagerlöf is depicting the search and comparing it to alchemy.
22. Milton, *Paradise Lost*, 147. On the phoenix and alchemy, see for instance Haeffner, *Dictionary of Alchemy*, 180, and C. Gilchrist, *The Elements of Alchemy* (Shaftesbury: Element, 1991), 6; see also A. E. Waite (ed.), *The Hermetic Museum, Volume I: The Sophic Hydrolith* (1893): 'The Philosopher's Stone is called the most ancient, secret or unknown, natural incomprehensible, heavenly, blessed sacred Stone of the Sages. It is described as ... the glorious Phoenix – the most precious of all treasures – the chief good of Nature.' C.-H. Tillhagen, *Fåglarna i folktron* (Stockholm: LTs förlag, 1978), 298f. (My translation of Lagerlöf, 'Karln', 752.)

'A wine that was drunk by the moon and the sun'
Per Faxneld

1. See, e.g., J. J. Thorsen, 'Wilhelm Freddie: frihedens store alkymist' [Wilhelm Freddie: The Great Alchemist of Liberty], in *Friheden er ikke til salg* (Copenhagen: Borgan, 1980); C. Simic, *Dime-Store Alchemy: The Art of Joseph Cornell* (New York: New York Review of Books, 1992).
2. P. Faxneld, *Det ockulta sekelskiftet: Esoteriska strömningar i Hilma af Klints tid* (Stockholm: Volante, 2020), 49.
3. W. Pater, *The Renaissance* (1873; Oxford: Oxford University Press, 1998), 75 (emphasis mine).
4. A. Breton, *Manifestoes of Surrealism*, trans. H. R. Lane and R. Seaver (Ann Arbor: University of Michigan Press, 1972), 178.
5. E. Maurer, 'Dada and Surrealism', in W. Rubin (ed.), *'Primitivism' in 20th Century Art: Affinity of the Tribal and the Modern*, vol. 2 (New York: Museum of Modern Art, 1984), 544.
6. Breton, *Manifestoes*, 174.
7. Ibid., 174–5.

8. A. Breton, *Arcanum 17*, trans. Z. Rogow (Copenhagen and Los Angeles: Green Integer, 2004), 60, 89–90, 118, 131, 159. Quote on 60.
9. K. Noheden, 'Magic Art between the Primitive and the Occult: Animal Sacrifice in Jan Švankmajer's Drawer Fetishes', in T. M. Bauduin and H. Johnsson (eds), *The Occult in Modernist Art, Literature, and Cinema* (London: Palgrave, 2017), 199–201.
10. A. Breton, *L'art magique* (Paris: Club français de l'art, 1957), 22–5, 49.
11. W. Atkin, 'Endless Metamorphosis: Surrealism and Alchemy', in V. Greene et al. (eds), *Surrealism and Magic: Enchanted Modernity* (New York: The Solomon R. Guggenheim Foundation, 2022), 137. Atkin also links Brauner's tarot-inspired paintings *The Surrealist* and *The Lovers* (both 1947) to alchemical symbolism, though the connection in this case seems a little less clear.
12. K. Seligmann, *The Mirror of Magic: A History of Magic in the Western World* (Rochester: Inner Traditions, 2018), 146.
13. Quoted in M. E. Warlick, *Max Ernst and Alchemy: A Magician in Search of Myth* (Austin: University of Texas Press, 2001), 1.
14. Ibid., 5.
15. Ibid., 125, 166, quote on 166.
16. Ibid., 176.
17. S. Aberth, *Leonora Carrington: Surrealism, Alchemy and Art* (Farnham: Lund Humphries, 2004/10), 7.
18. Ibid., 23.
19. Warlick, *Max Ernst*, 161.
20. L. Carrington, *Down Below*, trans. V. Llona (Chicago: Black Swan Press, 1983), 177.
21. K. Noheden, 'Leonora Carrington, Surrealism, and Initiation: Symbolic Death and Rebirth in *Little Francis* and Down Below', *Correspondences: Journal for the Study of Esotericism*, 2:1 (2014), 50–1.
22. P. Mabille, *Mirror of the Marvelous: The Classic Surrealist Work on Myth*, trans. J. Gladding (Rochester: Inner Traditions, 1998), 59.
23. Noheden, 'Leonora Carrington', 53–5, quote on 55. Cf. Warlick, *Max Ernst*, 166.
24. Noheden, 'Leonora Carrington', 58.
25. Aberth, *Leonora Carrington*, 64–6.
26. Ibid., 75.
27. The suggestion that there is a quote from the 1351 alchemical text *Asensus Nigrum* in Carrington's *AB EO QUOD* (1956) is from Aberth, *Leonora Carrington* (93), while W. J. Hanegraaff, 'A Visual World: Leonora Carrington and the Occult', *Abraxas: International Journal of Esoteric Studies*, 6 (2014), 109, points out that it is in fact derived from 'Marsilio Ficino's Latin translation of a Pythagorean passage, which had been quoted by Carl Gustav Jung in his book *Psychology and Alchemy*'.
28. Aberth, *Leonora Carrington*, 82.
29. Carrington quoted in ibid., 94.
30. On Carrington's ideas about the kitchen as a site of magic power, see ibid., 64–70.
31. Ibid., 59–64.
32. M. Nonaka, *Remedios Varo: The Mexican Years* (Mexico City: Editorial RM, 2020), 13 (for more on her occult interests, see 14–18).
33. Cf. ibid., 15.
34. A. Jodorowsky, *The Spiritual Journey of Alejandro Jodorowsky*, trans. J. Rowe (Rochester: Park Street Press, 2008), Chapter 3.
35. A. Susik, 'The Alchemy of Surrealist Presence in Alejandro Jodorowsky's "The Holy Mountain"', in K. Noheden and A. Susik (eds), *Surrealism and Film after 1945: Absolutely Modern Mysteries* (Manchester: Manchester University Press, 2021), 190.
36. L. Ivan-Zadeh, '*El Topo*: The Weirdest Western Ever Made', BBC Culture, 23 July, 2020, https://www.bbc.com/culture/article/20200723-el-topo-the-weirdest-western-ever-made.
37. Susik, 'The Alchemy', 187.
38. Cf. ibid., 186, 200–1.
39. Ibid., 189.
40. P. D. Ouspensky, *In Search of the Miraculous: Fragments of an Unknown Teaching* (New York: Harcourt Brace, 1977), 294.

41. D. Stermac and A. Jodorowsky, 'Two Visions', *Cinema Canada*, 13 (1974), 62.
42. Quoted in Susik, 'The Alchemy', 198.
43. Ibid., 191–2.
44. K. Noheden, 'Expo Aleby, 1949 – Wilhelm Freddie, Gösta Kriland and Surrealist Magic Art in Stockholm', in B. Hjartarson et al. (eds), *A Cultural History of the Avant-Garde in the Nordic Countries 1925–1950* (Leiden: Brill, 2019).
45. 'Några hållpunkter inför en poesins svenska historia', originally published in *Mannen på gatan*, 1991. Available online at the Surrealist Group's website: http://www.surrealist- gruppen.org/BJ_CME_nagra- hall.html (accessed 9 September, 2023).
46. Faxneld, *Det ockulta*, 190.
47. Noheden, 'Magic Art', 201–2.
48. Ibid., 202.
49. K. L. Johnson, *Jan Švankmajer* (Urbana: University of Illinois Press, 2017), 100–3.

EMBLEMA XIV. *De secretis Naturæ.* 65

Hic eſt Draco caudam ſuam devorans.

'This is the dragon eating its own tail.' Emblem 14 illustrated by Matthäus Merian the Elder. From *Atalanta fugiens* by Michael Maier, 1618.

Contributors

KURT ALMQVIST is CEO at the Axel and Margaret Ax:son Johnson Foundation for Public Benefit. During the years 2013–19, he initiated an extensive seminar series that accompanied the exhibition 'Hilma af Klint – Pioneer of Abstraction' worldwide. He became CEO of the non-profit foundation in 1999 and has subsequently developed the business by taking initiatives for Axess Magasin, Axess TV and the Engelsberg Seminar, among other things.

TOBIAS CHURTON is one of Britain's leading scholars in the field of Western esotericism. Holding a master's degree in theology from Brasenose College, Oxford, he was appointed honorary fellow and Faculty Lecturer in Western esotericism at Exeter University in 2005. Following Professor Gabriele Boccaccini's invitation in 2019 to participate in the Enoch Seminar in Florence, *The Lost Pillars of Enoch* was published in 2020. His comprehensive work on all aspects of Enochic tradition, *The Books of Enoch Revealed*, appeared in March 2025. Following studies on Gnostic traditions that included *The Gnostics* (1987), *Gnostic Philosophy* (2005), *Kiss of Death: The True History of the Gospel of Judas* (2008) and *Gnostic Mysteries of Sex* (2015), Churton's *The First Alchemists* – a comprehensive account of the origins of alchemy and its relation to Gnostic thought – was published in 2023. Tobias is also a filmmaker, poet, composer and author of acclaimed biographies of William Blake, Aleister Crowley, Elias Ashmole and G. I. Gurdjieff. Academic papers on Crowley, the Yezidis, Rosicrucianism, Freemasonry, the French Occult Revival and the Enochic tradition have appeared in anthologies published by E. J. Brill, Routledge and Oxford University Press. He has recently completed his 28th commissioned work, *The Celestial Realms, A Brief History of Heaven since the Dawn of Time*, for publication in 2026.

PER FAXNELD is Professor in the Study of Religions at Södertörn University. Specialising in alternative spirituality, secularity and esotericism, his earlier work (encompassing three monographs, three edited volumes and numerous articles and chapters) focused on esoteric art, popular culture and the intersections of religion and politics.

HJALMAR FORS heads the Hagströmer medico-historical library at Karolinska Institutet, Stockholm. He is a PhD (2003) and docent (2014) in the history of science. His primary research expertise concerns the history of empirical sciences – that is, alchemy, chemistry, mining knowledge, natural history, medicine and pharmacy – during the seventeenth and eighteenth centuries. He works with natural scientists in reproducing scientific experiments and craft practices (experimental history of science). Among his publications is acclaimed monograph *The Limits of Matter: Chemistry, Mining, and Enlightenment*, published by the University of Chicago Press in 2015. The Swedish translation is *Upplysningens element: Materia och världsbild under 1600- och 1700-talet* (2020).

PETER J. FORSHAW has a doctorate in intellectual history and is Associate Professor of Western Esotericism in the Early Modern Period at the Centre for History of Hermetic Philosophy at the University of Amsterdam. He was editor-in-chief of *Aries: Journal for the Study of Western Esotericism* for ten years (2010–20), has edited several books, including *Lux in Tenebris: The Visual and the Symbolic in Western Esotericism* (2016), and is author of *Occult: Decoding the Visual Culture of Mysticism, Magic & Divination* (2024) and a four-volume monograph, *The Mage's Images: Heinrich Khunrath in His Oratory and Laboratory* (2024–2025).

MATTIAS FYHR is an associate professor at Stockholm University, presently working at Örebro University. He has published extensively on the Gothic and horror in combination with the supernatural, for instance on how H. P. Lovecraft consciously behaved like a mystic, describing pseudo-memories of a past life, while adhering to a materialistic worldview.

GEORGIANA D HEDESAN is an early modern historian of science at the University of Oxford, specialising in the study of alchemy and alchemical medicine. Her first book, *An Alchemical Quest for Universal Knowledge: The 'Christian Philosophy' of Jan Baptist Van Helmont (1579–1644)*, was published in 2016 by Routledge.

JOHN M. MACMURPHY has degrees in psychology (UC Berkeley) and Western esotericism (University of Amsterdam/HHP). His work explores the relationship between various esoteric trends and theoretical schemes associated with practices such as meditation, sexuality, entheogens, altered states of consciousness and mysticism. His current research focuses on *Kabbalah Ma'asit* (Practical Kabbalah).

CHRISTOPHER MCINTOSH is a British-born writer and historian specialising in the esoteric traditions of the West. His recent books include *Beyond the North Wind* (2019), *Occult Russia* (2022) and *Occult Germany* (2024). With his wife, Dr Donate McIntosh, he produced a new translation of the Rosicrucian *Fama Fraternitatis* (2014). Since 1994 he has lived in northern Germany.

MARK S. MORRISSON is an associate dean and English professor at Penn State University and author of several monographs, including *Modern Alchemy: Occultism and the Emergence of Atomic Theory*. His new monograph, *Light on the Path: Advancing Occultism Through Esoteric Fiction, 1880-1940*, will be published by Oxford University Press in 2025.

FABRIZIO PREGADIO has taught in Italy, Germany, the USA and Canada. His research focuses on Taoist thought and religion, alchemy, self-cultivation and views of the human being. He edited the *Encyclopaedia of Taoism* (2008) and authored the *Dictionary of Taoist Internal Alchemy* (2024) and has also published numerous studies and translations of Taoist texts.

SALAM RASSI is Lecturer in Islam and Christian-Muslim Relations at the School of Divinity, University of Edinburgh. His main area of research is Christian–Muslim interactions across theology, philosophy and literature. Following the completion of his doctorate at the University of Oxford, he became a Mellon Foundation postdoctoral fellow at the American University of Beirut. He has also worked as a cataloguer of Syriac and Arabic manuscripts at the Hill Museum & Manuscript Library, Minnesota, and was a British Academy postdoctoral fellow at Oxford. His first book, entitled *Christian Thought in the Medieval Islamicate World*, was published by Oxford University Press in early 2022.

CRISTINA VIANO is directeur de recherche at the Centre national de la recherche scientifique, Centre Léon Robin, Sorbonne-Université, Paris. She works on the ancient doctrines of matter, emotions and causation, Aristotle, Alexandrian alchemy and the relationship with Greek philosophy. Among her publications are *La matière des choses. Le livre IV des* Météorologiques *d'Aristote, et son interprétation par Olympiodore* (published by Vrin, Paris, 2006) and the Italian translation with introduction and notes of Aristotle's *Rhetoric* (published by Laterza, Bari, 2021).

ANDREAS WINKLER is an assistant professor of Egyptology in the Department of Near and Middle Eastern Civilizations at the University of Toronto, specialising in history and philology. While his research spans the entirety of ancient Egyptian history, he focuses primarily on the Graeco-Roman period, particularly its socio-economic and intellectual history. His specific areas of work include ancient Egyptian astral sciences and the transfer of knowledge in the ancient Mediterranean world.

DAGMAR WUJASTYK is an associate professor in the Department of History, Classics, and Religion at the University of Alberta. She is an Indologist specialising in the history and literature of classical South Asia, including Indian medicine (Ayurveda), iatrochemistry (*rasaśāstra*) and yoga.

Image rights

p. 10	Photo © Christie's Images/Bridgeman Images
p. 12	Acquired by Henry Walters, 1915, The Walters Art Museum, Baltimore, 35.35 (CC0)
p. 14	National Museum of Asian Art, Smithsonian Institution/Arthur M. Sackler Collection/Bridgeman Images
p. 16	Wellcome Collection, London (PDM)
p.18	After Alexey I. Falev, Victor L. Kokorin. Determining role of cyclicity in cosmic and natural processes for formation of energy figures in Chinese Classical Zhen Jiu Therapy. *Cardiometry*; issue 9; Nov. 2016; p. 19. (CC BY 4.0)
p. 21	Wellcome Collection, London (CC BY 4.0)
p. 22	Courtesy of the author
p. 25	© Christie's Images/Bridgeman Images
p. 26	Courtesy of the author
p. 28	© GrandPalaisRmn (Musée du Louvre)/Hervé Lewandowski
p. 30	National Museum of Antiquities, Leiden
p. 32	National Library of Sweden/World Digital Library, Acc 2013/75, 14
p. 36	(left) National Library of Sweden/ World Digital Library, Acc. 2013/75, 15r (CC BY 4.0)
p. 36	(right) National Museum of Antiquities, Leiden, AMS 76 vel 1 (P. Leid. I 395), 25
pp. 40–41	Getty Images/Universal Images Group. Photo: Werner Forman
p. 44	Mondadori Portfolio/Archivio Dell'arte Luciano Pedicini/Luciano Pedicini/Bridgeman Images
p. 46	National Museum of Antiquities, Leiden, AMS 75 vel 1 (P. Leid. I 384)
p. 47	National Museum of Antiquities, Leiden, AMS 66 vel 1 (P. Leid. I 387)
p. 48	Kelsey Museum of Archaeology. University of Michigan, KM 10662
p. 49	Kelsey Museum of Archaeology. University of Michigan, KM 26801
p. 51	Kairoinfo4u via Flickr
p. 52	Syndics of Cambridge University Library, T-S Misc. 8.51.
p. 55	Diego Delso, via Wikipedia (CC BY-SA 4.0)
p. 57	Mondadori Portfolio/Archivio Dell'arte Luciano Pedicini/Luciano Pedicini/Bridgeman Images
p.60	Palazzo della Ragione, Padua. Photo © Superstock/Bridgeman Images
p. 62	J. Paul Getty Museum, Los Angeles, 83.AM.342 (CC0 1.0)
p.64	Biblioteca Nazionale Marciana, Venice, Ms. Gr. 299 (=584), fos. 188v and 193v. By permission of the Ministry of Culture – National Library Marciana
p. 66	MS Vaticanus Graecus 1134, fos. 1v-2r.
p. 67	Museu Nacional de Machado de Castro, Coimbra, Portugal. Photo: Manuel Cohen/© Scala, Florence
p. 72	Biblioteca Nazionale Marciana, Venice, Ms. Gr. 299 (=584), fo.8r. By permission of the Ministry of Culture – National Library Marciana
p. 75	(top and middle) © MAFDO, Jean-Pierre Brun
p. 75	(bottom) © MAFDO, Adam Bülow-Jacobsen
p. 80	U.S. National Library of Medicine (PDM)
p. 83	Sächsische Landesbibliothek – Staats- und Universitätsbibliothek (SLUB), Dresden (PDM 1.0)
p. 84	University of Wisconsin – Madison Libraries (CC BY 4.0)
p. 87	Leiden University Libraries, Vossianus MS Chym F 29, fo.98. (PDM)
p. 89	Bavarian State Library, Res/4 Alch. 51, 33
p. 92	Biblioteca Medicea Laurenziana, Cod. Ashburnham 1166, fo. 12r. Photo: Universal History Archive/UIG/ Bridgeman Images
p. 95	The British Library, London, Add. 25724, fo.18v. From the British Library archive/Bridgeman Images
p. 96	Bavarian State Library, Munich, Res/4 Alch. 51, 141
p. 98	Bibliothèque nationale de France, Paris, Arabe 5099, fos. 77v-78r
p. 104	Courtesy of the Topkapi Palace

Museum, Istanbul, Ahmet III 2075, fos. 2b-3a

p. 107 The British Library, London, Add. 25724, fo.36v. From the British Library archive/Bridgeman Images

p. 109 The British Library, London, Or. 13006, fo.5v, courtesy of Qatar Digital Library

p. 111 University of Pennsylvania Library, LJS 441, fo.85

p. 112 The British Library, London, Or. 12208, fo. 294. From the British Library archive/Bridgeman Images

p. 115 Bodleian Library MS. Ouseley Add. 166, fo. 16r. © Bodleian Libraries, University of Oxford

p. 116 National Crafts Museum, New Delhi/ PuranaStudy (CC BY-SA 4.0)

p. 119 Bhandarkar Institute, Poona © 2024 Bhandarkar Oriental Research Institute, MS P, fo. 35r

p. 121 Courtesy of Andrew Mason

p. 123 Lalbhai Dalpatbhai Institute of Indology, Ahmedabad, LDII MS Acc No 9442, fo.25

p. 125 Kislak Center for Special Collections, Rare Books and Manuscripts, University of Pennsylvania Libraries, Ms. Coll. 390 Item 763, fos. 1v–2r

p. 126 Wellcome Collection, London (PDM)

p. 128 Bavarian State Library, Res/4 Alch. 51, 57

p. 129 Berlin State Library, Ms. Or. Oct. 514 (Steinschneider 258), fo. 51r

p. 131 Bavarian State Library, Munich, Res/4 Alch. 51, titlepage

p. 132 Ohio State University Libraries (PDM)

p. 133 EFM/Bibliotheca Philosophica Hermetica Collection, Amsterdam, M 469

p. 134 Opensource

p. 135 Courtesy of Austrian National Library, https://onb.digital/result/1084B722

p. 137 Wellcome Collection, London (PDM)

p. 138 Courtesy of Kedem Auction House, Jerusalem

p. 141 Getty Research Institute, Los Angeles

p. 142 Museo di Palazzo Vecchio (Palazzo della Signoria), Florence/Luisa Ricciarini/Bridgeman Images p. 144 The Parker Library, Corpus Christi College, Cambridge, MS 395, fo. 61v

p. 146 Bodleian Libraries, University of Oxford, MS Digby 162, fo. 1r

p. 147 © Bodleian Libraries, University of Oxford, MS. e Mus. 155, p.591.

p. 148 University of Glasgow Library, Sp Coll Ferguson An-y.10, endpapers

p. 149 University of Glasgow Library, Sp Coll Ferguson An-y.10, fo.38r

p. 150 The Parker Library, Corpus Christi College, Cambridge, MS 395, fo. 50r

p. 151 Wellcome Collection, London (PDM)

p. 152 Wellcome Collection, London (PDM)

p. 154 The Parker Library, Corpus Christi College, Cambridge, MS 395, fos 61v-62r

p. 155 Courtesy of Science History Institute, Shelfmark Othmer MS 3, fo.1r

p. 156 Courtesy of Science History Institute, Shelfmark Othmer MS 3, fo.44r

p. 157 Wellcome Collection, London (PDM)

p. 158 Zurich Central Library, Ms. Rh. 172, fo. 3r (detail), https://www.e-codices.unifr.ch/en

p. 161 Kassel University Library, State Library and Murhard Library of the City of Kassel, Ms. Chem. 82, fos 25r, 26r, 27r, 29r

p. 162 Bavarian State Library, Munich, BSB Cgm 598, pp.90r (detail), 104v, 24v

p. 165 Zurich Central Library, Ms.Rh.172, fo.3r, https://www.e-codices.unifr.ch/en

pp. 166–167 Zurich Central Library, Ms.Rh.172, fo.10v, https://www.e-codices.unifr.ch/en

p. 169 University of Wisconsin Libraries (CC BY 4.0)

p. 172 Fototeca Gilardi/Bridgeman Images

p. 174 Courtesy of University of Glasgow Archives & Special Collections, Sp Coll Ferguson Al-e.66, titlepage

p. 175 Getty Research Institute, Los Angeles, 377017

p. 176 Baden Memorial Library, Karlsruhe, Cod. St. Peter perg. 92, iv

p. 177 State Library, Regensburg/Bavarian State Library, Munich, 999/Med.516

pp. 178–179 © Imagno/Getty Images

p. 180 Getty Research Institute, Los Angeles, 371996, titlepage and p.95

p. 181 Courtesy of University of Glasgow Archives & Special Collections, Sp Coll Ferguson, Ar-f.16, titlepage

p. 182 Bavarian State Library, Alch. 67#Beibd.2, titlepage
p. 183 Opensource, p.18
p. 184 Opensource, p.50
p. 187 Private Collection/Bridgeman Images
p. 188 Derby Museum & Art Gallery, 1883-152. © Derby Museums/Bridgeman Images
p. 190 Wellcome Library, London, 3560/D/1 (PDM)
p. 192 The British Library, London, 1039.f.20. (1.). From the British Library archive/ Bridgeman Images
p. 193 Wellcome Library, London, 2200305i (PDM)
p. 194 Library of Congress, Rare Book and Special Collections Division, Washington D.C., Incun. 1500.B78, titlepage and p.10
p. 195 BNF Bibliothèque de l'Arsenal, Paris, MS 7389 © BnF, Dist. GrandPalaisRmn/ image BnF
pp. 196–197 The Metropolitan Museum of Art, 41.44.1613. (PDM)
p. 198 Augsburg, State and City Library/ Bavarian State Library, 4 Med 908
p. 199 Bavarian State Library, Munich, Phys.m. 203
p. 201 Bavarian State Library, Munich, Med.g. 442m
p. 203 BNF Bibliothèque de l'Arsenal, Paris, 8-S-8231
p. 204 Zurich Central Library, SCH R 201, 1a
p. 206 Courtesy of Science History Institute, Philadelphia (PDM)
p. 208 Rijksmuseum, Amsterdam, SK-A-2342
p. 210 Courtesy of Science History Institute, Philadelphia (PDM)
p. 211 Musée de l'Histoire de France, Château de Versailles. Photo: © Fine Art Images/ Heritage Images/Getty Images
p. 213 Nationalmuseum, Stockholm, NMH 279/1892 TBC
pp. 214–215 Wellcome Collection, London, 37106i (PDM)
p. 219 Alchemy, Magic and Kabbalah, Foundation of the Works of C.G.Jung, Zurich (PDM)
p. 221 Wellcome Collection, London (PDM)
p. 224 New York Public Library (PDM)
p. 228 The Metropolitan Museum of Art, New York, 1977.10
p. 230 Wellcome Collection, London, 37207i
p. 232 (top) EFM/Bibliotheca Philosophica Hermetica Collection, Amsterdam
p. 232 (bottom left) Wellcome Collection, London, EPB/B/24863.v1
p. 232 (bottom right) Courtesy of University of Glasgow Archives & Special Collections, via Wellcome Collection, London
p. 233 Wellcome Collection, London, 3602i
p. 235 © 2007 Foundation of the Works of C.G. Jung. First published by W.W. Norton & Co.
p. 239 The Grace K. Babson Collection of the Works of Sir Isaac Newton at The Huntington Library, Art Museum, and Botanical Gardens, San Marino, mssBAB 2, fo.1r
p. 242 The Metropolitan Museum of Art, New York; The Friedsam Collection, Bequest of Michael Friedsam, 1931 (PDM)
p. 244 Courtesy of The Linda Hall Library of Science, Engineering & Technology (CC BY 4.0)
p. 246 Heidelberg University Library, Germany (CC BY-SA)
p. 248 Mary Evans Picture Library/Psychic News
p. 249 Wellcome Collection, London, 32971i
p. 250 After *Popular Science Monthly*, June 1923, vol. 102, issue 6, via Internet Archive (PDM)
p. 251 Wellcome Collection, London (PDM)
p. 252 Private collection
p. 253 Wikipedia (PDM)
p. 254 Alamy/ARCHIVIO GBB
p. 256 Private collection
p. 257 Heidelberg University Library, Germany (CC BY-SA)
p. 261 Photo: Dust Jackets L.L.C., San Francisco
p. 264 © 2007 Foundation of the Works of C.G. Jung
p. 267 Basel University Library, Portr BS Jung CG 1794, 3
p. 268 Süddeutsche Zeitung Photo/Alamy Stock Photo
p. 269 Getty Research Institute, Los Angeles
pp. 270–271 Getty Research Institute, Los Angeles
p. 272 General Collections, Library of Congress, Washington D.C., BL1900. L843 W5 1929

p. 273 Chinese Rare Book Collection, Library of Congress, Washington D.C., 2014514151
p. 274 © 2007 Foundation of the Works of C.G. Jung. First published by W.W. Norton & Co.
p. 275 Tim Nachum Gidal © The Israel Museum, Jerusalem
p. 277 DB/Ullstein bild via Getty Images
pp. 278, 279 Wellcome Library, London, 163/B/1
p. 280 Courtesy of the Strindberg Museum, Stockholm
p. 282–283 The Staatliche Museen zu Berlin, Kupferstichkabinett, inv. Botticelli, Inferno 18/Getty Images
p. 284 Science History Images/Alamy Stock Photo
p. 285 Courtesy of The Linda Hall Library of Science, Engineering & Technology (CC BY 4.0)
p. 287 Chronicle/Alamy Stock Photo
p. 288 Logic Images/Alamy Stock Photo
p. 289 National Library of Sweden, Collection of Manuscripts, Strindbergsrummet (CC BY 2.0)
pp. 290–291 piemags/CMB/Alamy Stock Photo
p. 292 Collection of Alejandra Varsoviano de Gruen, Mexico City. Index Fototeca/Bridgeman Images. © Remedios Varo/Bildupphovsrätt 2025
p. 294 Bavarian State Painting Collections, Alte Pinakothek, Munich, no. 1847 URL: https://www.sammlung.pinakothek.de/en/artwork/ZMLJykqxJv (CC BY-SA 4.0)
p. 296 Cyrille Cauvet/Musée d'art moderne et contemporain (MAMC) de Saint-Etienne Métropole. © Victor Brauner/Bildupphovsrätt 2025
p. 297 The Museum of Modern Art, New York, 828.1964.E. Digital image, The Museum of Modern Art, New York/Scala, Florence. © Max Ernst/Bildupphovsrätt 2025
p.298 Courtesy of Librairie-galerie l'Imaginaire, Brussels © Yves Tanguy/Bildupphovsrätt 2025
p. 299 Charles B. Goddard Center for Visual and Performing Arts, Ardmore, Oklahoma © Estate of Leonora Carrington/Bildupphovsrätt 2025
p. 301 ©athanor Ltd.
p. 303 © Photo 12/Alamy Stock Photo
p. 305 ©athanor Ltd.
p. 306 Photo by Fine Art Images/Heritage Images via Getty Images
p. 352 Photo by Fine Art Images/Heritage Images via Getty Images
p. 359 Photo by Fine Art Images/Heritage Images via Getty Images

EMBLEMA XXI. *De secretis Naturæ.* 93

Fac ex mare & fœmina circulum, inde quadrangulum, hinc triangulum, fac circulum & habebis lap. Philosophorum.

'Make a circle out of a man and a woman, out of this a square, out of this a triangle, make a circle, and you will have the Philosophers' Stone.' Emblem 21 illustrated by Matthäus Merian the Elder. From *Atalanta fugiens* by Michael Maier, 1618.

THE HISTORY OF ALCHEMY
Influences on Culture, Science and Society

Published by Bokförlaget Stolpe, Sweden, 2025

Edited by: Carl Philip Passmark

Text editor: Zoe Gullen
Picture editor: Sara Ayad
Translator: Clare Barnes (Kurt Almqvist)
Design: Patric Leo
Layout: Pontus Dahlström
Cover image: *The Alchemical World Landscape*,
from *Opus medico-chymicum* by Johann Daniel Mylius, 1618,
Heritage Image Partnership Ltd / Alamy Stock Photo.
Pre-press and print coordinator: Italgraf Media AB, Sweden
Print: PNB Print, Latvia, via Italgraf Media, 2025
First edition, first printing
ISBN: 978-91-89882-69-0

Bokförlaget Stolpe is a part of
Axel and Margaret Ax:son Johnson Foundation
for Public Benefit

AXEL AND MARGARET AX:SON JOHNSON
FOUNDATION FOR PUBLIC BENEFIT